Current Topics in Behavioral Neurosciences

Volume 56

Current Topics in Behavioral Neurosciences provides critical and comprehensive discussions of the most significant areas of behavioral neuroscience research, written by leading international authorities. Each volume in the series represents the most informative and contemporary account of its subject available, making it an unrivalled reference source. Each volume will be made available in both print and electronic form.

With the development of new methodologies for brain imaging, genetic and genomic analyses, molecular engineering of mutant animals, novel routes for drug delivery, and sophisticated cross-species behavioral assessments, it is now possible to study behavior relevant to psychiatric and neurological diseases and disorders on the physiological level. The Behavioral Neurosciences series focuses on translational medicine and cutting-edge technologies. Preclinical and clinical trials for the development of new diagnostics and therapeutics as well as prevention efforts are covered whenever possible. Special attention is also drawn on epigenetical aspects, especially in psychiatric disorders.

CTBN series is indexed in PubMed and Scopus.

Frederick S. Barrett • Katrin H. Preller
Editors

Disruptive Psychopharmacology

Springer

Editors
Frederick S. Barrett
Center for Psychedelic and Consciousness Research, Department of Psychiatry and Behavioral Sciences
Johns Hopkins University School of Medicine
Baltimore, MD, USA

Katrin H. Preller
Department of Psychiatry, Psychotherapy and Psychosomatics
Psychiatric University Hospital Zürich, University of Zurich
Zürich, Switzerland

ISSN 1866-3370 ISSN 1866-3389 (electronic)
Current Topics in Behavioral Neurosciences
ISBN 978-3-031-12183-8 ISBN 978-3-031-12184-5 (eBook)
https://doi.org/10.1007/978-3-031-12184-5

This Springer imprint is published by the registered company Springer Nature Switzerland AG
The registered company address is: Gewerbestrasse 11, 6330 Cham, Switzerland

Preface

Until recently, the modal treatment of many psychiatric disorders was defined by one or a combination of daily pharmacotherapy and various forms of structured or unstructured psychotherapy. While these approaches have been met with some success, many current psychiatric pharmacotherapies have modest effect sizes, undesirable side effects, and variable patient adherence. Many patients also remain refractory to treatment in many symptom domains. Hence, more effective therapies are urgently needed.

New therapeutic approaches that harness the power of psychedelic drugs are gaining traction as potential treatments for a wide range of indications. The structure and delivery of psychedelic therapies represent a sharp departure from more traditional models of psychotherapy and pharmacotherapy that have defined the treatment of psychiatric and other medical disorders. In the case of ketamine, patients may undergo weekly exposure to this psychoactive compound rather than chronic daily dosing that is encountered with standard pharmacotherapy (e.g., with a monoaminergic reuptake inhibitor). In the case of serotonergic psychedelics, a therapeutic approach may consist of careful volunteer preparation followed by 1–3 psilocybin administrations under closely monitored, controlled conditions. These departures from the standard psychiatric model may be critical to the success of psychedelic therapies. In this way, psychedelic therapies represent a type of *disruptive technology* that is emerging within the field of psychiatry and which may have the potential to revolutionize the treatment of psychiatric disorders.

The current volume consists of a series of monographs from thought-leaders across a range of areas that together provide a comprehensive snapshot of the state of the science of these disruptive psychopharmacological approaches. In the first section, the authors introduce perspectives on the approaches to determining appropriate dosing with psychedelics, conducting group psychedelic therapy, and the training of psychedelic therapists. The authors also discuss the literature on the effects of setting on psychedelic experiences. In the second section, targeted chapters review the growing literature supporting the use of ayahuasca, psilocybin, and ketamine for the treatment of mood disorders. The authors also address the use of psychedelic therapies for palliative care and the treatment of end-of-life distress. In the final section, the authors review emerging data and provide theoretical perspectives on the treatment of substance use disorders, inflammation, obsessive compulsive, trauma-related, neurodegenerative, headache, and chronic pain disorders.

Together, the chapters in this volume serve to summarize the state of the science of safety and efficacy of psychedelic therapies, introduce theoretical models and approaches to psychedelic therapies, and provide a roadmap to the disruptive psychopharmacology of psychedelic drugs.

We want to thank all authors and reviewers who have contributed their expertise to this volume. The list of reviewers includes in alphabetical order:

Dr. Brian Anderson
Dr. Alan Davis
Dr. Patrick Finan
Dr. Fernando Goes
Dr. Peter Hendricks
Dr. Marcus Herdener
Dr. Andrea Jungaberle
Dr. Henrik Jungaberle
Lea Mertens, M.Sc.
Dr. Stewart Mostofsky
Dr. Gerald Nestadt
Dr. David Nichols
Dr. Lorenzo Pasquini
Dr. Timo Schmidt
Dr. Luís Fernando Tófoli
Dr. Harriet de Wit

Additionally, we would like to thank the series editors, in particular Dr. Mark Geyer, for their support.

Baltimore, MD, USA Frederick S. Barrett
Zürich, Switzerland Katrin H. Preller

Contents

Part I
Models and Approaches

Dosing Psychedelics and MDMA

Matthias E. Liechti and Friederike Holze

Contents

Abstract Classic psychedelics, including psilocybin, lysergic acid diethylamide (LSD), dimethyltryptamine, and mescaline, and entactogens/empathogens, especially 3,4-methylenedioxymethamphetamine, have received renewed attention in psychiatric research and may be developed into medications for such indications as anxiety, depression, cluster headache, and posttraumatic stress disorder, among others. However, identifying proper doses is crucial. Controlled study data on dosing using well-characterized pharmaceutical formulations of the substances are scarce. The dose equivalence of different substances, dose-response effects, and subjective effects of different doses are of great interest and practically important for their clinical use in psychotherapy. Furthermore, the so-called microdosing of psychedelics has recently gained popularity, and the first placebo-controlled studies of LSD have been published. This chapter discusses different aspects of psychedelic dosing, including pharmaceutical aspects, definitions and characteristics of different doses, including microdoses, aspects of personalized dosing, and non-pharmacological factors, that can influence the response to psychedelics.

M. E. Liechti (✉) and F. Holze
Division of Clinical Pharmacology and Toxicology, Department of Biomedicine, University Hospital Basel, Basel, Switzerland

Department of Clinical Research, University Hospital Basel, Basel, Switzerland
e-mail: matthias.liechti@usb.ch

Curr Topics Behav Neurosci (2022) 56: 3–22
https://doi.org/10.1007/7854_2021_270
Published Online: 4 November 2021

Keywords Dosing · Ego-dissolution dose · Good effect dose · Microdosing · Personalized medicine · Pharmacogenetics

1 Introduction

Classic psychedelics, such as psilocybin, lysergic acid diethylamide (LSD), dimethyltryptamine (DMT), and mescaline, have seen a renaissance in psychiatric research. Notably, psilocybin, LSD, and DMT are currently under rigorous investigation for use as adjuncts in psychedelic-assisted therapy for several disorders, such as depression, anxiety, and substance use disorders (Gasser et al. 2015; Davis et al. 2020; Johnson et al. 2014; Carhart-Harris et al. 2016a; Griffiths et al. 2016; Bogenschutz et al. 2015). The focus of researchers goes beyond psychiatric disorders, exploring their potential use to treat cluster headaches and migraine (ClinicalTrials.gov: NCT03341689, NCT03781128). Furthermore, 3,4,-methylenedioxymethamphetamine (MDMA) is currently under investigation for the treatment of posttraumatic stress disorder (Mithoefer et al. 2010, 2019). Among the group of classic psychedelics, psilocybin is to date the most widely investigated substance in patients, with also several academic studies in healthy subjects (Hirschfeld and Schmidt 2021). Several modern Phase 1 and 2 studies have been conducted over recent years to investigate psilocybin dosing (Brown et al. 2017; Griffiths et al. 2011; Kraehenmann et al. 2017a; Preller and Vollenweider 2018; Madsen et al. 2019; Hasler et al. 2004; Hirschfeld and Schmidt 2021; Garcia-Romeu et al. 2021). However, despite increasing work in patients, scarce pharmacokinetic data on psilocybin are available (Kolaczynska et al. 2020; Madsen et al. 2019; Brown et al. 2017), with a lack of solid dose-finding studies using different doses of psilocybin within the same study (Hirschfeld and Schmidt 2021). The pharmacokinetics and dose-response characteristics of LSD are better defined (Holze et al. 2019, 2021a, b; Dolder et al. 2017). MDMA has been studied relatively extensively in the healthy population and in the therapeutic setting (Vizeli and Liechti 2017; Vollenweider et al. 1998, 1999; Parrott 2014; Mithoefer et al. 2010, 2018, 2019), including pharmacokinetics, safety pharmacology, and dosing (Studerus et al. 2021; Vizeli et al. 2017, 2019; Vizeli and Liechti 2017; Schmid et al. 2016).

2 Pharmaceutical Aspects

The majority of psychedelic substance use is non-medical and includes recreational, spiritual, or other types of non-controlled use. In these settings, the doses are generally not well-defined, and substance preparations may also not be stable over time compared with drugs that are dosed as pharmaceuticals. This means that the true dose is unknown in most cases. Additionally, natural products that contain psilocybin or DMT may be used, in which case the amount of active substance is also unknown. In the case of synthetic substances, such as LSD, often unclear is whether the reportedly used doses refer to the active substance (i.e., LSD base) or a typically used salt, such as LSD tartrate. However, even when psychedelics are used in research studies, the doses that are administered to subjects are often unknown or reported with insufficient details to be defined. For example, LSD may be used as LSD base or salt. If the tartrate salt is used, then it may be LSD 1:1 tartrate or LSD 2:1 tartrate and may contain additional amounts of crystal water, methanol solvate, or other solvents. Thus, correct dosing information requires that the type of substance that is used is indicated, together with the estimated or measured content of active substance. Specifically, pure LSD 1:1 and LSD 2:1 tartrate contains 68% and 81% active LSD base, respectively. Thus, a 1 μg dose unit of LSD base that is used by some researchers (Holze et al. 2019, 2020, 2021a, b; Hutten et al. 2020b; Carhart-Harris et al. 2016b; Yanakieva et al. 2020, 2019; Family et al. 2020) would correspond to 1.46 or 1.23 μg of LSD 1:1 or 2:1 tartrate salt, respectively, as used by other researchers (Bershad et al. 2019, 2020; Family et al. 2020; Yanakieva et al. 2019). However, the true identity of the salt and content of active substance may be unclear because the necessary tests to discern such information (e.g., high-performance liquid chromatography and nuclear magnetic resonance identification of the molecule that validly and analytically measures content uniformity in 10 samples) and stability data over the length of the full storage or study period may be unavailable. Furthermore, such substances as LSD and psilocybin may decompose over time, and the content of pharmaceutically active substance may thus be lower than initially thought (Holze et al. 2019). For example, LSD is inactivated to iso-LSD, depending on temperature, solvent, and pH, and may be unstable in certain formulations. Other stress factors, such as light, oxygen, and chlorine in tap water, may lead to decomposition of the LSD molecule. In fact, amounts of iso-LSD were detected in plasma from research subjects in one of our studies that used LSD in capsules (Mueller et al. 2017a, b; Dolder et al. 2016), indicating that approximately 30% of the LSD that was administered likely isomerized to inactive iso-LSD within the LSD capsules that were used (Steuer et al. 2017). Later pharmacokinetic studies that used validly defined doses and a novel liquid formulation of LSD (Holze et al. 2019, 2021b) indicated that the previous studies that used capsules had actually used 70 and 140 μg LSD base rather than the reported oral doses of 100 and 200 μg LSD base (Mueller et al. 2017a, b; Dolder et al. 2016; Preller et al. 2018, 2019; Barrett et al. 2018; Preller et al. 2017; Kraehenmann et al. 2017a, b; Schmid et al. 2015). Additionally, particularly in the case of not yet well-defined substance formulations, plasma concentrations of the psychoactive substance could provide a measure of the compound that actually arrives in the body, thereby accounting for dosing, bioavailability, and even interindividual differences in absorption, distribution, metabolism, and excretion. Plasma concentrations can also help compare formulations and substance exposure between different studies and different research groups and thus should be generated for each novel substance formulation. In the future, correct dose identification will likely benefit from the increasing use of investigational medicinal products that are produced according to Good Manufacturing Practice.

3 Psychedelic (Full) Doses

Psychedelics are psychoactive substances. The goal of substance-assisted psychotherapy is typically to induce a "good trip" or psychedelic experience that is characterized by mainly positively experienced subjective drug effects, including the induction of positive emotions and pleasurable alterations of the mind and perception. A positive acute response is associated with a more favorable long-term therapeutic outcome (Roseman et al. 2017; Griffiths et al. 2016; Ross et al. 2016). Doses that produce the full range of the psychedelic experience are termed full psychedelic effect doses, ego-dissolution doses, or experiential doses (Holze et al. 2021b), in contrast to "microdoses" or "minidoses" that induce no or only minimal acute subjective effects (Kuypers et al. 2019; Holze et al. 2021a, b). Negative subjective effects in the context of full doses (i.e., "bad trips") are sometimes called "challenging experiences" to indicate that they may have partly personal or therapeutic value (Barrett et al. 2016) rather than being exclusively untoward drug effects. Generally, negative subjective drug effects are less desired and likely more frequent in non-therapeutic settings and vulnerable individuals (Barrett et al. 2017). More pronounced acute negative drug effects, including anxiety or panic, are generally undesired. Negative effects may either occur at very high doses as suggested by a dose-response study that used LSD (Holze et al. 2021b) or be less dependent on the dose within the dose range that is currently used as indicated by an analysis of many studies that used psilocybin (Hirschfeld and Schmidt 2021).

The types of experiences that are induced by psychoactive substances depend on the specific substance and dose that are used. Different groups of substances have relatively similar subjective effects within the group, but clear distinctions manifest between groups, largely independent of the doses that are used. Specifically, effects of MDMA and MDMA-like entactogens or empathogens are distinct from classic serotonergic psychedelics, such as LSD, psilocybin, and mescaline, and also distinct from pure stimulants, such as D-amphetamine and methamphetamine (Holze et al. 2020). Already at relatively moderate doses, classic psychedelics clearly induce stronger alterations of mind, including greater ego-dissolution and perceptual alterations (e.g., changes in the perception of time, space, and the self) than relatively high doses of MDMA or pure stimulants. Classic psychedelics are also more likely to induce anxiety than MDMA (Holze et al. 2020), whereas MDMA is more likely than classic psychedelics to produce a state of predominant well-being, relaxation, talkativeness, and feelings of love and trust in the absence of fear ((Holze et al. 2020; Hysek et al. 2014); Fig. 1).

In addition to the substance that is used, the dose strength of the substance clearly alters both the intensity and quality of the acute experience, particularly in the case of classic psychedelics. For example, dose-acute response characteristics have been well-studied for LSD in healthy subjects (Holze et al. 2021b). Healthy participants

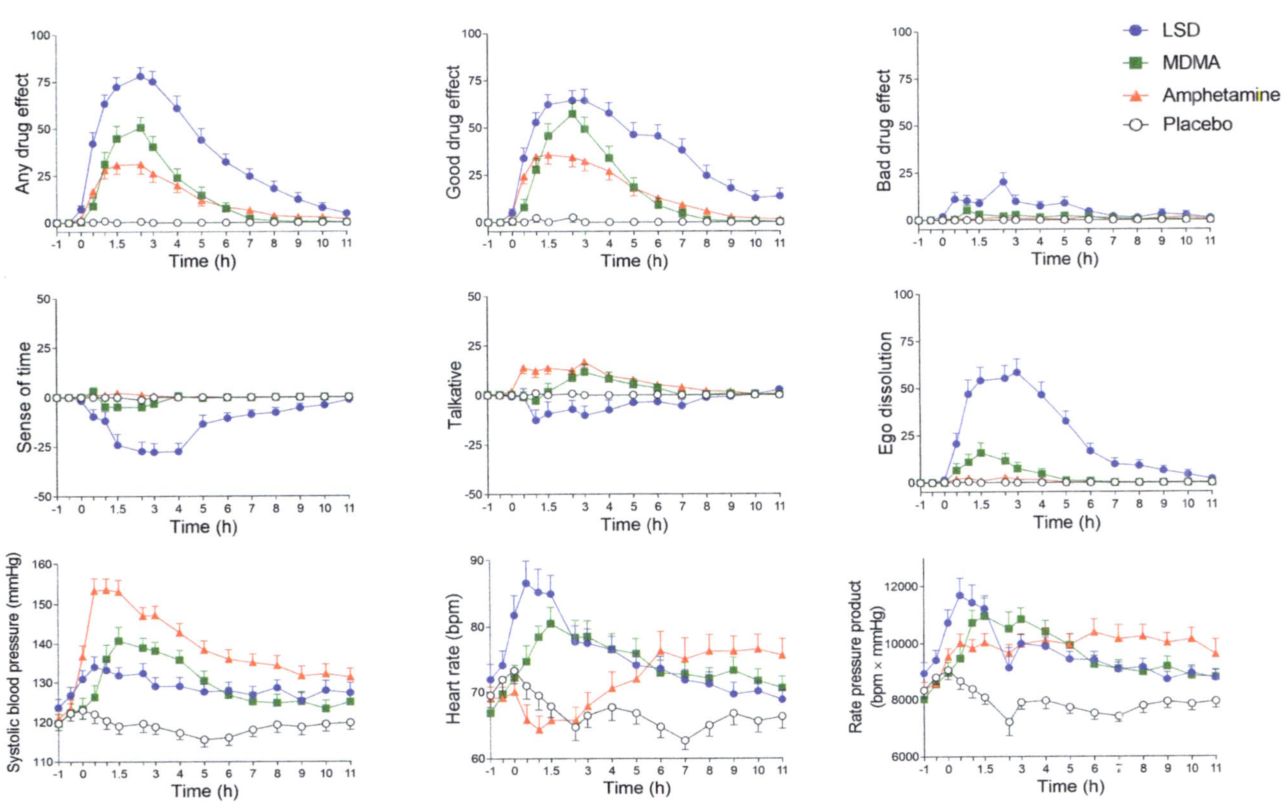

Fig. 1 Subjective and autonomic effects of lysergic acid diethylamide (LSD; 100 μg LSD base), 3,4,-methylenedioxymethamphetamine (MDMA; 125 mg MDMA hydrochloride), amphetamine (40 mg D-amphetamine sulfate), and corresponding placebo. The study used a double-blind within-subjects design. Doses of all active substances were equivalent with regard to autonomic stimulation reflected by the rate pressure product (RPP). The psychedelic LSD produced significantly greater ratings of "any drug effect," "good drug effect," "bad drug effect," and particularly "ego-dissolution" compared with the entactogen/

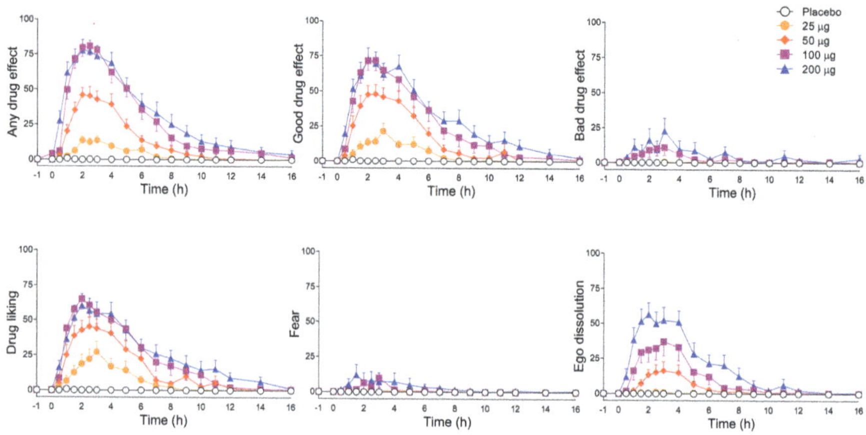

Fig. 2 Acute dose-dependent subjective effects of lysergic acid diethylamide (LSD). LSD (25–200 μg) or placebo was administered at $t = 0$ h in 16 healthy participants in a counter-balanced order using a double-blind cross-over study design (Holze et al. 2021b). LSD dose-dependently induced good drug effects, with a maximum effect at the 100 μg dose. The 200 μg dose of LSD did not further increase good drug effects or drug liking compared with the 100 μg dose, but it further increased ego-dissolution compared with the 100 μg dose. The data are expressed as the mean ± SEM. The figure was modified from Holze et al. (2021b). Table 1 shows corresponding maximal effects and effect duration values for "any drug effects"

were administered different single doses of 25, 50, 100, and 200 μg LSD (LSD base doses) or placebo on different treatment days using a random-order, counter-balanced, within-subjects design in a controlled setting. As expected, LSD produced subjective effects that increased in intensity with increasing doses. The dose-dependence of the LSD response is shown in Fig. 2. Typically, the relationship between the dose and effect of a drug is not linear but rather according to a maximum effect curve, in which subjective effects were minimal at the 25 μg dose. At the 50 μg dose, LSD produced approximately half-maximal effects that were overall positive, with high ratings of drug liking and no negative effects. However, ratings of psychedelic-typical ego-dissolution only amounted to approximately 20% of the scale maximum. At the 100 μg dose, LSD produced greater good drug effect ratings and only minimally greater drug liking than the 50 μg dose, but ratings of ego-dissolution further increased to approximately 40% of the scale maximum. Finally, the full 200 μg dose induced substantial ego-dissolution, with ratings in the range of 60% over several hours. However, good drug effects and drug liking

---

Fig. 1 (continued) empathogen MDMA and pure stimulant D-amphetamine. MDMA and D-amphetamine increased ratings of "talkative" compared with LSD. Only LSD altered "sense of time" compared with placebo. MDMA produced moderately greater ratings of "any drug effect," "good drug effect," and "ego-dissolution" compared with D-amphetamine. The data are expressed as mean ± SEM in 28 healthy volunteers. The figure was modified from Holze et al. (2020)

Table 1 Subjective effect intensity and duration after different doses of LSD

Effect	25 μg LSD	50 μg LSD	100 μg LSD	200 μg LSD
Maximal effect (%)	17 ± 11 (0–37)	46 ± 21 (7–86)	76 ± 21 (13–97)	76 ± 26 (7–98)
Time to maximal effect (h)	2.8 ± 0.8 (1.8–4.9)	2.4 ± 0.6 (1.6–3.3)	2.6 ± 1.1 (1.3–6.5)	2.2 ± 0.9 (1.3–5.1)
Time to onset (h)	1.0 ± 0.4 (0.6–1.8)	0.7 ± 0.2 (0.4–1.2)	0.6 ± 0.2 (0.4–0.9)	0.4 ± 0.2 (0.1–0.9)
Time to offset (h)	7.7 ± 2.1 (5.0–12.6)	8.1 ± 2.3 (4.9–13)	8.9 ± 3.0 (5.8–18)	11 ± 4.9 (6.3–22)
Effect duration (h)	6.7 ± 2.0 (4.3–12)	7.4 ± 2.3 (4.4–12)	8.3 ± 2.9 (5.0–18)	11 ± 5.0 (5.6–22)

Parameters are for "any drug effects" as predicted by pharmacokinetic-pharmacodynamic modeling. The threshold to determine times to onset and offset was set individually at 10% of the individual maximal response. Values are mean ± SD (range). Taken from Holze et al. (2021a, b)

Table 2 Dose comparison of different substances

	LSD[a] [μg]	Psilocybin [mg]	Mescaline[b] [mg]	DMT[c] [mg i.v. bolus]	MDMA[d] [mg]
Subthreshold or microdoses	<10	<2.5	<75	NA	NA
Low dose/minidose	20–50	5–10	100–200	15	25–50
Intermediate/good effect dose	100	20	500	25	125–200
High/ego-dissolution dose	200	30–40	1,000	30	No

Doses are considered approximately equivalent in terms of peak or overall response
[a]Dose given for LSD base. 100 μg of LSD base = 146 μg of LSD 1:1 tartrate
[b]Dose given for mescaline hydrochloride
[c]Dose given for DMT fumarate
[d]Dose given for MDMA hydrochloride. All doses are for oral administration in humans except for DMT as indicated

ratings were similar to the 100 μg LSD dose. Additionally, ratings of bad drug effects and fear, although low on average, increased in several participants. Altogether, the LSD dose-response curve showed a ceiling effect for subjective good drug effects, whereas ego-dissolution and anxiety increased further above 100 μg LSD base. Accordingly, a dose of 50–100 μg LSD base could be termed a good drug effect dose, producing a substantial part of the psychedelic experience in most subjects. A dose of 200 μg could be called an ego-dissolution dose, producing the full psychedelic experience in most subjects, including the overwhelming and existential experience of ego-dissolution but also being associated with a greater risk of producing anxiety and thus a bad trip. Additionally, higher doses produced longer acute effects as illustrated in Fig. 2 and Table 1. Importantly, these dose relationships were observed in healthy research subjects who at least partially had previous LSD experience in a controlled laboratory setting. These effects may be different in patients with psychiatric disorders or when LSD is used in an uncontrolled setting.

Few studies have validly compared different doses of psychedelic substances with each other to provide information on relative potency or efficacy. Table 2 presents an overview of different doses and relative potencies of the most frequently used substances. A 100 μg dose of LSD base corresponds to a 146 μg dose of LSD 1:1 tartrate or 123–133 μg dose of LSD 2:1 tartrate (without/with methanolate). A 10 mg dose of psilocybin corresponds to an 11 mg dose of psilocybin dehydrate. A 500 mg dose of mescaline hydrochloride corresponds to a 616 mg dose of mescaline sulfate (data from (Kuypers et al. 2019; Gallimore and Strassman 2016), and unpublished studies).

4 Personalized Dosing and Pharmacogenetics

Therapeutic effects of medications vary, and doses often need to be titrated individually. This is not different with psychoactive substances with regard to their subjective acute effects and potential longer-term therapeutic responses. Specifically, set (mental state) and setting (environment) influence the acute effects of psychedelics. Several studies have explored various factors that moderate the response to psychedelics and MDMA in a controlled setting (Studerus et al. 2012, 2021; Hirschfeld and Schmidt 2021). The most important determinant of the acute response to a psychoactive substance is clearly the dose of the substance or plasma concentration of the active substance (Holze et al. 2021a, b; Studerus et al. 2012, 2021; Hirschfeld and Schmidt 2021). Additionally, the substance concentration-time curve is closely linked to the effect-time curve during the acute effects of LSD (Holze et al. 2021a, b; Dolder et al. 2017). This means that an acute alteration of the mind is present as long as the substance is present in the body, and the intensity of subjective alterations changes with the plasma concentration within a given person. However, when substance concentrations are compared between individuals, the association between the substance concentration and subjective effect is weaker, and additional set (mental state) and setting (environment) variables are known to be at least partially relevant (Holze et al. 2021a, b; Studerus et al. 2012, 2021). For example, in a study of predictor variables of the acute response to psilocybin in healthy subjects, a few non-pharmacological variables contributed to its effects (Studerus et al. 2012). A high personality trait of absorption (i.e., openness to experiences and vivid imagery/fantasy) was positively associated with more pleasant and generally greater psilocybin effects, including mystical-type experiences and synesthesia (Studerus et al. 2012). Younger age was also associated with greater negative reactions and anxiety in response to psilocybin (Studerus et al. 2012). In contrast, older subjects reported less fear of loss of control in response to psilocybin (Studerus et al. 2012). Hallucinogen-naive subjects tended to report overall stronger psilocybin effects (Studerus et al. 2012). Subjects who sometimes smoked cannabis (i.e., more than once monthly) experienced more pleasurable effects and exhibited a trend toward less anxiety compared with subjects who rarely used cannabis (Studerus et al. 2012). Several studies also explored predictor variables of the response to MDMA in healthy volunteers (Vizeli et al. 2017; Vizeli and Liechti 2017; Schmid et al. 2016; Studerus et al. 2021). The most significant and relevant predictor variable of the response to MDMA was the plasma MDMA concentration, which was itself significantly and strongly dependent on the dose of MDMA per body weight of the person (Studerus et al. 2021). As a result, women exhibited significantly higher plasma MDMA concentrations than men when MDMA was dosed with similar total doses of MDMA and not accounting for the lower average body weight of women compared with men (Schmid et al. 2016; Vizeli and Liechti 2017). The plasma MDMA concentration was also significantly influenced by activity of the cytochrome CYP2D6 enzyme (Schmid et al. 2016; Studerus et al. 2021). CYP2D6 is a polymorphic enzyme and the main MDMA-metabolizing enzyme. Individuals with low CYP2D6 activity (defined genetically or phenotypically) have higher plasma MDMA levels than individuals with normal or high activity (Schmid et al. 2016). CYP2D6 poor metabolizers reach higher plasma concentrations of MDMA, reach these levels faster, and exhibit greater and more rapid increases in good drug effect, feelings of drug high, liking, and stimulation, and greater elevations of mean arterial blood pressure compared with individuals who are extensive or ultra-rapid metabolizers (Schmid et al. 2016). CYP2D6 poor metabolizer status may theoretically be a risk factor for acute cardiovascular toxicity and greater abuse liability. However, the effects on the pharmacokinetics of MDMA and response to MDMA are rather moderate and present only during the first 2 h (Schmid et al. 2016) because MDMA inhibits CYP2D6 (auto-inhibition) and rapidly turns all users into CYP2D6 poor metabolizers for approximately 10–14 days (Yang et al. 2006). LSD is also partly metabolized by CYP2D6 (Luethi et al. 2019) and poor metabolizers may similarly be at risk of greater exposure and effects compared with extensive metabolizers. Psilocin, the active metabolite of the prodrug psilocybin, is inactivated in the body to approximately equal degrees by glucuronidation/conjugation and formation of 4-hydroxy-indolic-3-acetic acid (Kolaczynska et al. 2020) and therefore polymorphisms in CYPs may be less relevant. Published data on the pharmacogenetics of LSD and psilocybin is currently lacking. There are also non-pharmacological predictors of the acute response to MDMA (Studerus et al. 2021). A recent large study reported that sex had no influence on the acute response to MDMA when adjusting the MDMA dose to body weight (Studerus et al. 2021), although an older study reported greater subjective effects in women compared with men (Liechti et al. 2001). Several studies that investigated LSD in healthy participants showed no influence of sex or bodyweight on plasma concentrations or subjective effect strength (Holze et al. 2019, 2021b; Dolder et al. 2015). A post hoc analysis of 288 participants from 10 studies that used fixed and body weight-adjusted doses of psilocybin similarly indicated no advantage of the weight-adjusted dosing of psilocybin over fixed dosing (Garcia-Romeu et al. 2021). These data indicate that fixed doses of LSD and psilocybin can be used, and dose adjustment to body weight is not needed for these psychedelics (Holze et al. 2019; Garcia-Romeu et al. 2021) but could be considered for MDMA (Studerus et al. 2021; Vizeli and Liechti 2017). Personality traits may also moderate the response to MDMA. Personality trait openness to novel experiences correlated with greater ratings of closeness and positively experienced derealization after MDMA administration (Studerus et al. 2021). In contrast, individuals who scored higher in neuroticism and trait anxiety were more likely to experience fear of ego-dissolution and feelings of impaired control and cognition in response to MDMA (Studerus et al. 2021). Whether a person had a few past experiences with MDMA (up to five times) or no experience at all did not relevantly alter the response to MDMA (Studerus et al. 2021). However, extensive past MDMA use may result in moderate tolerance to the acute effects of MDMA (Kirkpatrick et al. 2014). Responses to LSD were comparable, regardless of whether individuals had used it a few times in the past (up to 10 times) or not before administering a dose LSD in a controlled laboratory setting (Schmid et al. 2015; Holze et al. 2021b). Past drug use and pre-experiences with psychedelics only moderately affected psilocybin responses in healthy research subjects (Studerus et al. 2012). Greater well-being ratings immediately before MDMA administration had no significant effect on the response to MDMA (Studerus et al. 2021). However, feelings of anxiety or depression immediately before MDMA intake increased the likelihood of anxious reactions to MDMA. Nonetheless, anxiety scores in response to MDMA are rather small, and only 7% of subjects reported anxiety as an acute adverse effect to 125 mg MDMA (Vizeli and Liechti 2017). Thus, challenging experiences are considered less likely to occur after MDMA than after high-dose psychedelics. Mood before psilocybin administration also moderated the response to psilocybin (Studerus et al. 2012). Emotional excitability before psilocybin administration enhanced all aspects of the psilocybin-induced alteration of mind (Studerus et al. 2012). However, feeling more depressed before drug intake did not lead to more unpleasant experiences (Studerus et al. 2012). Importantly, most of these factors that moderate the response to psychedelics or MDMA have been investigated in healthy research subjects, and more data from clinical patients are needed. Notably, more positive acute experiences after psilocybin administration have been associated with greater long-term therapeutic benefits in patients (Roseman et al. 2017; Griffiths et al. 2016; Garcia-Romeu et al. 2015). Even in healthy subjects, positive acute responses to psychedelics, including LSD, have been shown to be linked to more positive long-term effects on well-being (Schmid and Liechti 2018; Griffiths et al. 2008). Thus, positive acute mood effects may partially mediate long-term beneficial effects or may at least be a measurable marker for such effects. Some individuals, possibly in particular patients with psychiatric disorders, may have no or only very weak subjective effects in response to even very high doses of LSD, but therapeutic effects may still be present (Müller et al. 2020). Finally, in addition to pharmacogenetic variables that influence drug metabolism, co-administered or previously used medications may also influence the response to psychoactive substances. Medications that are known or expected to alter the effects of hallucinogens include tricyclic antidepressants, lithium, serotonin uptake inhibitors, antipsychotics, and monoamine oxidase inhibitors (Johnson et al. 2008; Hintzen and Passie 2010). The concurrent or past use of serotonin or serotonin/norepinephrine uptake inhibitors also alters the acute response to MDMA

(Liechti et al. 2000; Hysek et al. 2012; Farre et al. 2007) and may also influence longer-term responses in patients (Feduccia et al. 2021). Research continues to be conducted to achieve a better understanding of drug–drug interactions with psychedelics and MDMA, particularly the ways in which recent antidepressant use may influence the therapeutic outcome.

5 Microdosing

Microdosing psychedelics has recently gained popularity, referring to the (repeated multiple-dose) use of a sub-perceptual dose of a classic psychedelic, such as LSD and psilocybin (Kuypers et al. 2019). There are many anecdotal reports and books that describe the beneficial effects of microdosing (Anderson et al. 2019; Passie 2019; Polito and Stevenson 2019; Prochazkova et al. 2018). A self-blinding study found greater well-being after microdosing LSD and psilocybin, but empty capsules similarly improved mood (Szigeti et al. 2021). A few placebo-controlled studies investigated effects of single microdose administration of LSD (Family et al. 2020; Yanakieva et al. 2019; Bershad et al. 2019, 2020; Hutten et al. 2020b); (Hasler et al. 2004). Placebo, 5, 10, or 20 μg LSD (possibly LSD base (Yanakieva et al. 2020)) was randomly administered to groups of older adults aged 55–75 years (12/dose group) (Family et al. 2020; Yanakieva et al. 2019). LSD produced a significant acute subjective drug effect at these low doses without relevantly altering perception or subjective concentration (Yanakieva et al. 2019). However, LSD altered the perception of time already at a dose of 10 μg LSD tartrate. LSD was well tolerated and did not impair cognition or balance. LSD did not produce more adverse events than placebo, although LSD increased the frequency of headaches (Family et al. 2020). Another study administered placebo, 6.5, 13 or 26 μg of LSD tartrate (possibly 4.5–17.8 μg LSD base) in a randomized order at 1-week intervals to healthy young adults aged 18–40 years (Bershad et al. 2019). LSD produced dose-dependent subjective effects. Specifically, LSD (13 and 26 μg) significantly increased ratings of "feel drug." LSD (26 μg) increased ratings of "feel high," "like drug," and "dislike drug." LSD increased "experience of unity" and "blissful state" at 13 and 26 μg, and "impaired control and cognition" at 26 μg (Bershad et al. 2019). LSD had no relevant effects on cognition or vital signs at these low doses (Bershad et al. 2019). A functional magnetic resonance imaging study that compared 13 μg LSD tartrate and placebo in healthy subjects showed an acute alteration of brain connectivity in the limbic system that was comparable to higher doses of LSD and psilocybin (Bershad et al. 2020). Another study administered placebo, 5, 10, and 20 μg of LSD base also using a random-order cross-over design in healthy young adults aged 18–40 years (Holze et al. 2021a, b; Hutten et al. 2020a; Ramaekers et al. 2020; Hutten et al. 2020b). The 10 μg dose of LSD significantly increased ratings of "under the influence" and "good drug effect" compared with placebo (Holze et al. 2021a; Hutten et al. 2020b). These effects began an average of 1.1 h after 10 μg LSD administration, peaked at 2.5 h, and ended at 5.1 h (Holze et al. 2021a). The 20 μg dose of LSD significantly increased ratings of "under the influence," "good drug effects," and "bad drug effects" (Holze et al. 2021a). Negative effects also manifested as an increase in confusion and anxiety (20 μg) (Hutten et al. 2020b). In this study, LSD (20 μg) attenuated experimental pain (Ramaekers et al. 2020). LSD also increased circulating levels of brain derived neurotrophic factor (BDNF) at 4 h (5 μg) and at 6 h (5 and 20 μg) after administration (Hutten et al. 2020a) similar to a full 200 μg dose (Holze et al. 2021b). This finding indicates neuroplastic effects of LSD consistent with preclinical studies (Ly et al. 2018) and even at low doses. LSD concentrations dose-proportionally increased at doses as low as 5–20 μg and decreased with a half-life of 3 h (Holze et al. 2021a). Notably, the relatively short half-life of LSD of 3 h indicates that LSD does not accumulate in the body with repeated administration (e.g., during microdosing when small doses of LSD are used repeatedly), even when used at 24-h intervals (Holze et al. 2021a). The plasma concentration-time curve of LSD was consistent with its within-subject effect-time curve as documented with the PK-PD modeling even at very low doses of LSD (Holze et al. 2021a), thus confirming the results with high doses (Holze et al. 2019, 2021b; Dolder et al. 2017). This means that the subjective effects of LSD relatively closely mirror LSD plasma concentrations in healthy subjects. Psychotropic effects of LSD are generally present as long as LSD is present in the body. Accordingly, the long residence time of LSD at the 5-HT_{2A} receptor observed in vitro (Kim et al. 2020; Wacker et al. 2017) is not required to explain the longer duration of action of LSD in vivo compared with psilocybin. The terminal elimination half-life of LSD and unconjugated psilocin are 4 and 2.5 h (Holze et al. 2021b; Kolaczynska et al. 2020), respectively, and very consistent with the effect half-lives and durations of action of the two substances. While LSD concentrations and its effects are closely linked within subjects as evidenced by the good PK-PD model fit, greater variance in the effects of LSD is observed between individuals. However, the variance in plasma concentrations between subjects at a given dose was surprisingly small even when using microdoses of LSD (Holze et al. 2021a), indicated by the coefficients of variation for the C_{max} values of 30–36%. Similar low variability in plasma has previously been reported with the same formulation of LSD base when used at high doses (Holze et al. 2019). In contrast, higher variation was seen with older and less stable formulations that were used in older studies (Dolder et al. 2015, 2017) and would be expected with non-controlled recreational products. This observation indicates that more consistent exposure to the drug is produced with well-defined formulations of LSD as used in some modern research studies (Holze et al. 2019, 2021a) which may then likely result in more consistent and predictable effects compared with past and less well-characterized preparations. Taken together, the available data indicate that the threshold dose for subjective effects of LSD is approximately 10 μg LSD base equivalent (Bershad et al. 2019; Holze et al. 2021a). A 10 μg LSD dose produced small subjective good drug effects, whereas the effects of 5 μg LSD base equivalent were not distinguishable from placebo. A 20 μg LSD base equivalent clearly produced subjective effects, good drug effects, and even unspecified bad drug effects (Fig. 3, Table 3). Additionally, a 20 μg LSD base equivalent produced weak psychedelic-typical alterations of consciousness

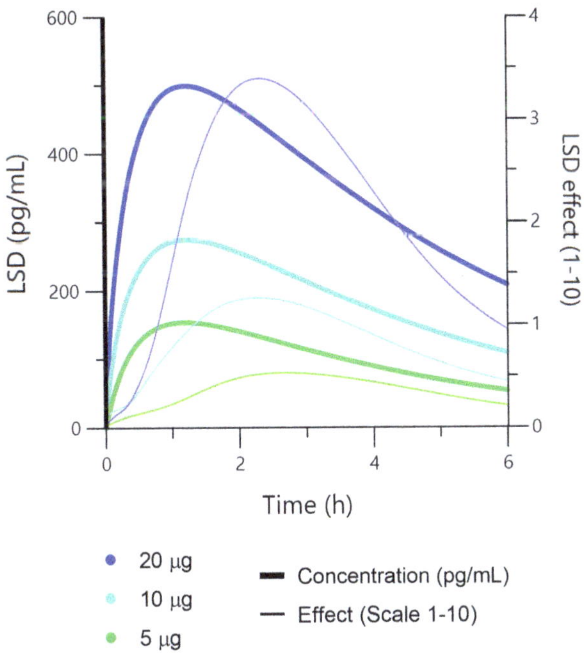

Fig. 3 Acute subjective effects and plasma concentrations of lysergic acid diethylamide (LSD) microdoses predicted using pharmacokinetic-pharmacodynamic modeling. LSD (5–20 μg) or placebo was administered at $t = 0$ h in 23 healthy participants in a counter-balanced order using a double-blind cross-over study design (Holze et al. 2021a). The left y-axis shows dose-proportional LSD plasma concentrations. The right y-axis shows the subjective overall effect, assessed using a visual analog scale for "under the influence," from 1 ("not at all") on the left to 10 ("extreme") on the right. The increase in subjective drug effect was dose-proportional and consistent with the dose-proportional pharmacokinetics, with clearly perceptible effects at 20 μg LSD base. The maximal subjective effect occurred approximately 1 h after maximal plasma LSD concentrations were reached. Table 1 shows corresponding maximal effects and effect duration values for the subjective response

(Bershad et al. 2019; Hutten et al. 2020b). Thus, 20 μg LSD base equivalent can be considered a "minidose" (Bershad et al. 2019; Holze et al. 2021a). While microdosing is reportedly used to improve mental health and enhance cognition (Lea et al. 2019, 2020; Passie 2019; Fadiman and Korb 2019; Cameron et al. 2020), presently, no controlled study data is available on any beneficial effects of repeated administration of LSD microdoses or on its therapeutic use. It is also unclear whether there is tolerance to the effects of repeated administration of low doses of psychedelics. Furthermore, effects of various low doses of psilocybin need to be determined.

Preliminary studies indicate that the effects of psilocybin can be subjectively perceived at doses of 3 mg/70 kg (Hasler et al. 2004). Whether repeated microdosing with psychedelics is safe and exerts beneficial effects remains to be studied.

Table 3 Subjective effect intensity and duration after LSD micro- and minidoses

Effect	5 μg LSD	10 μg LSD	20 μg LSD
Maximal effect (%)	0.57 ± 1.2 (0–4.3)	1.4 ± 1.6 (0–5.4)	3.6 ± 2.0 (0.44–7.5)
Time to maximal effect (h)	1.5 ± 1.2 (0–3.7)	2.5 ± 1.6 (0–6)	2.3 ± 0.84 (1.2–4.6)
Time to onset (h)	0.71 ± 0.58 (0.25–1.6)[a]	1.1 ± 0.52 (0.35–2.3)[b]	0.85 ± 0.41 (0.10–1.7)
Time to offset (h)	5.4 ± 0.57 (4.6–5.9)[a]	5.1 ± 0.94 (2.9–6.0)[b]	5.2 ± 0.62 (4.2–6.0)
Effect duration (h)	4.7 ± 0.96 (3.7–5.6)[a]	4.0 ± 0.97 (2.3–5.6)[b]	4.3 ± 0.57 (3.2–5.0)

Parameters are for the visual analogue scale "under the influence" as predicted by the pharmacokinetic-pharmacodynamic link model. The threshold to determine times to onset was set individually at 25% of the individual maximal response. Values are mean ± SD (range). Adapted from Holze et al. (2021a, b)

[a]For 4 subjects. Ratings of other subjects were too low to define onset and offset

[b]For 13 subjects. Ratings of other subjects were too low to define onset and offset

6 Summary

This chapter discussed different aspects of psychedelic dosing, including pharmaceutical properties, some characteristics of high doses and microdoses, personalized dosing, and personality and other factors, that can influence the response to psychedelics. More pharmacokinetic data are needed, particularly for psilocybin, to allow better comparisons of different studies and preparations. Pharmacogenetic data on psilocybin are also lacking. Measures of acute subjective effects are currently mostly used as a proxy for exposure to psilocybin or other psychedelics in the absence of pharmacokinetic measures (Hirschfeld and Schmidt 2021). Furthermore, more data on the dose equivalence of different psychedelics is needed. Such data could be generated by directly comparing acute effects of psychedelics using well-characterized formulations and doses within the same studies.

Acknowledgement The authors acknowledge the assistance of Michael Arends for proofreading the manuscript.

References

Anderson T, Petranker R, Rosenbaum D, Weissman CR, Dinh-Williams LA, Hui K, Hapke E, Farb NAS (2019) Microdosing psychedelics: personality, mental health, and creativity differences in microdosers. Psychopharmacology (Berl) 236(2):731–740

Barrett FS, Bradstreet MP, Leoutsakos JS, Johnson MW, Griffiths RR (2016) The challenging experience questionnaire: characterization of challenging experiences with psilocybin mushrooms. J Psychopharmacol 30(12):1279–1295

Barrett FS, Johnson MW, Griffiths RR (2017) Neuroticism is associated with challenging experiences with psilocybin mushrooms. Pers Individ Differ 117:155–160

Barrett FS, Preller KH, Herdener M, Janata P, Vollenweider FX (2018) Serotonin 2A receptor signaling underlies LSD-induced alteration of the neural response to dynamic changes in music. Cereb Cortex 28(11):3939–3950

Bershad AK, Schepers ST, Bremmer MP, Lee R, de Wit H (2019) Acute subjective and behavioral effects of microdoses of lysergic acid diethylamide in healthy human volunteers. Biol Psychiatry 86(10):792–800

Bershad AK, Preller KH, Lee R, Keedy S, Wren-Jarvis J, Bremmer MP, de Wit H (2020) Preliminary report on the effects of a low dose of LSD on resting-state amygdala functional connectivity. Biol Psychiatry Cogn Neurosci Neuroimaging 5(4):461–467

Bogenschutz MP, Forcehimes AA, Pommy JA, Wilcox CE, Barbosa PC, Strassman RJ (2015) Psilocybin-assisted treatment for alcohol dependence: a proof-of-concept study. J Psychopharmacol 29(3):289–299

Brown RT, Nicholas CR, Cozzi NV, Gassman MC, Cooper KM, Muller D, Thomas CD, Hetzel SJ, Henriquez KM, Ribaudo AS, Hutson PR (2017) Pharmacokinetics of escalating doses of oral psilocybin in healthy adults. Clin Pharmacokinet 56(12):1543–1554

Cameron LP, Nazarian A, Olson DE (2020) Psychedelic microdosing: prevalence and subjective effects. J Psychoactive Drugs 52(2):113–122

Carhart-Harris RL, Bolstridge M, Rucker J, Day CM, Erritzoe D, Kaelen M, Bloomfield M, Rickard JA, Forbes B, Feilding A, Taylor D, Pilling S, Curran VH, Nutt DJ (2016a) Psilocybin with psychological support for treatment-resistant depression: an open-label feasibility study. Lancet Psychiatry 3:619–627

Carhart-Harris RL, Muthukumaraswamy S, Roseman L, Kaelen M, Droog W, Murphy K, Tagliazucchi E, Schenberg EE, Nest T, Orban C, Leech R, Williams LT, Williams TM, Bolstridge M, Sessa B, McGonigle J, Sereno MI, Nichols D, Hellyer PJ, Hobden P, Evans J, Singh KD, Wise RG, Curran HV, Feilding A, Nutt DJ (2016b) Neural correlates of the LSD experience revealed by multimodal neuroimaging. Proc Natl Acad Sci U S A 113:4853–4858

Davis AK, Barrett FS, May DG, Cosimano MP, Sepeda ND, Johnson MW, Finan PH, Griffiths RR (2020) Effects of psilocybin-assisted therapy on major depressive disorder: a randomized clinical trial. JAMA Psychiat. https://doi.org/10.1001/jamapsychiatry.2020.3285

Dolder PC, Schmid Y, Haschke M, Rentsch KM, Liechti ME (2015) Pharmacokinetics and concentration-effect relationship of oral LSD in humans. Int J Neuropsychopharmacol 19: pyv072

Dolder PC, Schmid Y, Mueller F, Borgwardt S, Liechti ME (2016) LSD acutely impairs fear recognition and enhances emotional empathy and sociality. Neuropsychopharmacology 41:2638–2646

Dolder PC, Schmid Y, Steuer AE, Kraemer T, Rentsch KM, Hammann F, Liechti ME (2017) Pharmacokinetics and pharmacodynamics of lysergic acid diethylamide in healthy subjects. Clin Pharmacokinet 56:1219–1230

Fadiman J, Korb S (2019) Might microdosing psychedelics be safe and beneficial? An initial exploration. J Psychoactive Drugs 51(2):118–122

Family N, Maillet EL, Williams LTJ, Krediet E, Carhart-Harris RL, Williams TM, Nichols CD, Goble DJ, Raz S (2020) Safety, tolerability, pharmacokinetics, and pharmacodynamics of low dose lysergic acid diethylamide (LSD) in healthy older volunteers. Psychopharmacology (Berl) 237(3):841–853

Farre M, Abanades S, Roset PN, Peiro AM, Torrens M, O'Mathuna B, Segura M, de la Torre R (2007) Pharmacological interaction between 3,4-methylenedioxymethamphetamine (ecstasy) and paroxetine: pharmacological effects and pharmacokinetics. J Pharmacol Exp Ther 323 (3):954–962

Feduccia AA, Jerome L, Mithoefer MC, Holland J (2021) Discontinuation of medications classified as reuptake inhibitors affects treatment response of MDMA-assisted psychotherapy. Psychopharmacology (Berl) 238(2):581–588

Gallimore AR, Strassman RJ (2016) A model for the application of target-controlled intravenous infusion for prolonged immersive DMT psychedelic experience. Front Pharmacol 7:211

Garcia-Romeu A, Griffiths RR, Johnson MW (2015) Psilocybin-occasioned mystical experiences in the treatment of tobacco addiction. Curr Drug Abuse Rev 7(3):157–164

Garcia-Romeu A, Barrett FS, Carbonaro TM, Johnson MW, Griffiths RR (2021) Optimal dosing for psilocybin pharmacotherapy: considering weight-adjusted and fixed dosing approaches. J Psychopharmacol. https://doi.org/10.1177/0269881121991822

Gasser P, Kirchner K, Passie T (2015) LSD-assisted psychotherapy for anxiety associated with a life-threatening disease: a qualitative study of acute and sustained subjective effects. J Psychopharmacol 29(1):57–68

Griffiths R, Richards W, Johnson M, McCann U, Jesse R (2008) Mystical-type experiences occasioned by psilocybin mediate the attribution of personal meaning and spiritual significance 14 months later. J Psychopharmacol 22(6):621–632

Griffiths RR, Johnson MW, Richards WA, Richards BD, McCann U, Jesse R (2011) Psilocybin occasioned mystical-type experiences: immediate and persisting dose-related effects. Psychopharmacology (Berl) 218(4):649–665

Griffiths RR, Johnson MW, Carducci MA, Umbricht A, Richards WA, Richards BD, Cosimano MP, Klinedinst MA (2016) Psilocybin produces substantial and sustained decreases in depression and anxiety in patients with life-threatening cancer: a randomized double-blind trial. J Psychopharmacol 30(12):1181–1197

Hasler F, Grimberg U, Benz MA, Huber T, Vollenweider FX (2004) Acute psychological and physiological effects of psilocybin in healthy humans: a double-blind, placebo-controlled dose-effect study. Psychopharmacology (Berl) 172(2):145–156

Hintzen A, Passie T (2010) The pharmacology of LSD: a critical review. Oxford University Press, Oxford

Hirschfeld T, Schmidt TT (2021) Dose-response relationships of psilocybin-induced subjective experiences in humans. J Psychopharmacol. https://doi.org/10.1177/0269881121992676

Holze F, Duthaler U, Vizeli P, Muller F, Borgwardt S, Liechti ME (2019) Pharmacokinetics and subjective effects of a novel oral LSD formulation in healthy subjects. Br J Clin Pharmacol 85:1474–1483

Holze F, Vizeli P, Muller F, Ley L, Duerig R, Varghese N, Eckert A, Borgwardt S, Liechti ME (2020) Distinct acute effects of LSD, MDMA, and D-amphetamine in healthy subjects. Neuropsychopharmacology 45(3):462–471

Holze F, Liechti ME, Hutten N, Mason NL, Dolder PC, Theunissen EL, Duthaler U, Feilding A, Ramaekers JG, Kuypers KPC (2021a) Pharmacokinetics and pharmacodynamics of lysergic acid diethylamide microdoses in healthy participants. Clin Pharmacol Ther 109(3):658–666

Holze F, Vizeli P, Ley L, Muller F, Dolder P, Stocker M, Duthaler U, Varghese N, Eckert A, Borgwardt S, Liechti ME (2021b) Acute dose-dependent effects of lysergic acid diethylamide in a double-blind placebo-controlled study in healthy subjects. Neuropsychopharmacology 46 (3):537–544

Hutten N, Mason NL, Dolder P, Theunissen EL, Holze F, Liechti ME, Varghese N, Eckert A, Feilding A, Ramaekers JG, Kuypers KP (2020a) Low dose LSD acutely increases BDNF blood plasma levels in healthy volunteers. ACS Pharmacol Transl Sci. https://doi.org/10.1021/acsptsci.0c00099

Hutten N, Mason NL, Dolder PC, Theunissen EL, Holze F, Liechti ME, Feilding A, Ramaekers JG, Kuypers KPC (2020b) Mood and cognition after administration of low LSD doses in healthy volunteers: a placebo controlled dose-effect finding study. Eur Neuropsychopharmacol 41:81–91

Hysek CM, Simmler LD, Nicola V, Vischer N, Donzelli M, Krähenbühl S, Grouzmann E, Hoener MC, Liechti ME (2012) Duloxetine inhibits effects of MDMA ("ecstasy") in vitro and in humans in a randomized placebo-controlled laboratory study. PLoS One 7:e36476

Hysek CM, Schmid Y, Simmler LD, Domes G, Heinrichs M, Eisenegger C, Preller KH, Quednow BB, Liechti ME (2014) MDMA enhances emotional empathy and prosocial behavior. Soc Cogn Affect Neurosci 9:1645–1652

Johnson M, Richards W, Griffiths R (2008) Human hallucinogen research: guidelines for safety. J Psychopharmacol 22(6):603–620

Johnson MW, Garcia-Romeu A, Cosimano MP, Griffiths RR (2014) Pilot study of the 5-HT_{2A}R agonist psilocybin in the treatment of tobacco addiction. J Psychopharmacol 28(11):983–992

Kim K, Che T, Panova O, DiBerto JF, Lyu J, Krumm BE, Wacker D, Robertson MJ, Seven AB, Nichols DE, Shoichet BK, Skiniotis G, Roth BL (2020) Structure of a hallucinogen-activated Gq-coupled 5-HT_{2A} serotonin receptor. Cell 182(6):1574–1588 e1519

Kirkpatrick MG, Baggott MJ, Mendelson JE, Galloway GP, Liechti ME, Hysek CM, de Wit H (2014) MDMA effects consistent across laboratories. Psychopharmacology (Berl) 231 (19):3899–3905

Kolaczynska KE, Liechti ME, Duthaler U (2020) Development and validation of an LC-MS/MS method for the bioanalysis of psilocybin's main metabolites, psilocin and 4-hydroxyindole-3-acetic acid, in human plasma. J Chromatogr B Analyt Technol Biomed Life Sci 1164:122486

Kraehenmann R, Pokorny D, Aicher H, Preller KH, Pokorny T, Bosch OG, Seifritz E, Vollenweider FX (2017a) LSD increases primary process thinking via serotonin 2A receptor activation. Front Pharmacol 8:814

Kraehenmann R, Pokorny D, Vollenweider L, Preller KH, Pokorny T, Seifritz E, Vollenweider FX (2017b) Dreamlike effects of LSD on waking imagery in humans depend on serotonin 2A receptor activation. Psychopharmacology (Berl) 234(13):2031–2046

Kuypers KP, Ng L, Erritzoe D, Knudsen GM, Nichols CD, Nichols DE, Pani L, Soula A, Nutt D (2019) Microdosing psychedelics: more questions than answers? An overview and suggestions for future research. J Psychopharmacol 33:1039–1057

Lea T, Amada N, Jungaberle H (2019) Psychedelic microdosing: a subreddit analysis. J Psychoactive Drugs:1–12

Lea T, Amada N, Jungaberle H, Schecke H, Klein M (2020) Microdosing psychedelics: motivations, subjective effects and harm reduction. Int J Drug Policy 75:102600

Liechti ME, Baumann C, Gamma A, Vollenweider FX (2000) Acute psychological effects of 3,4-methylenedioxymethamphetamine (MDMA, "Ecstasy") are attenuated by the serotonin uptake inhibitor citalopram. Neuropsychopharmacology 22(5):513–521

Liechti ME, Gamma A, Vollenweider FX (2001) Gender differences in the subjective effects of MDMA. Psychopharmacology (Berl) 154(2):161–168

Luethi D, Hoener MC, Krahenbuhl S, Liechti ME, Duthaler U (2019) Cytochrome P450 enzymes contribute to the metabolism of LSD to nor-LSD and 2-oxo-3-hydroxy-LSD: implications for clinical LSD use. Biochem Pharmacol 164:129–138

Ly C, Greb AC, Cameron LP, Wong JM, Barragan EV, Wilson PC, Burbach KF, Soltanzadeh Zarandi S, Sood A, Paddy MR, Duim WC, Dennis MY, McAllister AK, Ori-McKenney KM, Gray JA, Olson DE (2018) Psychedelics promote structural and functional neural plasticity. Cell Rep 23(11):3170–3182

Madsen MK, Fisher PM, Burmester D, Dyssegaard A, Stenbaek DS, Kristiansen S, Johansen SS, Lehel S, Linnet K, Svarer C, Erritzoe D, Ozenne B, Knudsen GM (2019) Psychedelic effects of psilocybin correlate with serotonin 2A receptor occupancy and plasma psilocin levels. Neuropsychopharmacology 44(7):1328–1334

Mithoefer MC, Wagner MT, Mithoefer AT, Jerome I, Doblin R (2010) The safety and efficacy of ±3,4-methylenedioxymethamphetamine-assisted psychotherapy in subjects with chronic, treatment-resistant posttraumatic stress disorder: the first randomized controlled pilot study. J Psychopharmacol 25:439–452

Mithoefer MC, Mithoefer AT, Feduccia AA, Jerome L, Wagner M, Wymer J, Holland J, Hamilton S, Yazar-Klosinski B, Emerson A, Doblin R (2018) 3,4-Methylenedioxymethamphetamine (MDMA)-assisted psychotherapy for post-traumatic stress disorder in military veterans, firefighters, and police officers: a randomised, double-blind, dose-response, phase 2 clinical trial. Lancet Psychiatry 5(6):486–497

Mithoefer MC, Feduccia AA, Jerome L, Mithoefer A, Wagner M, Walsh Z, Hamilton S, Yazar-Klosinski B, Emerson A, Doblin R (2019) MDMA-assisted psychotherapy for treatment of

PTSD: study design and rationale for phase 3 trials based on pooled analysis of six phase 2 randomized controlled trials. Psychopharmacology (Berl) 236(9):2735–2745

Mueller F, Lenz C, Dolder PC, Harder S, Schmid Y, Lang UE, Liechti ME, Borgwardt S (2017a) Acute effects of LSD on amygdala activity during processing of fearful stimuli in healthy subjects. Transl Psychiatry 7:e1084

Mueller F, Lenz C, Dolder PC, Lang UE, Schmidt A, Liechti ME, Borgwardt S (2017b) Increased thalamic resting-state connectivity as a core driver of LSD-induced hallucinations. Acta Psychiatr Scand 136:648–657

Müller F, Mühlhauser M, Holze F, Lang UE, Walter M, Liechti ME, Borgwardt S (2020) Treatment of a complex personality disorder using repeated doses of LSD: a case report on significant improvements in the absence of acute drug effects. Front Psych 11:573953

Parrott AC (2014) The potential dangers of using MDMA for psychotherapy. J Psychoactive Drugs 46(1):37–43. https://doi.org/10.1080/02791072.2014.873690

Passie T (2019) The science of microdosing psychedelics. Psychedelic Press, London

Polito V, Stevenson RJ (2019) A systematic study of microdosing psychedelics. PLoS One 14(2): e0211023

Preller KH, Vollenweider FX (2018) Phenomenology, structure, and dynamic of psychedelic states. Curr Top Behav Neurosci 36:221–256

Preller KH, Herdener M, Pokorny T, Planzer A, Kraehenmann R, Stämpfli P, Liechti ME, Seifritz E, Vollenweider FX (2017) The fabric of meaning and subjective effects in LSD-induced states depend on serotonin 2A receptor activation. Curr Biol 27:451–457

Preller KH, Burt JB, Ji JL, Schleifer CH, Adkinson BD, Stampfli P, Seifritz E, Repovs G, Krystal JH, Murray JD, Vollenweider FX, Anticevic A (2018) Changes in global and thalamic brain connectivity in LSD-induced altered states of consciousness are attributable to the 5-HT2A receptor. Elife 7:e35082

Preller KH, Razi A, Zeidman P, Stampfli P, Friston KJ, Vollenweider FX (2019) Effective connectivity changes in LSD-induced altered states of consciousness in humans. Proc Natl Acad Sci U S A 116(7):2743–2748

Prochazkova L, Lippelt DP, Colzato LS, Kuchar M, Sjoerds Z, Hommel B (2018) Exploring the effect of microdosing psychedelics on creativity in an open-label natural setting. Psychopharmacology (Berl) 235(12):3401–3413

Ramaekers JG, Hutten N, Mason NL, Dolder P, Theunissen EL, Holze F, Liechti ME, Feilding A, Kuypers KP (2020) A low dose of lysergic acid diethylamide decreases pain perception in healthy volunteers. J Psychopharmacol. https://doi.org/10.1177/0269881120940937

Roseman L, Nutt DJ, Carhart-Harris RL (2017) Quality of acute psychedelic experience predicts therapeutic efficacy of psilocybin for treatment-resistant depression. Front Pharmacol 8:974

Ross S, Bossis A, Guss J, Agin-Liebes G, Malone T, Cohen B, Mennenga SE, Belser A, Kalliontzi K, Babb J, Su Z, Corby P, Schmidt BL (2016) Rapid and sustained symptom reduction following psilocybin treatment for anxiety and depression in patients with life-threatening cancer: a randomized controlled trial. J Psychopharmacol 30(12):1165–1180

Schmid Y, Liechti ME (2018) Long-lasting subjective effects of LSD in normal subjects. Psychopharmacology (Berl) 235(2):535–545

Schmid Y, Enzler F, Gasser P, Grouzmann E, Preller KH, Vollenweider FX, Brenneisen R, Mueller F, Borgwardt S, Liechti ME (2015) Acute effects of lysergic acid diethylamide in healthy subjects. Biol Psychiatry 78(8):544–553

Schmid Y, Vizeli P, Hysek CM, Prestin K, Meyer ZU, Schwabedissen HE, Liechti ME (2016) CYP2D6 function moderates the pharmacokinetics and pharmacodynamics of 3,4-methylene-dioxymethamphetamine in a controlled study in healthy individuals. Pharmacogenet Genomics 26(8):397–401

Steuer AE, Poetzsch M, Stock L, Eisenbeiss L, Schmid Y, Liechti ME, Kraemer T (2017) Development and validation of an ultra-fast and sensitive microflow liquid chromatography-tandem mass spectrometry (MFLC-MS/MS) method for quantification of LSD and its metabolites in plasma and application to a controlled LSD administration study in humans. Drug Test Anal 9:788–797

Studerus E, Gamma A, Kometer M, Vollenweider FX (2012) Prediction of psilocybin response in healthy volunteers. PLoS One 7(2):e30800

Studerus E, Vizeli P, Harder S, Liechti ME (2021) Prediction of MDMA response in healthy humans: a pooled analysis of placebo-controlled studies. J Psychopharmacol. https://doi.org/10.1177/0269881121998322

Szigeti B, Kartner L, Blemings A, Rosas F, Feilding A, Nutt DJ, Carhart-Harris RL, Erritzoe D (2021) Self-blinding citizen science to explore psychedelic microdosing. Elife 10:e62878

Vizeli P, Liechti ME (2017) Safety pharmacology of acute MDMA administration in healthy subjects. J Psychopharmacol 31(5):576–588

Vizeli P, Schmid Y, Prestin K, Meyer ZU, Schwabedissen HE, Liechti ME (2017) Pharmacogenetics of ecstasy: CYP1A2, CYP2C19, and CYP2B6 polymorphisms moderate pharmacokinetics of MDMA in healthy subjects. Eur Neuropsychopharmacol 27:232–238

Vizeli P, Meyer ZU, Schwabedissen HE, Liechti ME (2019) Role of serotonin transporter and receptor gene variations in the acute effects of MDMA in healthy subjects. ACS Chem Nerosci 10(7):3120–3131

Vollenweider FX, Gamma A, Liechti ME, Huber T (1998) Psychological and cardiovascular effects and short-term sequelae of MDMA ("ecstasy") in MDMA-naive healthy volunteers. Neuropsychopharmacology 19(4):241–251

Vollenweider FX, Gamma A, Liechti M, Huber T (1999) Is a single dose of MDMA harmless? Neuropsychopharmacology 21(4):598–600

Wacker D, Wang S, McCorvy JD, Betz RM, Venkatakrishnan AJ, Levit A, Lansu K, Schools ZL, Che T, Nichols DE, Shoichet BK, Dror RO, Roth BL (2017) Crystal structure of an LSD-bound human serotonin receptor. Cell 168(3):377–389 e312

Yanakieva S, Polychroni N, Family N, Williams LTJ, Luke DP, Terhune DB (2019) The effects of microdose LSD on time perception: a randomised, double-blind, placebo-controlled trial. Psychopharmacology (Berl) 236(4):1159–1170

Yanakieva S, Polychroni N, Family N, Williams LTJ, Luke DP, Terhune DB (2020) Correction to: the effects of microdose LSD on time perception: a randomised, double-blind, placebo-controlled trial. Psychopharmacology (Berl) 237(12):3803

Yang J, Jamei M, Heydari A, Yeo KR, de la Torre R, Farre M, Tucker GT, Rostami-Hodjegan A (2006) Implications of mechanism-based inhibition of CYP2D6 for the pharmacokinetics and toxicity of MDMA. J Psychopharmacol 20(6):842–849

Psychedelic Group Therapy

Peter Gasser

Contents

Abstract Gatherings in groups are a ubiquitous phenomenon throughout human history. This is true for everyday social tasks as well as for healing and spiritual purposes. In psychotherapy, group treatment started soon after developing psychoanalytic treatment procedures. For psychedelic therapy however, individual treatment guided by one or sometimes even two therapists is the most common and widespread treatment model for clinical research and therapy thus far. Since the foundation of the Swiss Medical Society for Psycholytic Therapy (Schweizerische Ärztegesellschaft für psycholytische Therapie, SÄPT) in 1985 in Switzerland, we however had the opportunity to conduct psychedelic group treatment in specific settings, which the following article describes.

Keywords Psychedelics · Psychedelics assisted therapy · Group psychotherapy · Psychedelic therapy · Group therapy · Limited use of scheduled substances

In this article, the terms "psycholytic" and "psychedelic" are used as synonyms. In addition, I chose the term "patient" for the person coming to treatment. It is the common labelling in medicine, lent from Latin and means "the one who has patience", which in fact is necessary for most psychological treatments. I care about both genders including description, although this is not always visible in English. In general, both genders are included.

P. Gasser (✉)
Private Practice, Solothurn, Switzerland

Curr Topics Behav Neurosci (2022) 56: 23–34
https://doi.org/10.1007/7854_2021_268
Published Online: 29 January 2022

1 Introduction

Group therapy is a well-investigated field of psychotherapy (Rutan 1993; Yalom 1995) and psychoanalytic group therapy has been practiced for over one hundred years now. Psychoanalytic groups, humanistic encounter groups (e.g., theme-centred interaction) and self-help groups (e.g., "Alcoholics Anonymous") have shown their usefulness and efficacy over decades (Barkowski et al. 2020; McDermut et al. 2001).

But, does this apply for psychedelic therapy groups as well? What is common to, and what is different from, groups where no altered states of consciousness occur and no substances are used as catalysts for the psychotherapeutic process are given?

When colleagues with different academic backgrounds hear that we conduct psychedelic group therapy in Switzerland, they normally react curious, puzzled, or astonished. Often the first and doubtful question in response to that information is whether that can work at all. They ask: "What about disturbing other participants when emotional or cathartic expression happens?", "How to provide safety for all?", "How can one support individual inner processes in a group setting?" There is a long list of critical and curious questions that must be addressed and of course answered in a scientific or at least satisfying and comprehensible way.

Up until today there is a lack of evidence-based data which can show safety and efficacy of psychedelic group therapy, and thus the reader looking for hard scientific data might be disappointed after having read this chapter.

What I put together in this article is a track record of our experiences from 2016–2020 when we conducted psychedelic group therapy in Switzerland as part of the so-called compassionate use program, i.e. the psychedelics assisted therapy (LSD; MDMA and Psilocybin) based on individual permissions by the Swiss Federal Office for Public Health (Bundesamt für Gesundheit, BAG) (Gasser 2017).

To the best of my knowledge, this is currently the only legal psychedelic group psychotherapy in the world in the sense that the administration of the psychedelic substance happens within the group.

2 Administration of Mind-Altering Substances in Groups

In ancient cultures, the ingestion of mind-altering substances has a long tradition (Eliade 1974; Harner 1990). Indigenous cultures incorporated rituals with the intake of these substances for healing and religious purposes with no strong distinction between the two. We know from anthropologic research that there were several forms of rituals. At times, the healer (shaman, curandero or curandera) was the only person who ingested the mind-altering substance and the sick person with or without his family were soberly present while the healing ritual went on. It is believed that certain levels of consciousness or layers of the cosmology of that society only can be entered by the Shaman (Homan 2011). However, often the community gathered for healing or religious rituals and all participants including the shaman took the sacred substances such as psilocybin, ayahuasca, mescaline and others (Liggenstorfer and Raetsch 1996). It is obvious that such rituals served functions such as community building and community preservation. Group rituals are known in indigenous cultures in North and South America, Siberia and other parts of the world (Eliade 1974).

Recreational group gatherings and self-administration of psychedelic drugs started at times when the substances were not banned yet and continued to happen in underground settings for purposes of the so-called psychonautic exploration of the realms of psychedelic experiences as well as therapeutic and healing purposes. Books like “The Secret Chief” by M. J. Stolaroff (1997) and “Therapy with Substance” by F. Meckel Fischer (2015) show the existence and the structure of such underground group meetings.

Religious groups like the Native American Church, Santo Daime and Uniao do Vegetal have been given permission to practice their rituals with peyote or ayahuasca in Brazil, in the USA and the Netherlands mostly based on religious freedom laws. In Peru the use of ayahuasca is legitimate as a traditional indigenous medicine. Meetings and rituals of these communities happen in groups, since they are part of a common spiritual experience involving the ingestion of a sacred substance, chanting and celebrating the common rituals. Religious rituals are centred around a communion and connection with the divine, while psychotherapy is focused on individual improvement by reduction of suffering from psychological distress, psychosomatic symptoms or resolving personal conflictual situations.

Trope and collaborators (2019) published a systematic review of scientific publications with psychedelic group therapy. This shows that administration of psychedelic substances in a group setting even in western medicine started to a little extent in the 1950s and stopped in the late 1960s like almost all human research and therapy with psychedelics due to the worldwide ban of LSD, Psilocybin and others.

3 Groups in Psychotherapy

Many psychological symptoms and difficulties are born out of dysfunctional relationships, and it is obvious to therapists that these symptoms and difficulties need helpful, good relationships to improve or heal. Grawe et al. (1994) undertook a comprehensive meta-analysis of psychotherapy research showing that the relationship between therapist and patient is the first among five main factors of efficacy in psychotherapy. More recent publications (Wampold 2015) confirm the value of the therapeutic relationship among common beneficial factors for psychotherapy.

Yalom (1995) sees one of the great values of group therapy in “interpersonal learning”, the first of 12 general categories of efficacy factors in group psychotherapy.

However, even though the usefulness and therapeutic value of group therapy and group encounter have been shown in many studies, patients often hesitate to accept a group therapy as their proposed treatment method. Yalom (2005) dedicated a whole novel "The Schopenhauer Cure" to the long lasting ambivalence of the protagonist to participate in a group therapy.

When I introduce the group setting to my patients, their hesitation to join a psychedelic group setting is often emotional, expressed by concerns like "I already have to deal with my own problems, why on earth should I listen to ten other patients having problems themselves?" Other typical questions are "Will that pull me down?", "Isn't it contagious when I learn about the suffering of all the other participants?", "What happens when I have to cry or shout in the presence of others?" And as well the other way round: "What happens to me when I am amidst a deep inner process and someone else starts crying loudly?"

Of course, these questions are completely understandable and I have to find ways to answer them that satisfy the patients so that she or he can accept to participate in a psychedelic group treatment.

The following anecdote is a part of a patient's personal report after her first workshop on Nov 8th, 2018 with MDMA (125 mg): "*… I returned home in the evening* (after the first preparatory meeting), *full of doubt if the decision to engage in this thing* (i.e. the group therapy with MDMA) *was good. The many participants which I don't know and the atmosphere that I experienced as burdensome, all this was discouraging and I wondered if a psycholytic therapy could be counterproductive for me…*". The next day when taking MDMA, she had both pleasant and challenging moments. On the same evening, she had a difficult argument with her partner which kept her awake at night. She described in her report: "*… I wasn't in the mood at all to participate in the upcoming sharing. I felt burdened after the argument with T. and I was tired and depressed. …*". But then, after the sharing in the group: "*… How fortunate that there was a sharing round. To learn what and how the others in the group experienced and to listen to the feedback they gave each other was extremely valuable. I would not have wanted to miss their mutual sharing of experiences, nor my own one. I am proud and grateful that I participated in this workshop*". This report following her first experience of psychedelic group therapy highlights the radical change in how she experienced the group situation, and how her trust was built. She continued the therapy with five more MDMA experiences and being a member of the group was no longer questioned.

The fear of being inhibited by listening to all the problems of others and the threat of being intimidated when showing their personal failures, problems and conflicts within a group very often needs a longer time for providing information and addressing the fears, so that the patient can finally agree to participate. Normally patients don't need as much time as Yalom's protagonist in the novel, who spent a lot of time to withdraw from interpersonal experiences by explaining this with his philosophical rationale.

The first group experiences often are trust building so that the possibilities for interpersonal learning may become evident as a direct experience not for patients but also for trainees of a learning group.

The following is an excerpt from a personal report of a colleague who participated in a SÄPT training group for therapists from 1989–1992:

First Psycholytic training weekend. November, 17th – 19th, 1989. MDMA 125 mg.

Although I dealt with being open to everything that would occur to me and I tried not to have any expectations . . . I was quite afraid when entering the room. There were far too many people and they all seemed to know each other, only I didn't know anybody. I felt excluded and somehow wrong here. . . . I had pictures of religious sects, whose members always seem to be so suspiciously nice and cheerful. I really felt bad and I decided that this would be the last weekend.

It is difficult to describe the next day, because it was an un-describable experience. Suddenly there were no limits, not inside nor outside. There was no more inner chat, no self-controlling ego. There was pure being. I felt a floating calmness within myself . . . and a connectedness with every single person, and together as a whole in that room.

Unsurprisingly, this colleague remained in the group and completed all 3 years of training.

Benefits of psychedelic group therapy and group therapy in general are the learning of empathic contact and prosocial behaviour, the sharing of real life matters with others and learning from each other.

I learned that some members of the ongoing psychedelic group therapy meet in between the group sessions. Unlike for psychoanalytic groups, where such meetings are discouraged and seen as a resistance to bring all the important topics into the therapy process, I mostly see the wish to meet for members of the psychedelic group as a part of the integration process. Having the strong experience of connectedness with other participants and meeting them in ordinary life can be a step of normalizing a far-out experience in a non-ordinary state of consciousness. The psychedelic experience is a journey into inner realms and sometimes into spiritual dimensions of human existence. These deep encounters have to be brought back and translated into everyday life, like Christian Scharfetter stated (Scharfetter 1997): "*I am not so much interested in what kind of spiritual experiences people have, I am more interested in what they have done with them*".

The group provides the opportunity for one of the first steps of integration of the experience, because there are living beings around with whom it is possible to get into contact, to share, to hold, to understand and being understood. However, in my own groups, I discourage too much contact during the acute effect of the substance for the first 5 h approximately. This allows an undisturbed individual process. But after that, a gentle and precautious contact (always asking the other person if it is okay to approach) can be of great help and be deeply relaxing.

4 Psychedelic Group Therapy in Switzerland

In Switzerland, we already have a tradition of conducting therapy with mind-altering drugs in group settings. Two of the founding members of SÄPT were trained in the individual one-on-one treatment setting approach with Stanislav Grof in Czechoslovakia (Juraj Styk) or with Hanscarl Leuner in Germany (Peter Baumann). But they applied a group setting approach immediately in 1988 when they received their special permission for treatments with MDMA or LSD (Jungaberle and Verres

2008). Benz (1989) interviewed the members of SÄPT that worked with MDMA and LSD for his dissertation. In four of six different settings he describes that the therapists worked with psychedelic group therapy. So, the vast majority of the treatments done by SÄPT from 1988 until 1993 (nearly 200 patients in total) were done in groups. They showed surprisingly good outcomes as stated in a follow-up investigation (Gasser 1996). It is to say that the psychedelic group therapy sessions were incorporated for a majority of patients in an ongoing individual talking psychotherapy. On average the patients underwent seven sessions with MDMA and/or LSD within 3 years and during this time 70 individual talking psychotherapy sessions were held.

The 3-year training that SÄPT offered for therapists from 1989 until 1992 with 12 weekend workshops was all done in a group setting.

In addition to the general counterindications for psychedelic treatment such as risk for psychotic decompensation, severe personality disorder, acute psychological and especially suicidal crisis, there are a few more restrictions to psychedelic group therapy for people who are overburdened with the group situation. As part of my individual treatment permissions (so-called compassionate use) for psychedelic treatment, I treat a male patient with Asperger autism. He feels highly overwhelmed when he is around more than one other person and stated that he would not be able to come to such a group. Another patient is visually impaired (secondary blindness after retina impairment). She said that she would not have enough sense of security and orientation when together with so many others in the same room while being under the influence of LSD. During the preparation phase of the treatment, it is to determine whether participation in a group would be a permanent obstacle for a therapeutic process, even harmful. Or a difficult challenge only and worth of being confronted.

Another concern that people often raise is whether such groups tend to be chaotic and non- manageable. I have no reporting about such escalations in the cases I have overseen in a therapeutically oriented and legally regulated setting, i.e. the population of about 200 patients of the publications of Gasser (1996) and Schmid et al. (2020).

However, in an illegal context which is by nature less controlled and where maybe new substances are administered, the risk for incidents in my opinion is higher.

During an underground group therapy workshop in Berlin in 2009, two participants died and two others were several weeks in hospital treatment due to the therapist combining two substances (MDMA and Methcathinone) and applying a 10 times overdosage of MDMA (information from the therapist involved, during a personal conversation).

In Handeloh, Germany, in 2015, a massive overdosage of an unknown new substance happened during an underground group gathering of health practitioners. In a large-scale ambulance operation 27 therapists had to be treated in emergency at the location where the workshop happened.

4.1 Group Therapy in the Current Program for Limited Medical Use of Scheduled Substances

After 2014 the Swiss Federal Office for Public Health (Bundesamt fuer Gesundheit, BAG) has given individual permission for patients to be treated with LSD or MDMA (and from 2020 on also for Psilocybin) given that they already underwent other psychotherapies and/or pharmacological therapies with insufficient outcomes. The term "compassionate use" is lent from oncology where individual treatments with new compounds that are not licensed yet can be individually permitted by the authorities.

When my colleague Peter Oehen (a psychiatrist-psychotherapist and also member of SÄPT) and I started with these so-called compassionate use treatments, we initially treated patients in an individual setting like we did in our pilot studies (Oehen et al. 2013; Gasser et al. 2014). When I first requested permission to treat the patients in groups, the authorities expressed concerns about safety and manageability of difficult situations in group treatment. We agreed that first I would do a small group of three patients and report after the session.

Because that group went well, the BAG agreed to let me continue in groups, allowing myself to reflect on setting, size, frequency and structure of the groups corresponding to my therapeutic reflections and methods.

Schmid et al. (2020) evaluated the data from patients receiving LSD within a psychedelic group therapy and compared it with data from research with healthy volunteers receiving LSD in an individual setting. She wanted to know whether acute effects of LSD differ between healthy volunteers and patients coming to treatment and also between participants taking the substance in a group or in an individual setting. She demonstrated that the acute drug effects in a group setting are quite comparable for categories of the altered states questionnaire (5-ASC) between group therapy with patients and individual setting with healthy volunteers. See Fig. 1 for detailed comparison (Table 1).

Fig. 1 A group session with LSD and MDMA (left) and the next day's sharing round (right) are part of a 3 days' workshop offered in the current research of psychedelic group therapy in Switzerland

Table 1 Effects of lysergic acid diethylamide (LSD) on the first completed five dimensions of Altered States of Consciousness (5D-ASC) scale in patients in the compassionate use group therapy ($n = 11$) and in healthy volunteers ($n = 40$). Overall, LSD produced similar effects in patients in the compassionate use group therapy and healthy volunteers. LSD predominantly increased ratings of OB and VR in all subjects. However, increases in VR were significantly greater in healthy subjects receiving 200 µg LSD compared with patients ($p < 0.05$). Ratings of "audio-visual synaesthesia", "changed meaning of percepts" and "blissful state" were higher after 200 µg LSD compared with patients ($p < 0.05$), but also compared with 100 µg LSD in healthy subjects ($p < 0.01$). (Abbreviations: ASC: Altered States of Consciousness (summation score); OB: Oceanic Boundlessness; AED: Anxious Ego Dissolution; VR: Visionary Restructuring; AA: Auditory Alterations; VIR: Vigilance Reduction)

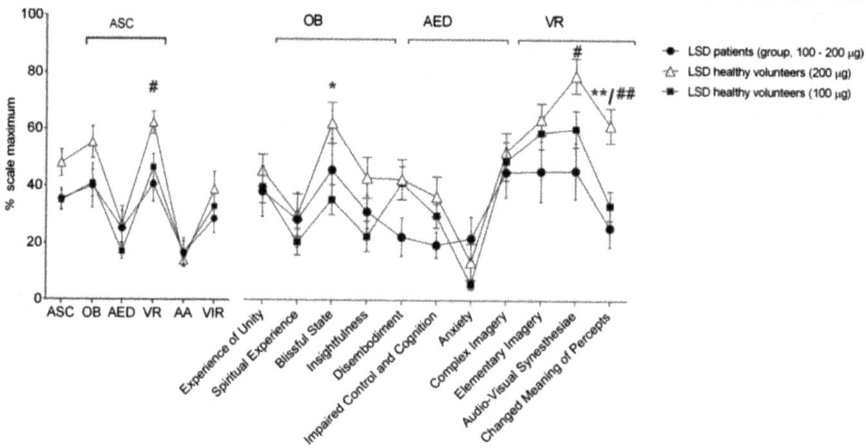

After 2016, Peter Oehen and I decided to do the LSD and MDMA group sessions together and we asked a female co-therapist to work with us in these workshops. This allowed us to always have three therapists and both genders in the group. From 2016 until February 2020, we ran four group workshops per year. During that time we directed groups of 5–13 patients and we included patients using LSD and MDMA in the same group. We encouraged a quiet meditative setting. For this setting the patients with MDMA were instructed to stay calm and mindful even when the main action of MDMA is fading out (approx. 6 h). The action of LSD is longer (8–12 h).

For our own training, Oehen and I were participants in the SÄPT training program from 1989 to 1992. In this context we appreciated workshops that lasted 3 days, with the day of the drug experience on the second day. That allows to have completely separate time slots for preparation and integration processes.

For our compassionate use psychedelic group therapy in Switzerland we usually met on Wednesday evening at 7:30 pm (day 1). For about 2½ h we did basic mindfulness or breathing exercises, nonverbal contact exercises with other participants or guided imagery. Then we sat in a circle and all participants shared thoughts with the group regarding their current personal situation as well as their fears and expectations concerning the next day. Questions about the following day, dosing, setting, etc. were answered and finally they listened to one or two pieces of music in a relaxed way laying on the mat the same way they would do the next day. Around 10 pm they were released and went home or to the hotel.

On Thursday (day 2), we met at 9:30 am. Everyone found a place where she or he would feel comfortable for the whole day, and oral intake of the substance happened at 10 am approximately. We administered LSD in doses varying from 100 μg to 200 μg with very few exceptions from that dosing (when there was a lack of effect with 200 μg) and MDMA as a standard dosage of 125 mg. Until now, we had no treatments with Psilocybin for limited medical use in Switzerland because this substance was not available. This will change in 2021, so we will be able to offer treatments with Psilocybin as well.

Every participant has approval for one specific substance only. It is possible to replace one substance with another over the course of the treatment (e.g. starting with MDMA for two or three sessions to process traumatic experiences and to allow to feel confident with oneself and the group and then changing to LSD for existential issues). This requires a further report to BAG explaining the reason for the change.

On the day of the session with the substance, we varied between music and a quiet meditative atmosphere, with about one-third of music and two-thirds of silence. Participants did not wear headphones or eyeshades, so they listened to the music from the hi-fi system and regulated visible contact to the outer world by opening or closing their eyes. The therapists went to the mats where the patients were asked how their actual state was and offered a guided process, short talks or touch, such as a gentle hold of the hand or putting the hand on the shoulder (i.e. soothing, caring and reassuring touch but no structured body work to induce specific processes) and guiding them through their processes of anxiety, psychic pain, grief, anger, etc. When strong emotional processes appeared, longer phases of close presence or physical contact were necessary.

For every workshop, we alternated the role of the leading therapist who moderated the sharing rounds, played the selected music, and sometimes gave advice to the whole group (like bringing attention inwards again, connecting with breath, etc.).

During the first few hours of the drug action (usually until 3 pm approximately) there was not (and should not be) much interaction between the patients. Extended interaction would disturb the inner process in the opening and plateau phase of the experience.

Around 4 pm, the participants with MDMA slowly came back to a more everyday-like state of consciousness. We advised them to remain silent, with the main attention directed inwards and to move cautiously in the room. When approaching other people, they should ask them if it feels right to be in touch for a short while. In the last hours of the experience, to be connected to real and present persons can be of help in integrating and transforming the spiritual and/or regressive experiences towards everyday behaviour.

Around 6 pm, we had a picnic in the room while sitting on the floor. At 7 pm, patients were allowed to leave the room and be brought home or to their hotel (no one is allowed to drive on that evening). Alternatively, they could choose to stay in the room for another 2–3 h and relax and leave then. Often one of the participants played some music, some people danced smoothly or laid on the mat talking or relaxing.

On Friday (day 3), we met at 9:30 am for sharing in the group. Everyone talked in detail about their experiences of the previous day. This was an important part of the workshop. We called it an integration step, i.e. participants had to find words for the rich, overwhelming, associative, often ineffable, spiritual or psychodynamic processes they experienced. Integration is not only recounting and reporting to the others in the group, but also – and maybe even more so – a step towards understanding and incorporating the drug session. There was also an important group dynamic that usually showed up on this third day, when participants could look to others who either had similar difficulties or experienced a certain situation or music in a different way. They also gave feedback or received such from other participants.

At the end of that sharing, around 12.30 pm, all patients were given advice to treat themselves carefully over the course of the next days, since they were still open in perception and emotion and thus vulnerable. They were asked to write a personal report of the workshop on the same or the next day and to bring this to the next individual psychotherapy session usually happening the following week.

The scheduling for the next group workshop was done individually depending on the therapeutic process. Most of the patients did three or four experiences with psychedelics during the year.

In 2020 we had to interrupt the group treatment due to COVID-19 pandemics, also Peter Oehen retired by the end of that year. I have planned to continue the group treatment as soon as the pandemic restrictions allow, within the same structure as described here.

5 Discussion and Conclusion

Administration of mind-altering substances in groups has a long tradition in human history (Furst 1972).

Underground group therapy, i.e. illegal consumption of scheduled substances in groups is happening to some extent in many places of the world (Stolaroff 1997; Meckel Fischer 2015).

Group therapy or group learning in psychological treatment or support settings have also proven their usefulness over the last decades (McDermut et al. 2001; Barkowski et al. 2020).

The research of the risks and benefits of psychedelic group therapy has hardly started in recent times and scientific data that has been published until now is from studies done in the 1950s and 1960s (Trope et al. 2019).

This article describes and summarizes experiences and anecdotal reports of group therapy conducted by myself and colleagues in Switzerland. These constitute mostly positive findings and encouraging reports about psychedelic group therapy. However, their level in terms of scientific evidence is still poor.

The program for restricted medical use of psychedelics in Switzerland, based on individual approval of treatments is a precious possibility to treat individually, i.e. outside standardized research protocols. It gives hope and perspective to severely ill people in the sense of "compassionate use" treatments.

The Swiss authorities are open enough to give permission to different kinds of psychological distress and psychiatric diseases. Thus, we can get preliminary data and gather experiences with problems that are poorly researched yet, e.g. obsessive compulsive disorder, cluster headache, gambling, etc. or treatment of PTSD (Posttraumatic Stress Disorder) with LSD or establishing a group therapy setting to allow to test the usefulness of psychedelic group therapy.

Another consideration is the economic aspect of psychedelic group therapy. It is obvious that a treatment that requires personal guidance and surveillance by one or even two therapists for at least 6 h (Psilocybin, MDMA) or at least 9 h (LSD) will be an expensive treatment. A group setting allows a better cost effectiveness ratio since the therapists' fees can be distributed among multiple patients. This advantage of course will be of importance only after having shown scientifically that this treatment setting is safe and effective.

In order to learn more about the safety and efficacy of psychedelic group therapy, we have to start with clinical research based on today's methodology in order to learn more about safety and efficacy of this setting.

For now, the placebo-controlled randomized trial is the only way to develop "scheduled narcotics with no medical and scientific value" (this is the legal term they are labelled, although they are not narcotic at all in the way they act) into prescribable medicine. This is the golden standard of drug research. However, we already know that the placebo control condition is a major problem when doing research with highly psychoactive drugs. It is very probable that there will be unblinding for the patient as well as for the therapist. This is highlighted by the pilot study with LSD-assisted therapy (Gasser et al. 2014) in which the patients and therapists had to guess after the session whether the participant was taking 200 μg LSD or the active placebo 20 μg LSD. The patients' guesses were correct in 23 of 24 sessions and the therapists were correct in 24 of 24 sessions.

To look for alternatives to the placebo-controlled trial has another, even more important aspect. The evaluation of risk and benefit of psychedelic group therapy will be psychotherapy research with the methods and standards of psychotherapy as well as drug research because it is really both not just one or the other. Sound methodological approaches still need to be developed to reveal the value and the difficulties of psychedelic group therapy.

References

Barkowski S, Schwartze D, Strauss B, Burlingame GM, Rosendahl J (2020) Efficacy of group psychotherapy for anxiety disorders: a systematic review and meta-analysis. Psychother Res 30 (8):965–982. https://doi.org/10.1080/10503307.2020.1729440

Benz E (1989) Halluzinogen-unterstützte Psychotherapie. Erhebung bei der Schweizerischen Ärztegesellschaft für Psycholytische Therapie. Universität Zürich. Dissertation

Eliade M (1974) Schamanismus und archaische Ekstasetechnik. Suhrkamp, Frankfurt

Furst PT (1972) Flesh of the gods: the ritual use of hallucinogens. Praeger Publishers, New York

Gasser P (1996) Die Psycholytische Therapie in der Schweiz von 1988 – 1993. Schweiz Arch Neurol Psych 147:59–65

Gasser P (2017) Research update: psychedelic group therapy in Switzerland. MAPS Bulletin 27 (1):28–29

Gasser P, Holstein D, Michel Y, Doblin R, Yazar-Klosinski B, Passie T, Brenneisen R (2014) Safety and efficacy of lysergic acid diethylamide-assisted psychotherapy for anxiety associated with life-threatening diseases. Nerv Ment Dis 202(7):513–520. https://doi.org/10.1097/NMD.0000000000000113

Grawe K, Donati R, Bernauer F (1994) Psychotherapie im Wandel. Von der Konfession zur Profession. Hogrefe, Göttingen

Harner M (1990) The way of the shaman. Harper One, San Francisco

Homan JE (2011) Charlatans, seekers, and shamans: the ayahuasca boom in western Peruvian Amazonia. Master thesis, University of Kansas

Jungaberle H, Verres R (2008) Regeln und Standards in der substanzunterstützten Psychotherapie (SPT). In: Jungaberle H, Gasser P, Weinhold J, Verres R (eds) Therapie mit Psychoaktiven Substanzen. Praxis und Kritik der Psychotherapie mit LSD, Psilocybin und MDMA. Huber, Bern, pp 41–109

Liggenstorfer R, Raetsch C (eds) (1996) Maria Sabina, Botin der heiligen Pilze. Nachtschatten Verlag, Solothurn

McDermut W, Miller IW, Brown RA (2001) The efficacy of group psychotherapy for depression: a meta-analysis and review of the empirical research. In: Database of abstracts of reviews of effects (DARE): quality-assessed reviews [internet]. Centre for Reviews and Dissemination, York, 1995. https://www.ncbi.nlm.nih.gov/books/NBK68475/

Meckel Fischer F (2015) Therapy with substance. Muswell Hill Press, London

Oehen P, Traber R, Widmer V, Schnyder U (2013) A randomized, controlled pilot study of MDMA (±3,4-Methylenedioxymethamphetamine)-assisted psychotherapy for treatment of resistant, chronic post-traumatic stress disorder (PTSD). Psychopharmacology 27(1):40–52. https://doi.org/10.1177/0269881112464827

Rutan JS (1993) Psychoanalytic group psychotherapy. In: Kaplan HI, Sadock BJ (eds) Comprehensive group psychotherapy. Lippincott Williams and Wilkins, Philadelphia, pp 138–146

Scharfetter C (1997) Der spirituelle Weg und seine Gefahren. Enke, Stuttgart

Schmid Y, Gasser P, Oehen P, Liechti ME (2020) Acute subjective effects in LSD- and MDMA-assisted psychotherapy. J Psychopharmacol 35:362–374. https://doi.org/10.1177/0269881120959604

Stolaroff MJ (1997) The secret chief: conversations with a pioneer of the underground psychedelic therapy movement. Multidisciplinary Association for Psychedelic Studies (MAPS), Charlotte

Trope A, Anderson BT, Hooker AR, Glick G, Stauffer C, Woolley JD (2019) Psychedelic-assisted group therapy: a systematic review. J Psychoactive Drugs 51(2):174–188. https://doi.org/10.1080/02791072.2019.1593559

Wampold BE (2015) How important are the common factors in psychotherapy? An update. World Psychiatry 14(3):270–277. https://doi.org/10.1002/wps.20238

Yalom ID (1995) The theory and practice of group therapy. Basic Books, New York. German edition: Yalom ID (1996) Theorie und Praxis der Gruppenpsychotherapie. Klett-Cotta, Stuttgart

Yalom ID (2005) The Schopenhauer cure. Harper and Collins, New York

Effects of Setting on Psychedelic Experiences, Therapies, and Outcomes: A Rapid Scoping Review of the Literature

Tasha L. Golden, Susan Magsamen, Clara C. Sandu, Shuyang Lin, Grace Marie Roebuck, Kathy M. Shi, and Frederick S. Barrett

Contents

T. L. Golden (✉), S. Magsamen, C. C. Sandu, S. Lin, and G. M. Roebuck
International Arts + Mind Lab, Center for Applied Neuroaesthetics, Johns Hopkins University School of Medicine, Baltimore, MD, USA
e-mail: tasha.golden@artsandmindlab.org

K. M. Shi and F. S. Barrett (✉)
Department of Psychiatry and Behavioral Sciences, Center for Psychedelic and Consciousness Research, Johns Hopkins University School of Medicine, Baltimore, MD, USA
e-mail: fbarrett@jhmi.edu

Curr Topics Behav Neurosci (2022) 56: 35–70
https://doi.org/10.1007/7854_2021_298
Published Online: 10 February 2022

Abstract The health and well-being impacts of art and aesthetic experiences have been rigorously studied by a range of disciplines, including cognitive neuroscience, psychiatry, public health, and translational clinical research. These experiences, encompassed in the concepts of *set* and *setting*, have long been claimed to be pivotal in determining the acute and enduring effects of psychedelic experiences. Responding to the field's longstanding emphasis on the role and value of setting, a rapid scoping review was undertaken to identify the extent to which effects of setting and aesthetics on psychedelic experiences and therapies have been explicitly studied. It offers an analysis of the strengths and limitations of the extant literature and discusses evidentiary gaps as well as evidentiary opportunities for the field. The 43 included studies indicate apparent consensus regarding the importance of setting in psychedelic therapies, as well as consistent interest in theorizing about these effects. However, this consensus has yet to generate consistent, prospective, rigorous tests of setting and its complexities. As a result, the field continues to lack understanding or agreement regarding the effects of various specific elements of setting, the mechanisms by which they affect outcomes, for whom these effects occur, under what circumstances, given what conditions, and other critical factors. Further studies of setting and aesthetics in the context of psychedelic therapies are likely to not only improve these therapies and their delivery, but also inform considerations of setting and aesthetics for non-psychedelic interventions.

Keywords Aesthetics · Arts · Arts and health · Ceremonial setting · Classic psychedelics · Enriched environment · Music · Naturalistic setting · Neuroaesthetics · Plant based medicine · Psychedelic therapy · Psychedelics · Ritual setting · Scoping review · Setting · Sociocultural setting

1 Introduction

1.1 Setting, Aesthetics, and Health

For millennia, arts and aesthetic experiences have played key roles in humans' physical, social, and spiritual health. Storytelling, music making, and other arts-based rituals were critical in supporting social relationships, group cohesion, and bonding among Neanderthal populations (Boyd 2018; Fancourt 2017, p. 6). In 25,000 BC, carved figurines such as the Paleolithic "Venus of Willendorf" promoted fertility (Kuiper 2018), and cave paintings supported shamanic rituals and provided hunting insights or "hunting magic" (Fancourt 2017; Powell 2019). Thousands of years later, around 1,000 BC, the Atharvaveda medical text encouraged people to "enjoy soft sounds, pleasant sights, and tastes" after eating, to aid digestion (Fancourt 2017, p. 7). Greek philosophers saw the natural link between art and everyday life as a cure for body and mind (Mason 2016); Pythagoras and Plato were among the first to institutionalize the role of the arts in health, prescribing music and harmonic melodies as medicine to heal or purify the mind, body, and soul (Mason 2016, p. 8). Over time, explorations, uses, and studies of art's effects on health have continued to develop.

In the last two decades, impacts of art and aesthetic experiences on health and well-being have been rigorously studied by a range of disciplines, including cognitive neuroscience, psychiatry, public health, and translational clinical research (Golden et al. 2021; Fancourt and Finn 2019; Boyce et al. 2018; Staricoff 2004). A full review of this literature is beyond the scope of this chapter; however, as brief examples, researchers have found that exposure to visual arts and art-making can decrease negative emotions such as depression, grief, and fatigue, while improving social connections and self-worth among trauma, chronic illness, and cancer patients (Stuckey and Nobel 2010; Elimimian et al. 2020; Kaimal et al. 2016). Playing patient-selected music has been shown to reduce pain and ease anxiety (Kilic et al. 2015); in a separate study, listening to music helped reduce the amount of time that patients in acute care settings were maintained on mechanical ventilation (Liang et al. 2016). Numerous studies have also demonstrated the effects of aesthetics and the built environment on health (WHO 2021; Yin et al. 2020; Kondo et al. 2018; Lankston et al. 2010), leading hospitals, clinics, and educational settings to make increased efforts to consider and incorporate art, sound, biophilic design, and green spaces (Lankston et al. 2010; Iyendo 2016; Scott 2020; Franklin 2018; Lambert et al. 2017). In 2019, the World Health Organization (WHO) published an extensive scoping review regarding roles of art in addressing individual, community, and population health (Fancourt and Finn 2019), confirming a growing global interest in understanding and applying art and aesthetics for mental and physical well-being. In short, humans have long understood the value of setting and aesthetics for supporting health and well-being; as the scientific evidence base develops, this value is increasingly translating into intentional shifts in health care, educational, and community settings.

1.2 Applications in Psychedelic Experiences and Psychedelic Therapies

Echoing the above findings, the *setting* of psychedelic experiences and therapies has long been claimed to be pivotal in determining outcomes. Traditional shamanistic rituals to enhance the healing experience of psychedelic medicines have been reported to contain such setting manipulations as "icaros (ritual songs), whistles, smoke blowing, and sucking" (Hartogsohn 2017; Beyer 2010; Dobkin de Rios 1975; Pinch and Bijker 1984). More recently, in the midst of the boom of psychedelic research in psychiatry between 1950 and the mid-1960s (Grinspoon 1981), the importance of context was captured by the phrase "set and setting." First coined by Timothy Leary in 1961 (Hartogsohn 2017; Leary 1961), set and setting refers to the mind*set* – the mental state or expectancy with which someone comes into psychedelic therapy, and the *setting*: the environment or context in which that person has the psychedelic experience (Metzner and Leary 1967). Over time, the role of "set and setting" has become so fundamental to psychedelic therapies that it is difficult to find an article about psychedelic therapies that does not mention these variables.

In particular, the extensive attention given to setting has generated clinical guidance for psychedelic therapy settings. For example, the current model for treatment with classic psychedelics involves a warm room with pleasing artwork and a sofa to recline on (Johnson et al. 2008). Recommendations have indicated that the patient should have privacy, but also be accompanied by two therapists or guides with whom the patient has already built rapport. Patients typically wear eyeshades and listen to music from curated playlists (Garcia-Romeu and Richards 2018). Of course, the broad concepts of setting and aesthetics extend beyond these specific recommendations to include relationships with facilitators or therapists, group vs. individual participation, the built environment beyond the specific therapy room, naturalistic settings, sociocultural and religious norms, and more. Such concepts and dimensions of experience also have the potential to impact both therapeutic and non-therapeutic experiences outside the context of psychedelic therapy, rendering their study within this field a potential model for other applications. However, despite the field's expectation that setting can support positive outcomes in psychedelic therapy, and despite the resulting attention given to therapy settings, the extent to which effects of setting have been empirically tested remains unclear.

1.3 Scoping Review

Responding to the field's longstanding emphasis on the role and value of setting, and recognizing the breadth of research conducted regarding effects of aesthetics on health and well-being, a rapid scoping review was undertaken to identify the extent to which effects of setting and aesthetics on psychedelic experiences and therapies have been explicitly studied. Published evidence was gathered regarding effects of setting on psychedelic experiences and their outcomes, with an emphasis on four questions:

- What aspects of setting have been studied? with what drugs?
- *How* has setting been studied? using what study designs, measures?

- What were the initial findings?
- What gaps and densities exist in this area of research?

The goals of this review were to consolidate existing knowledge, identify evidentiary gaps and opportunities, and lay the groundwork for continued research regarding effects of setting and aesthetics on psychedelic experiences and therapies, with the ultimate aim of advancing knowledge and improving the delivery of psychedelic therapies.

2 Methods

2.1 Design

Completion of this review followed the Joanna Briggs Institute's (JBI) Methodology for JBI Scoping Reviews (Peters et al. 2020). The general purpose of a scoping review is "to map rapidly the key concepts underpinning a research area and the main sources and types of evidence available" (Mays et al. 2001, p. 194), with the aim of examining "the extent, range, and nature of a research activity" and "identify [ing] research gaps" (Arksey and O'Malley 2005, n.p.). Scoping reviews differ from systematic reviews in that they do not include the critical appraisal of included studies; they are instead undertaken regarding emerging areas of knowledge, when the information needed to develop systematic reviews may not yet be available. Scoping reviews are similar to systematic reviews in their employment of systematic literature searches and rigorous screening and data extraction processes.

In June 2021, this review's research team searched for similar reviews or protocols using the JBI Database of Systematic Reviews and Implementation Reports, Cochrane Database of Systematic Reviews, BioMed Central Systematic Reviews, and PROSPERO: International Prospective Register of Systematic Reviews. One systematic review is in progress related to "extra-pharmacological factors" in "classical psychedelic-assisted experiences" (Axon et al. 2021); however, such a review can include varieties of therapeutic approaches or interventions that are used in psychedelic therapy, which is beyond the scope of the current review. No reviews were found to have gathered all published evidence related specifically or exclusively to setting and psychedelics.

2.2 Protocol Registration and Reporting

A protocol for this review, designed according to the JBI scoping review protocol guidelines (Peters et al. 2020), was registered in July 2021 with the Open Science Foundation (OSF) under the title, "Studies of the Effects of Setting and Aesthetics on Psychedelic Experiences, Therapies, and Outcomes: A Rapid Scoping Review of the

Literature." Reporting has been completed in accordance with the PRISMA-ScR Reporting Guidelines.

2.3 *Definitions and Classifications*

In this review, psychedelics were limited to "classic psychedelics," including psilocybin, ayahuasca, peyote, LSD, mescaline, and DMT (N,N-dimethyltryptamine). Multiple terms are utilized for these compounds, and studies were not excluded based upon use of alternative terms.

This review understood setting and aesthetics to refer to any aspects of one's environment during psychedelic use that were examined or discussed in relation to outcomes or reports of the experience.

2.4 *Inclusion and Exclusion Criteria*

This scoping review sought to consolidate all published research regarding effects of setting on psychedelic experiences and therapies; as a result, it did not limit results by date or geographic location. The PICOS framework (Amir-Behghadami and Janati 2020) was utilized to identify inclusion criteria:

- Population (P): Humans of all ages
- Intervention (I): Any uses of classic psychedelics that were studied, reported, or discussed in conjunction with information about effects of the use-setting(s)
- Comparator (C): Alternate settings, such as differing: music, physical environment(s), participation (e.g., group vs. individual), rituals, facilitator(s), cultural norms, etc.
- Outcome (O): Any outcomes related to effects of settings on psychedelic experiences or therapies, where outcomes may be a function of the acute experience
- Study design (S): All study designs were eligible, including randomized controlled trials (RCTs), within-subjects designs, qualitative studies, case reports, systematic reviews with and without meta-analyses, etc. Theoretical articles were also included if they centered effects of setting on psychedelic experiences or therapies.

Studies that were not available in English ($n = 4$) were excluded from this scoping review. In addition, one full text remained inaccessible, and was thus excluded without being screened (Chung 1983).

2.5 *Literature Search*

Comprehensive literature searches were conducted by an experienced health sciences librarian in PubMed, PsychINFO, and Embase, utilizing a search strategy that had been iteratively developed by the librarian and lead researchers. The research team reviewed the outcomes of this iterative search in order to recommend for inclusion any additional articles with which the team was already familiar but which had not been identified by the initial search. When searches and additions were complete, the librarian exported results to EndNote, de-duplicated them, and uploaded them to the Covidence platform (https://www.covidence.org/home) in preparation for screening by the research team. The search strategy is available as supplemental material.[1]

2.6 *Data Collection and Analysis*

Blinded pairs of research team members conducted title and abstract screening in Covidence. Full-text screening followed, and a shared document was utilized to collect data that was extracted from included studies. Meetings preceded each level of screening to ensure preparation and consistency across reviewers. Inter-rater discrepancies were resolved via discussion and/or by the engagement of a third reviewer.

2.7 *Critical Appraisal*

Unlike systematic reviews, scoping reviews are designed to generate "an overview of existing evidence regardless of methodological quality or risk of bias" (Tricco et al. 2018). Consistent with this purpose, the sources included in this review were not critically appraised.

3 Results

3.1 *Summary*

This review's formal literature search identified 1,160 articles for potential inclusion; seven additional articles were then recommended by the team for screening. After all

[1] https://mfr.osf.io/render?url=https://osf.io/zxkdh/?direct%26mode=render%26action=download%26mode=render

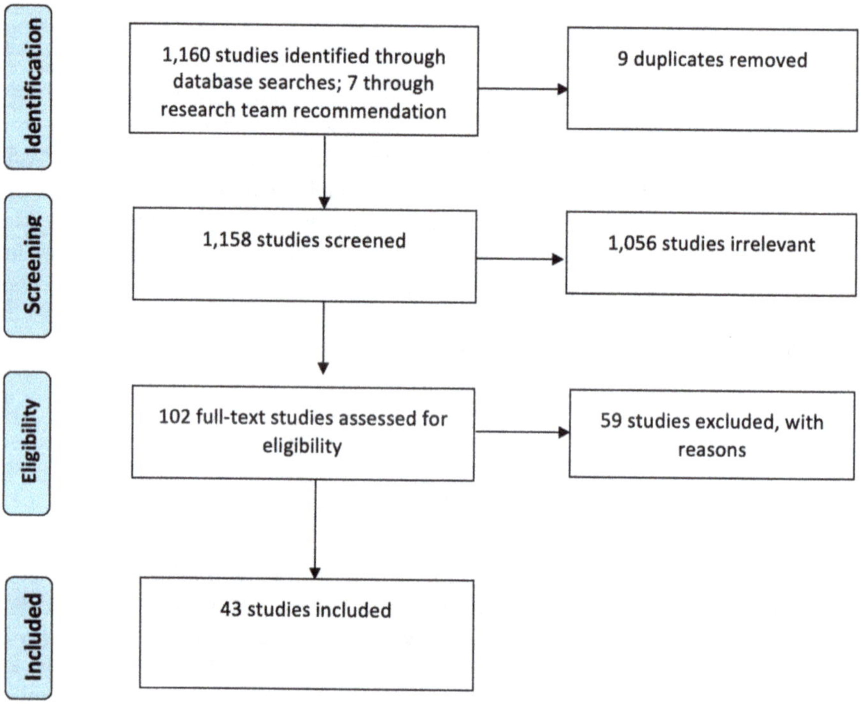

Fig. 1 PRISMA flow diagram

screening had been completed, 43 articles were found to meet the inclusion criteria; their data are represented in the present analysis (see Fig. 1). A list of all included articles is available as supplemental material.[2] Of these, only 29 reported primary empirical data, with 14 of these reports representing data from controlled laboratory studies, eight reports representing large-scale online surveys with retrospective ratings, and seven studies reporting on observational data from naturalistic settings (e.g., ritual psychedelic use, outdoor festivals). To prevent duplicative reporting (e.g., double-reporting of data that was reported both in a primary empirical publication and in a review), only primary empirical studies were included when calculating some descriptive statistics.

Data were coded to identify and quantify the psychedelic compounds studied, locations in which they were studied, study aims and designs, populations, demographics, and setting components under study, comparators (for both drugs and settings), dose and duration, outcomes measured, instruments and measures used, and reported findings. Descriptive statistics were generated for each of these coded

[2] https://mfr.osf.io/render?url=https://osf.io/vczj2/?direct%26mode=render%26action=download%26mode=render

Table 1 Primary empirical studies involving elements of setting in psychedelic research

								Compound under study							
	Location (country)	Study population	% Female	% White	Mean age [min–max]	Sample size	Study design	Psilocybin	LSD	Ayahuasca	DMT	5-MeO-DMT	Peyote	Mescaline	Other/synthetic
Eisner and Cohen (1958)	U.S.	Patients	36	NR	NR [21–60]	22	Quasi-experimental (pre/post test, no control)		X						
Sandison (1959)	U.K.	Patients	NR	NR	NR [NR-NR]	NR	Cohort		X						
Chandler and Hartman (1960)	U.S.	Patients	44	NR	38 [15–62]	110	Within-subjects		X						
Leary et al. (1963)	U.S.	Healthy	36	NR	[NR-NR]	98	Quasi-experimental (pre/post test, no control)	X							
Faillace and Szara (1968)	U.S.	Healthy	<50	>50	NR [20–45]	19	RCT								Xd
Gaston and Eagle (1970)	U.S.	Patients	NR	NR	NR [NR-NR]	59	NRCS		X						
Eagle (1972)	U.S.	Patients	0	100	47 [34–59]	32	Case report		X						

(continued)

Table 1 (continued)

								Compound under study							
	Location (country)	Study population	% Female	% White	Mean age [min–max]	Sample size	Study design	Psilocybin	LSD	Ayahuasca	DMT	5-MeO-DMT	Peyote	Mescaline	Other/ synthetic
Schlicht et al. (1972)	U.K.	Other	NR	NR	NR [16–late 20s]	34	Qualitative study		X						
Zinberg et al. (1975)	U.S.	Participants with substance use history	25	84	NR [14–70]	57	Qualitative study								X
Barbosa et al. (2005)	Brazil	Other	57	NR	36 [18–56]	28	Qualitative study			X					
Kaelen et al. (2016)	U.K.	Healthy	10	NR	34 [26–47]	10	Within-subjects		X						
Carbonaro et al. (2016)	U.S.	Online survey respondents	22	89	30 [18–79]	1993	Other	X							
Kaelen et al. (2018)	U.K.	Healthy	17	NR	33 [22–47]	12	Within-subjects		X						
Preller et al. (2017)	Switzerland	Healthy	23	NR	26 [20–34]	22	RCT		X						
Barrett et al. (2017)	U.S.	Psychedelic therapists	30	NR	NR [25–75+]	10	Qualitative study	X							

Belser et al. (2017)	U.S.	Patients	46	92	50 [18–76]	13	Qualitative study	X							
Bøhling (2017)	Denmark	Other	50	NR	NR [NR-NR]	100	Qualitative study	X	X						
Kaelen et al. (2018)	U.K.	Patients	NR	NR	NR [NR-NR]	19	Quasi-experimental (pre/post test, no control)	X							
Vitos (2017)	Australia	Other	NR	NR	28 [NR-NR]	17	Qualitative study		X						
Haijen et al. (2018)	a	Online survey respondents	25	NR	29 [18–40]	g	Quasi-experimental (pre/post test, no control)	X	X	X	X	X	X	X	X
Smigielski et al. (2019)	Switzerland	Other	NR	NR	NR [NR-NR]	39	RCT	X							
Olson et al. (2020)	U.S.	Healthy	45	NR	22 [NR-NR]	33	Quasi-experimental (pre/post test, no control)								Xe
Sepeda et al. (2020)	U.S.	Online survey respondents	32	8	36 [21-NR]	578	Other					X			
Strickland et al. (2020)	U.S.	Patients	20	90	51 [37–65]	10	Within-subjects	X							
Kettner et al. (2021)	b	Healthy	44	91		h	Other	X	X	X	X	X	X	X	X

(continued)

Table 1 (continued)

								Compound under study								
	Location (country)	Study population	% Female	% White	Mean age [min–max]	Sample size	Study design	Psilocybin	LSD	Ayahuasca	DMT	5-MeO-DMT	Peyote	Mescaline	Other/ synthetic
					44 [32–57]										
Pallavicini et al. (2021)	Argentina	Healthy	20	NR	33 [21–65]	35	Quasi-experimental (pre/ post test, no control)				X				
Perkins et al. (2021)	c	Online survey respondents	50	NR	39 [18-NR]	6,877	Other			X					
Roseman et al. (2021)	Israel, Palestine	Participants in religious rituals	42f	NR	NR [28–59]	31	Qualitative study			X					
Uthaug et al. (2021)	The Netherlands	Healthy	60	NR	50 [30–41]	30	RCT			X					

				Outcome measures included									
Dosage	Drug comparator	Facilitator	Element of setting under study	Subjective effects	Acute effects on cognition	Acute effects on emotion	Acute neural effects	Substance use	Psychiatric outcomes	Well-being/ quality of life	Other self-report behavior	Spiritual outcomes	Personality
Varied	NA	Therapist (s)	Multiple components						X				
NR	Insulin	Therapist (s)	Group/social setting						X	X			
	NA			X					X	X			

25-150 µg		Therapist (s)	Multiple components										
Varied (4–100 mg)	NA	Researcher (s)	Multiple components	X									
1 mg/kg	DPT	Therapist (s)	Physical environment						X				
500 mg	NA	Therapist (s)	Music	X									
100 mcg	Dose change	Therapist (s)	Music					X	X	X			
NR	NA	Self	Other	X									
NR	NA	Self	Expectations	X									
NR	Other	Religious leader(s)	Religious ceremony	X					X		X		
Varied (40–80 µg)	Placebo	Researcher (s)	Music	X		X							
NA	NA	Various	Multiple components	X						X		X	
75 mg	Placebo	Researcher (s)	Music	X	X		X						
100 mg	Placebo	Researcher (s)	Music	X	X	X							
NA	NA	NA	Music										
0.3 mg/kg	Other	Researcher (s)	Music						X	X			
NR	NA	NR	Multiple components	X		X							
10 mg	Dose change	Researcher (s)	Music	X					X	X			

(continued)

Table 1 (continued)

				Outcome measures included									
Dosage	Drug comparator	Facilitator	Element of setting under study	Subjective effects	Acute effects on cognition	Acute effects on emotion	Acute neural effects	Substance use	Psychiatric outcomes	Well-being/ quality of life	Other self-report behavior	Spiritual outcomes	Personality
NR	NA	Self	Physical environment	X							X		
Varied (100–> 300 µg LSD)	NA	NR	Physical environment	X						X			X
315 µg/ kg	Placebo	Researcher (s)	Religious ceremony	X							X		X
Placebo	NA	Researcher (s)	Expectations	X									
NR	NA	NR	Structured vs unstructured environment	X									
20 mg	Dose change	Researcher (s)	Music	X				X					
NA	NA	NR	Group/social setting	X					X	X			X
40 mg	NA	Varied	Naturalistic setting	X			X						
NA	NA	NR	Religious ceremony	X		X			X	X		X	
"2–3 cups"	NA	Varied	Group/social setting							X		X	

14.1–20.1 mg DMT 39–56 mg harmine 2.7–3.9 mg harmaline	Placebo	Researcher (s)	Naturalistic setting	X					X	X			

NR not reported, *NRCS* Non-randomized Controlled Study, *RCT* Randomized Controlled Trial

[a] Participants' responses are collected online from more than 50 countries, the majority are from the U.S. (30.4%). Authors are in U.K. and U.S.

[b] Participants' responses are collected online from more than 50 countries, the majority are from the U.S. (. . .8%). Authors are from the U.K.

[c] Participants' responses are collected online from more than 40 countries

[d] DET (N,N-diethyltryptamine)

[e] Placebo (no actual drug administered)

[f] 1 person out of 31 was non-binary. 13 were women; others were men

[g] Sample sizes for the five time points were $N = 654$, $N = 535$, $N = 379$, $N = 315$, and $N = 212$ respectively

[h] N1 (baseline) = 819; N2 (pre-session) = 582; N3 (post-session) = 533; N4 (post-retreat) = 432; N5 (4-week endpoint) = 399

dimensions. Table 1 offers summaries of several variables; it is also available as supplemental material.[3]

3.2 Study Dates and Locations

Studies ranged in publication date from 1958 to 2021 and were found to have originated or been conducted in 10 countries, including the U.S., Brazil, Australia, Israel, Palestine, the Netherlands, the U.K., Argentina, Denmark, and Switzerland. 10 primary studies were conducted in clinical or lab settings, while 17 took place in non-clinical or naturalistic environments. Two primary studies included both types of locations, and the location for the final two – including a scale development paper and an analysis of music-theoretic features of psilocybin session stimuli – was identified as "other."

3.3 Populations

Among all of the 43 identified articles, the most common population reported was healthy volunteers ($n = 8$; 18.6%). Patients who had a mental health diagnosis and/or were receiving therapy comprised the next largest group ($n = 6$; 13.9%). Additional populations included individuals who had a history of using psychedelics ($n = 5$), participants in religious rituals ($n = 4$), and individuals currently using substances ($n = 3$). For eight studies, the population was identified as "other," which included meditators (Smigielski et al. 2019), survey respondents selected for their use of a substance in structured vs unstructured contexts (Sepeda et al. 2020), attendees at parties or outdoor festivals (Vitos 2017; Schlicht et al. 1972), and other unique participant attributes. The remaining nine articles did not identify a study population. It should be noted that while study populations are distinctly recruited and reported, they are not necessarily independent of one another; i.e., individuals with histories of psychedelic use could also be healthy individuals, and meditators or festival attendees could include individuals with a mental health diagnosis or receiving therapy.

Variations were noted in the reporting of demographics (see Table 1). Of the 29 primary empirical studies, six (21%) did not report participants' sex. Of the remaining studies, 21 (72.4%) were coded as "mixed group," indicating that the study involved both males and females. One study involved only male participants; none were documented as involving only female participants.

[3] https://mfr.osf.io/render?url=https://osf.io/v5yhz/?direct%26mode=render%26action=download%26mode=render

Race and ethnicity were inconsistently reported; 18 primary studies (62%) did not report the race or ethnicity of their participants. Nine of the 10 remaining studies indicated that multiple races or ethnicities were represented; where the breakdown within studies was reported, representation was predominantly white. The final study involved only white participants.

Eight primary studies (27.6%) failed to report participant ages. Among studies reporting age data, ages ranged from 14 to 76. 16 studies (55.1%) reported mean ages, generating an overall mean age of 36.68.

3.4 *Sample Sizes and Study Designs*

Twenty eight of the twenty nine primary studies reported sample sizes, indicating a sample size range of 10–100 participants across experimental laboratory or clinic studies, and a range of 10–6,877 participants across internet survey studies. Two studies offered multiple sample sizes for multiple timepoints when data were collected. The mean sample size for all reporting primary studies was 395.7, with a median sample size of 31.5.

Many study designs were represented, with qualitative ($n = 7$; 16.3%) or quasi-experimental ($n = 6$; 14%) being most prevalent. Five studies (11.6%) were randomized controlled trials (RCTs), four were within-subjects (9.3%), and five were coded as "other." This review noted one each (2.3%) of non-randomized controlled trials, case reports, systematic reviews, and cohort studies. The remaining 12 articles were theoretical or conceptual. Additional information can be found in Tables 1 and 2.

3.5 *Study Purposes*

The included studies represent a wide range of aims and research questions. They predominantly addressed novel hypotheses and proposed new approaches and inquiries rather than establishing programs of foundational research followed by confirmatory or additive studies. While all included articles considered effects of setting or aesthetics on the psychedelic experience, they varied widely in the particular aspects of setting that were reported. These included (but were not limited to) effects of religious, ritual, or ceremonial settings (e.g., Perkins et al. 2021; Uthaug et al. 2021); the effects of contextual factors on expectancy and psychedelic outcomes (Olson et al. 2020); naturalistic or recreational settings vs. clinical or laboratory environments (e.g., Pallavicini et al. 2021; Bøhling 2017; Vitos 2017); effects of social norms (e.g., Zinberg et al. 1975; Lyttle and Montagne 1992) and race (Neitzke-Spruill 2019); group vs. individual experiences (e.g., Kettner et al. 2021; Roseman et al. 2021); structured vs. unstructured environments (Sepeda 2020); facilitator rapport and experience (Phelps 2017); and effects of music (e.g.,

Table 2 Characteristics of included studies

Sex (primary studies only)	*N*	%	Study location	*N*	%	Psychedelics	*N*	%	Drug comparators	*N*	%
Mixed group	22	75.86	Naturalistic	19	44.19	LSD	21	30.43	Not applicable	32	74.42
Not reported	6	20.69	Other	12	27.91	Psilocybin	16	23.19	Placebo (unspecified)	6	13.95
Male	1	3.45	Clinical	10	23.26	Ayahuasca	12	17.39	Dose change	2	4.65
Female	0	0	Combination	2	4.65	DMT	6	8.7	Insulin	1	2.33
N/A	0	0	**Total**	**43**	**100**	Other/synthetic	6	8.7	DPT	1	2.33
Total	**29**	**100**				Peyote	5	7.25	Other	1	2.33
						Mescaline	3	4.35	Not reported	0	0
						Total	**69**	**100**	**Total**	**43**	**100**
Population	*N*	%	**Setting components**	*N*	%	**Race/ethnicity (primary studies only)**	*N*	%	**Facilitator**	*N*	%
Not applicable	10	23.26	Music	15	34.88	Not reported	18	62.07	Therapist(s)	11	25.58
Healthy volunteers	8	18.6	Multiple components	8	18.6	Multiple represented	10	34.48	Not applicable	8	18.6
Other	7	16.28	Religious/ritual setting	5	11.63	White	1	3.45	Not reported	6	13.95
Patients diagnosed or receiving therapy	6	13.95	Other	4	9.3	American Indian/Alaskan native	0	0	Self	5	11.63
Participants with use history	5	11.63	Physical environment	3	6.98	Asian	0	0	Researcher(s)	4	9.3
Participants in religious rituals	4	9.3	Group/social setting	3	6.98	Black or African American	0	0	Religious leader (s)	4	9.3
Patients using substances	3	6.98	Expectations	3	6.98	Mixed race	0	0	Various	3	6.98

Various	0	0	Naturalistic setting	2	4.65	Native Hawaiian/Pacific islander	0	0	Music therapist	1	2.33
Not reported	0	0	Not reported	0	0	Not applicable	0	0	Trained guides	1	2.33
Total	**43**	**100**	Not applicable	0	0	**Total**	**29**	**100**	**Total**	**43**	**100**
			Total	**43**	**100**						

Barrett et al. 2017; Kaelen et al. 2018; Strickland et al. 2020; Gaston and Eagle 1970).

While some studies were designed specifically to test, explore, or discuss setting and its effects on subjective aspects of the psychedelic experience (e.g., Strickland et al. 2020; Perkins et al. 2021; Uthaug et al. 2021; Roseman et al. 2021; Sepeda et al. 2020), others were developed to test effects of psychedelics on biological markers and therapeutic outcomes. The latter were included in this study because they either offered information about setting as part of their description of study methods and ultimate findings (e.g., Chandler and Hartman 1960; Belser et al. 2017; Haijen et al. 2018), or they conducted qualitative studies of setting as an addition to an existing drug study (e.g., Kaelen et al. 2018). More generally, while some studies sought to understand the extent to which setting may distinguish a "good trip" from one involving adverse experiences (e.g., Schlicht et al. 1972), others aimed to learn more about setting in order to optimize psychedelic therapies (e.g., Barrett et al. 2017) or to understand the mechanisms by which they work (e.g., Kettner et al. 2021).

3.6 Psychedelics Under Study

Across all 43 included studies, LSD was the most studied or discussed psychedelic compound ($n = 21$), followed by psilocybin ($n = 16$) and ayahuasca ($n = 12$). Six studies involved DMT; five involved peyote, and three involved mescaline. Six studies examined other classic psychedelic compounds. Because 13 studies involved more than one psychedelic, the total number of studies reported here (69) is greater than the sum of included studies (43).

3.7 Drug Comparators

In primary studies that involved administration of psychedelics, 9 studies included the comparison of different drug conditions within the study. In these studies, the study drugs were compared to placebo ($n = 4$), both placebo and pretreatment with ketanserin (a serotonin 2A receptor antagonist; $n = 1$), insulin ($n = 1$), niacin ($n = 1$), or varied doses of the study drug ($n = 1$). No studies involved a systematic comparison of the interaction between drug condition and setting conditions.

3.8 Settings and Aesthetics Addressed

Across all 43 included studies, music was the most studied or discussed setting or aesthetic ($n = 15$), followed by articles that addressed multiple setting components

($n = 7$), religious or ritual settings ($n = 5$), "other" setting elements ($n = 4$), the physical environment in which an experience took place ($n = 3$), group or social context ($n = 3$), a person's expectations of the experience (which might relate more to "set" than "setting"; $n = 3$), and naturalistic settings ($n = 2$). "Other" included therapist attributes or competencies; access to comfortable rest, lavatories, and hydration; racial socialization; and structured vs. unstructured settings.

3.8.1 Comparators

Many primary studies documented effects of a single setting, using qualitative, quasi-experimental, or survey approaches. As a result, fewer than a quarter of the primary studies ($n = 7$) involved setting comparators. These comparators included alternate music types (or silence), unstructured contexts (compared to structured), and pleasant settings with facilitator rapport (compared to austere settings and minimal emotional support).

3.9 Facilitators

The authors have used the term "facilitator" to refer to the individual or individuals administering the drug and its experience. Facilitators have been identified by the literature as an aspect of psychedelic "setting;" because of this, facilitator identities were documented for all articles in this scoping review, then coded according to the list in Table 2.

3.10 Outcomes

Outcomes for primary studies were initially documented verbatim from each text. These were developed into a taxonomy of 10 outcome domains by which each study was categorized (see Table 1). These included:

- Subjective effects
- Acute effects on cognition
- Acute effects on emotion
- Acute neural effects
- Substance use
- Psychiatric outcomes

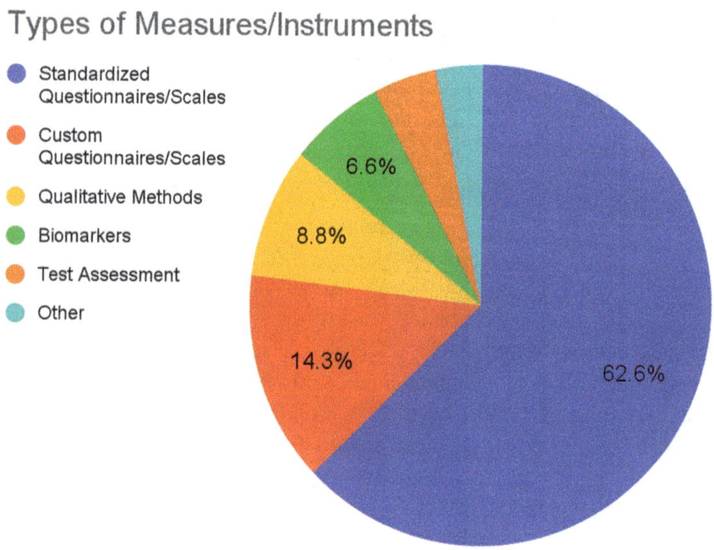

Fig. 2 Types of measurements/instruments

- Well-being/Quality of life
- Other self-report behaviors
- Spiritual outcomes

Across primary studies, subjective effects were the most common outcome measure (reported in 22 studies, or 51.2% of all studies), followed by psychiatric outcomes and well-being/quality of life (reported in 11 studies, or 25.6% of all studies), with few studies reporting on outcomes related to acute effects on emotion ($n = 4$; 9.3%), spiritual outcomes, personality, or other self-report behaviors (each $n = 3$; 7%). Lastly, effects on substance use, acute effects on cognition, and acute neural effects were each reported in two studies (4.7%). Note that domains are not mutually exclusive, and some primary papers report on multiple domains.

3.10.1 Outcome Measures/Instruments

Across the 29 primary studies, a total of 90 unique measures or instruments were used, with many measures being used across multiple studies. A complete list of measures/instruments, along with usage counts, is available as Supplemental Material.[4] These were also grouped according to type, with the most prominent by far being Standardized Questionnaires/Scales – as seen in Fig. 2.

[4]https://mfr.osf.io/render?url=https://osf.io/yzrvk/?direct%26mode=render%26action=download%26mode=render

3.11 Associations Between Setting and Psychedelic Experiences

Given the breadth of study purposes and measured outcomes, reported findings varied widely. Generally, some studies and theoretical articles appear or claim to confirm or elaborate upon the hypothesis that setting and aesthetics affect or even determine outcomes of psychedelic experiences, while other studies seem to assume these hypotheses are true. However, many studies do not provide rigorous or empirical tests of the relationship between setting and acute or enduring effects of psychedelics. Reviewing all reported findings in each article is beyond the scope of the current article; however, this section offers an overview of the findings regarding the relationship between psychedelic experiences and setting.

3.11.1 General Settings

An early study used music, photographs of loved ones, reflections in a mirror, and art therapy during and towards the end of LSD administration sessions, suggesting that these interventions could encourage relaxation or otherwise aid the therapeutic process (Eisner and Cohen 1958). A later study reported that preparation and a pleasant and supportive environment was more conducive to a positive psychedelic experience (i.e., lower scores on a measure of psychotic-like symptoms) than an austere environment accompanied by no preparation (Faillace and Szara 1968). A recent uncontrolled study (Olson et al. 2020) and an ethnographic analysis of the experiences of partygoers (Vitos 2017) suggested that psychedelic environments, such as spaces featuring specific artwork, color-changing lights, psychedelic trance music, and other individuals tripping (or appearing to), might elicit psychedelic experiences in some participants (as determined by self-report measures), even without psychedelic consumption. In addition, a description and summary of psychedelic trip reports suggests that participants can have similar experiences in uncontrolled or recreational environments as are encountered in controlled laboratory studies (Bøhling 2017). However, the foregoing studies (with the possible exception of Faillace and Szara 1968) did not provide empirical controls or rigorous tests of the hypotheses that elements of setting had any influence on psychedelic experiences. Also, these studies are unable to address which (if any) of their multiple elements of setting had an effect on psychedelic experiences.

3.11.2 Social/Cultural Settings

Positive social relationships and interactions, as well as one's perception of support and safety during psychedelic use, have been claimed to increase the likelihood of positive experiences and beneficial effects (Eisner 1997; Leary et al. 1963), and collective psychedelic use by participants with sociocultural differences appeared to support feelings of unity and understanding (Roseman et al. 2021). Conversely, lack of social support and safety measures were suggested as contributing to psychedelic users seeking medical attention during an outdoor festival (Schlicht et al. 1972); however, this is a highly selective sample representing only those encountering challenging experiences. Additionally, some form of group therapy was suggested to be potentially beneficial to participants undergoing psychedelic therapy (Sandison 1959, Leary et al. 1963). Generally, the studies above have suggested, assumed, or tried to assert that environmental sociocultural context can influence psychedelic experiences and outcomes. However, they did not provide empirical controls or rigorous testing of these hypotheses.

3.11.3 Ritual or Ceremonial Settings

A recent controlled trial raised the possibility that a shared ceremonial environment may evoke psychotropic effects without psychedelic consumption (Uthaug et al. 2021). These comport with results from a survey study that suggest that elements of setting associated with religious ritual use of psychedelics may support intended outcomes such as increased self-insight, spiritual growth, and mystical experiences (Perkins et al. 2021). Together, these suggest that ritual or ceremonial settings may contain elements that contribute to the psychedelic experience; however, it is unclear which element or elements of the ritual or ceremonial environment in these studies are important.

An observational study investigated the experiences of two separate ceremonial groups, the Santo Daime (whose ceremonies lasted from 4 to 12 h and were characterized by active engagement of congregants and study participants in singing and dancing in this study) and Uniao do Vegetal (whose ceremonies always lasted 4 h and were characterized by passive observation of singing and dancing by study participants in this study) (Barbosa et al. 2005). Participants differed in the degree of awe, peace, and psychological insight that they experienced during the two different ceremonies (Barbosa et al. 2005), suggesting that direct engagement in singing and dancing, and/or ceremony time, may contribute to psychedelic experiences. It may also be that a social or group context is an important factor in the ritual or ceremonial context. Survey studies have provided evidence that social and emotional support (Carbonaro et al. 2016) and feeling comfortable in the setting (Haijen et al. 2018) may reduce the intensity and/or duration of challenging experiences (e.g., "bad trips") with psychedelic drugs, and may improve enduring positive outcomes (Haijen et al. 2018). They have also indicated that a structured environment for psychedelic use may confer both acute and enduring positive benefits over unstructured environments (Sepeda et al. 2020), and that social connection during psychedelic experiences can increase the likelihood of enduring increases in well-being (Kettner et al. 2021). Contrary to this, a description and summary of anonymous psychedelic trip reports asserts that social interaction is less important than the experience of individual pleasure gained from various activities during psychedelic experiences (Bøhling 2017). While such naturalistic retrospective and prospective sources can yield rich and seemingly compelling information, it is important to note that survey studies and anonymous trip reports must be approached with caution, given (1) known biases in retrospective reports, including recall and sampling biases (Raphael 1987; Coughlin 1990; Sato and Kawahara 2011), which can vary as a function of different cohorts (Ebner-Priemer et al. 2006), and (2) the ultimately uncontrolled nature of the experiences under study.

Study findings additionally raise the possibility that the act of setting an intention, or engaging intentionally in a community or psychedelic experience, confers benefit. A prospective survey of psychedelic drug use found evidence that intentions to connect with nature or to have a spiritual and/or therapeutic experience predicted stronger acute mystical experiences and greater overall well-being, compared to experiences that occurred in absence of prior intentions (Haijen et al. 2018). A study of ayahuasca use in a natural setting also demonstrated that the degree of "clear intentions" before a ceremony predicted the degree of participants' "mystical experience;" however, questions related directly to "setting" in that study did not correlate either with subjective or neurobiological measures (Pallavicini et al. 2021). Additionally, intentions may more reasonably relate to set rather than setting, though they are a likely important component of engaging in ritual or ceremonial use of psychedelic drugs.

Effects of psilocybin when administered to long-term meditators during a meditation retreat have been reported (Smigielski et al. 2019), and results on self-report questionnaires were compared post-hoc to other published data on psychedelic experiences in healthy non-meditators outside the context of a meditation retreat. However, there was no direct comparison condition within the study or within the given population for the meditation retreat context. Thus the impact of the meditation retreat setting overall, or of any particular elements of the meditation retreat setting, remains unclear; it may be that participants in the studied population respond differently to psilocybin than participants in other psychedelic studies.

3.11.4 Music as Setting

Several studies have examined effects of music on psychedelic experiences and associated brain function. Early studies involving treatment of alcohol use suggested that music that is familiar to a patient (in one case, compared to randomly selected music, music that is unfamiliar and disliked, or "no music" conditions) and music related to the themes of religion or romantic love may be most beneficial during psychedelic therapy sessions (Gaston and Eagle 1970; Eagle 1972). However, these studies were not well-controlled. In one case, patients were not randomly or equally assigned to different music intervention conditions (Gaston and Eagle 1970), and in the other, music selection was not systematically manipulated to test hypotheses related to familiarity or thematic content (Eagle 1972).

Qualitative interviews of patients who had undergone psilocybin therapy for depression indicated that liking of, resonance with, and openness to music presented during therapeutic sessions predicted therapeutic outcomes (Kaelen et al. 2018).

While this does not represent a systematic manipulation of music within the study (all participants heard the same playlist), and while the identified variables may relate more closely to set (e.g., the participant's internal experience of the music) rather than explicit setting, this study identifies important vectors to consider for the selection of music during psychedelic experiences. Another qualitative study of patients in a clinical trial for the treatment of depression and anxiety related to a late-stage cancer diagnosis emphasized vivid descriptions of the importance of music during psychedelic therapy sessions, as recounted by study volunteers. However, this study included no control conditions or any test of hypotheses related to the role or function of music, or to the type of music that was supportive or helpful (Belser et al. 2017). These findings all comport with descriptions of a collection of anonymous psychedelic trip reports, which include emphasis on the role of music in supporting or modulating the emotional component of psychedelic experiences (Bøhling 2017).

More recently, a small-sample, open-label pilot study showed no drawback to deviating from a standardized western art-music playlist and using a playlist primarily constructed over overtone music (Strickland et al. 2020). Though this study did not include assessment or consideration of participants' qualitative experience of these playlists, it does challenge the notion that western art music, or any particular playlist, holds a privileged place in terms of being able to support psychedelic experiences. An additional survey study and mixed methods analysis identified some common acoustic and music-theoretical features of musical stimuli that seemed consistent across musical stimuli that were recommended to support peak experiences by therapists who used psychedelic interventions. However, this study did not in any way assess direct patient information or the outcomes of actual psychedelic therapy sessions (Barrett et al. 2017).

Functional neuroimaging studies have provided evidence for a biological basis of the effects of music on psychedelic experiences. One such study showed that personally meaningful music (compared to music that was not personally meaningful) more strongly engaged somatosensory brain regions and brain regions involved in self-referential processing during the effects of LSD compared to placebo and a serotonin 2A antagonist pretreatment (Preller et al. 2017). Secondary analyses of these data showed that both music cognition and anterior memory networks in the brain more closely tracked the time-varying tonal structure of music when participants were listening to personally meaningful music, and when participants were experiencing the effects of LSD (Barrett et al. 2017). Another study demonstrated that music increased the coupling of memory-related brain regions with visual cortex during the acute effects of LSD, and this coupling was associated with music-evoked imagery during the LSD experience, providing direct evidence for a mechanism underlying vivid visualizations specifically with music listening during psychedelic experiences (Kaelen et al. 2018). Together, these reports provide support for a biological basis of the role of psychedelic drugs in increasing meaning making, engaging vivid subjective experiences, and engaging personally-relevant content during a psychedelic experience, with specific effects resulting from music listening.

4 Discussion

4.1 *Summary*

This scoping review found that effects of setting on classic psychedelic therapies and experiences have been studied in multiple countries since 1958, among varied populations. Aims, study designs, outcomes, and measures have varied considerably; however, a common goal appears to be the identification and examination of setting-effects in order to (1) better understand the benefits and risks associated with psychedelic use and treatments, and (2) to improve treatment outcomes and the delivery of psychedelic therapies.

4.2 *Evidentiary Gaps*

Despite widespread recognition of setting as a critical aspect of psychedelic experiences and outcomes, this scoping review found that empirical studies of the effects and the optimization of setting remain highly limited, with very few studies rigorously testing hypotheses regarding the interaction between setting and either acute or enduring effects of psychedelics. Studies that directly examined the effects of setting or aesthetics on psychedelic outcomes appear primarily to have set out to test the existence or extent of these effects. Articles that addressed setting as a *secondary* aspect of study design or analysis also appear to have essentially confirmed or acknowledged that some element of setting affected study outcomes, though with few prospective controls. As a result, accumulating evidence underscores the importance of setting, but provides limited direct evidence for that importance. It also does not yet provide a robust understanding of the mechanisms by which setting affects psychedelic experiences (music being a potential exception, to a small extent); nor does it typically disaggregate elements of setting to help operationalize findings for optimal care delivery.

For example, despite consistent interest in effects of music on psychedelic experiences, few studies have compared one music type to another, or sought to explain why some playlists appear to be more conducive to positive results, or for whom. Similarly, while pleasant, safe physical environments have become standard for psychedelic therapies, the many elements comprising such environments – artwork, furniture, objects, textures, colors, temperature, scent, familiarity with the space, *etc.* – have not been empirically investigated for their individual contributions. In addition, the field does not yet have access to evidence that would allow selection of these elements based upon specific participants, conditions, or outcomes.

Contributing to a lack of granularity is a tendency to utilize embellished language when discussing experiences related to psychedelics, including elements of setting. For example, Dobkin De Rios (1975) noted that "music replaces with its own implicit structure a set of banisters and pathways through which the drug user in a non-Western ritual setting negotiates his way" (p. 64). While this hypothesis is intriguing, it is not clearly testable, and does not accommodate other applications or purposes of music suggested by other studies. To be sure, many aspects of psychedelic experiences challenge one's ability to describe them, and creative language can help ensure that diverse and nuanced experiences are included in data collection (see Golden 2020). However, creative language regarding psychedelic therapies may be most useful when combined with granular details and operationalizable, measurable constructs that allow interpretation in the context of concrete application. While many included articles make claims regarding the effects of social settings, physical environments, music, festivals, general naturalistic contexts, and more, study designs and variations in reporting often preclude a full exploration or understanding of these effects, including their replication and iteration.

In sum, it has been claimed for some time that the "black box" of setting and aesthetics contributes to psychedelic experiences and outcomes. Existing studies appear to repeatedly test the existence or extent of this contribution, in varied ways. Over time, researchers have separated the black box of setting into multiple black boxes: music, ritual or ceremony, group settings, etc. However, the existing evidence base does not provide enough insight into these phenomena to inform the establishment of evidence-based best practices, individualized or population-based approaches, or mechanistic understandings. In short, it is clear from the literature that various aspects of setting may contribute to positive results, but which settings are best, and for whom? under what conditions? when? for what indications? These questions remain unanswered.

4.3 Evidentiary Opportunities

That said, this scoping review does reveal *categories of setting* for which important groundwork has been laid for further investigation and application. These include music/sound, religious and ritual contexts, group vs. individual experiences, sociocultural expectations or norms, and the physical environment. These categories reflect what seems to be the center of gravity of ideas for the aspects of setting that may be most important and/or most influential in psychedelic experiences. Subsequently, these may be the areas in which optimization of setting has the greatest impact on both acute and enduring effects of psychedelic drugs. General intuition and available evidence converge on the finding that optimal musical settings, appropriate rituals, optimized environments, and alignment of one's sociocultural expectations with these contextual elements may provide for the greatest optimization of psychedelic effects; however, as stated above, the actual details (the "what" and "how") remain elusive.

The existing literature illuminates the value of studying how idiosyncratic preferences and participant agency can shape the effects of music and environment on psychedelic experiences. Given that elements of setting appear to have differential effects on study participants, it is clear that methods to personalize or customize settings may yield optimal results, though further research is needed to determine the specifics of this customization. In line with this, the literature also suggests that despite the popularity of the clinic-room-based model designed for individual therapeutic application, group and naturalistic settings may be ideal for particular individuals, populations, or outcomes. As a result, research regarding optimization of setting for psychedelic therapies will do well to include considerations of extra-clinical environments. Finally, although "setting" has commonly referred to immediate contexts or environments for individual participants, this review's findings indicate that increased attention to overarching settings such as sociocultural norms or religious or cultural narratives – and their differences among varied participants – may improve outcomes.

It is clear that researching setting more extensively or more granularly involves considerable challenges. As Hartogsohn (2017) argued, "[i]ntegrating variables of set and setting into clinical drug research would entail great complications for a pharmaceutical industry bent on randomized controlled trials (RCTs) and with limited patience for injecting fuzzy social and cultural elements into its considerations" (n.p.). However, as social ecological frameworks indicate (Brofenbrenner 1979; McLeroy et al. 1988; Stineman and Streim 2010; Golden and Earp 2012; Golden and Wendel 2020), individual medical decisions and health behaviors – including drug use and associated experiences – cannot be separated from their intrapersonal, interpersonal, community, political, and sociocultural contexts. Though standard approaches to pharmacological research have historically emphasized designs that neglect or exclude such variables, additional approaches are necessary to inform improved understanding and delivery of psychedelic therapies.

In addition, it should be acknowledged that existing models and practices for psychedelic therapies have delivered consistently promising results; as a result, the modification and testing of more (or more specific) setting variables would require researchers to risk negative or neutral outcomes in exchange for the potential of generating increased understanding and improved future outcomes. This exchange may be a difficult sell, but it must be acknowledged that some individuals do not respond to psychedelic therapy, while others seem to not respond fully. The alternative to testing hypotheses regarding the optimization of setting may be stagnation and continued inability to customize or optimize care.

Having long valued the effects of setting in treatment and delivery of care, the field of psychedelic research offers a significant model for the larger health care field – reminding researchers and practitioners that the elimination or neglect of contextual variables precludes accurate and optimized understandings of how, when, for whom, and under what conditions a given intervention "works." Further empirical studies of setting components and their effects within the field of psychedelic research will develop this model for the potential benefit of health research at large. Findings are likely to have application not only to health research, practice, and care delivery, but also to the betterment of well people more broadly. Empirically guided decisions regarding the therapeutic impact of setting may have import in the design of wellness and retreat centers, as well as public, private, corporate, government, education, and other institutional settings. In other words, improvements to setting research within the context of psychedelic therapy may eventually generate meaningful shifts in expectations or priorities regarding a variety of environments.

4.4 Reporting

Beyond evidentiary gaps, this scoping review identified gaps in reporting that limit the synthesis of evidence across studies, and thereby put limits on continued improvements in research and practice. For example, reporting of participant demographics was inconsistent, perpetuating questions regarding who has been studied and for whom psychedelic therapies or experiences – and particular settings – may be most helpful. In addition, setting is a complex and subjective phenomenon, entailing multiple potential factors and components. As a result, descriptions of settings and setting-components varied considerably in their specificity and ultimate replicability. Greater use of reporting guidelines may help ensure that evidence can be synthesized and applied over time for the optimization of treatment and care. Reporting guidelines offer "minimum lists of information needed to ensure a manuscript can be, for example, understood by a reader, replicated by a researcher, used by a doctor to make a clinical decision, and included in a systematic review" (EQUATOR 2018, n.p.). Standardized reporting practices can support evidence synthesis as well as translation of research into practice, policy, and improved research.

Of course, the issue of underreporting critical information in research is not unique to psychedelics or related setting/aesthetics research. For example, non-pharmacological interventions have been found to provide adequate documentation of interventions about half as often as pharmacological interventions (29% vs. 67%; see Glasziou et al. 2008), and one comprehensive study found suboptimal adherence to reporting guidelines in 87.9% of included studies – compelling the authors to identify "the need for...the scientific community to encourage the use of reporting guidelines" (Jin et al. 2018, n.p.). That said, the field of psychedelics research may consider whether the particular complexities and challenges of studying settings and aesthetics necessitate new guidelines. For example, Hartogsohn (2017) noted that the ability to understand effects of setting on psychedelic experiences requires, at the least, "a detailed description of the set and setting conditions in which research took place, including reference to such parameters as subject selection, researcher expectations, subject expectation, preparation, and physical setting" (n.p.). If new guidelines are necessary, these are best developed via highly interdisciplinary collaborations involving all interested parties and end users of evidence. To assist in this effort, the EQUATOR Network offers an extensive yet accessible toolkit (www.equator-network.org).

4.5 *Theory to Practice*

Finally, this scoping review illuminates the prominence of theoretical articles in literature related to setting and psychedelic therapies. Many of these articles curate and discuss literature, histories, and therapist observations, offering collective knowledge formed over decades. Diverse knowledges and knowledge-generation practices are of critical value to the field; the data, theories, and narratives collected by practitioners and users over time can and should inform continued clinical research and practice. However, many theories and standard practices remain unexamined via empirical research. When researchers fail to rigorously test well-known theories and assumptions, they do not thereby honor or value practitioners' accumulated knowledge; they instead relegate it to limited application. The field's many theoretical articles suggest rich opportunities to develop empirical studies that test existing theories, generate translational research, and propel continued improvement.

5 Limitations

The search strategy utilized for this review limited results to those available through the identified databases, and the rapid nature of the review precluded hand-searches of the literature. It is therefore conceivable that additional empirical studies of setting have been conducted but were inaccessible via the selected databases. Similarly, as this review was limited to articles available in English, the four studies in other languages may have offered new information. In addition, setting and context are broad concepts that are variously defined and understood. As a result, despite the comprehensive nature of the search strategy, it is possible that relevant studies may have been missed due to disparities in terms or descriptions. There also seems to be the potential for some ambiguity regarding whether an aspect of psychedelic therapy is best considered under the heading of "set" or "setting," when the two have the opportunity to interact. Finally, extraction and coding of data for this review involved adapting theoretical, review, and study articles to the extraction tool. Though multiple research team members reviewed the extracted data, it is possible that other research teams may have coded or reported them differently. Notably, it is unlikely that potential alterations due to the above limitations would have changed the overarching conclusions of the review.

6 Conclusion

This scoping review was designed to support the field's understanding and further study of the effects of setting on psychedelic therapies by gathering and organizing existing evidence regarding these effects. To the authors' knowledge, it is the first scoping review undertaken with this goal, and it provides a critical foundation for continued study of the complexities of setting, aesthetics, contextual factors, and their impacts. The 43 included studies indicate apparent consensus regarding the importance of setting in psychedelic therapies, as well as consistent interest in theorizing about these effects. However, this consensus has yet to generate consistent, rigorous tests of setting and its complexities. As a result, the field continues to lack understanding or agreement regarding the effects of various specific elements of setting, the mechanisms by which they affect outcomes, for whom these effects occur, under what circumstances, given what conditions, and other critical factors.

This review also illuminated a disparity of knowledge in this field, in that the knowledge driving current practice is disproportionately derived from clinical practice and guidance rather than empirical study. Though decades of applied practice provide clear value to the field, multiple robust forms of knowledge and knowledge generation are needed to support evidence-based interventions and their continued development and optimization. This scoping review indicates extensive opportunities for further theory-driven, practice-based research, which in turn will continue to inform evidence-based practice.

Throughout human history, setting and aesthetics have been recognized as critical aspects of human experience, including human health and well-being. Psychedelic therapies and their study have underscored this importance, and they model the integration of extra-drug variables into research and practice. Further studies of setting and aesthetics in the context of psychedelic therapies are likely to not only improve these therapies and their delivery, but also inform considerations of setting and aesthetics for non-psychedelic interventions.

References

Amir-Behghadami M, Janati A (2020) Population, intervention, comparison, outcomes and study (picos) design as a framework to formulate eligibility criteria in systematic reviews. Emerg Med J 37(6):387–387. https://doi.org/10.1136/emermed-2020-2095

Arksey H, O'Malley L (2005) Scoping studies: towards a methodological framework. Int J Soc Res Methodol 8(1):19–32. https://doi-org.echo.louisville.edu. https://doi.org/10.1080/1364557032000119616

Axon R, Whaley S, Marupuru S (2021) Extra-pharmacological factors predictive of clinical and humanistic outcomes associated with classical psychedelic assisted experiences: a systematic review. CRD42021234760. https://www.crd.york.ac.uk/prospero/display_record.php?ID=CRD42021234760

Barbosa PC, Giglio JS, Dalgalarrondo P (2005) Altered states of consciousness and short-term psychological after-effects induced by the first time ritual use of ayahuasca in an urban context in Brazil. J Psychoactive Drugs 37(2):193–201. https://doi.org/10.1080/02791072.2005.10399801

Barrett FS, Robbins H, Smooke D, Brown JL, Griffiths RR (2017) Qualitative and quantitative features of music reported to support peak mystical experiences during psychedelic therapy sessions. Front Psychol 8:1238

Belser AB, Agin-Liebes G, Swift TC, Terrana S, Devenot N, Friedman HL, Guss J, Bossis A, Ross S (2017) Patient experiences of psilocybin-assisted psychotherapy: an interpretative phenomenological analysis. Journal of Humanistic Psychology 57(4):354–388. https://doi.org/10.1177/0022167817706884
Beyer SV (2010) Singing to the plants: a guide to Mestizo Shamanism in the Upper Amazon, Reprint edn. University of New Mexico Press, Albuquerque
Bøhling F (2017) Psychedelic pleasures: an affective understanding of the joys of tripping. Int J Drug Policy 49:133–143. https://doi.org/10.1016/j.drugpo.2017.07.017
Boyce M, Bungay H, Munn Giddings C, Wilson C (2018) The impact of the arts in healthcare on patients and service users: a critical review. Health Soc Care Community 26(4):458–473. https://doi.org/10.1111/hsc.12502
Boyd B (2018) The evolution of stories: from mimesis to language, from fact to fiction. WIREs Cogn Sci 9(1444). https://doi.org/10.1002/wcs.1444
Brofenbrenner U (1979) The ecology of human development. Harvard University Press, pp 1–352
Carbonaro TM, Bradstreet MP, Barrett FS, MacLean KA, Jesse R, Johnson MW, Griffiths RR (2016) Survey study of challenging experiences after ingesting psilocybin mushrooms: acute and enduring positive and negative consequences. J Psychopharmacol 30(12):1268–1278. https://doi.org/10.1177/0269881116662634
Chandler AL, Hartman MA (1960) Lysergic acid diethylamide (LSD-25) as a facilitating agent in psychotherapy. AMA Arch Gen Psychiatry 2:286–299. https://doi.org/10.1001/archpsyc.1960.03590090042008
Chung IM (1983 Sep) Bad trips. Aust Fam Physician 12(9):689–690. PMID: 6651634
Coughlin SS (1990) Recall bias in epidemiologic studies. J Clin Epidemiol 43(1):87–91. https://doi.org/10.1016/0895-4356(90)90060-3
Dobkin De Rios M (1975) Man, culture and hallucinogens: an overview. In: Rubin V (ed) Cannabis and culture. Mouton Publishers, The Hague, pp 401–416
Eagle CT (1972) Music and LSD: an empirical study. J Music Ther 9(1):23–36. https://doi.org/10.1093/jmt/9.1.23
Ebner-Priemer UW, Kuo J, Welch SS, Thielgen T, Witte S, Bohus M, Linehan MM (2006) A valence-dependent group-specific recall bias of retrospective self-reports: a study of borderline personality disorder in everyday life. J Nerv Ment Dis 194(10):774–779. https://doi.org/10.1097/01.nmd.0000239900.46595.72
Eisner B (1997) Set, setting, and matrix. J Psychoactive Drugs 29(2):213–216. https://doi.org/10.1080/02791072.1997.10400190
Eisner BG, Cohen S (1958) Psychotherapy with lysergic acid diethylamide. J Nerv Ment Dis 127(6):528–539. https://doi.org/10.1097/00005053-195812000-00006
Elimimian EB, Elson L, Stone E et al (2020) A pilot study of improved psychological distress with art therapy in patients with cancer undergoing chemotherapy. BMC Cancer 20(899). https://doi.org/10.1186/s12885-020-07380-5
Equator (2018) Centre for statistics in medicine (CSM), NDORMS, University of Oxford
Faillace LA, Szara S (1968) Hallucinogenic drugs: influence of mental set and setting. Dis Nerv Syst 29(2):124–126
Fancourt D (2017) Arts in health: designing and researching interventions, vol 1. Oxford University Press, pp 3–8
Fancourt D, Finn S (2019) What is the evidence on the role of the arts in improving health and well-being? WHO Regional Office for Europe, Copenhagen, p 67. https://www.ncbi.nlm.nih.gov/books/NBK553773/
Franklin N (2018) Biophilic design the key to improving mental health, productivity and stress levels. Workplace Insights
Garcia-Romeu A, Richards WA (2018) Current perspectives on psychedelic therapy: use of serotonergic hallucinogens in clinical interventions. Int Rev Psychiatry 30:291–316. https://doi.org/10.1080/09540261.2018.1486289
Gaston ET, Eagle CT Jr (1970) The function of music in LSD therapy for alcoholic patients. J Music Ther 7(1):3–19

Glasziou P, Meats E, Heneghan C, Shepperd S (2008) What is missing from descriptions of treatment in trials and reviews? BMJ 336:1472–1474. https://doi.org/10.1136/bmj.39590.732037.47

Golden T (2020) Innovation and equity in public health research: testing arts-based methods for trauma-informed, culturally-responsive inquiry. [Doctoral Dissertation, University of Louisville.] Think IR Electronic Theses and Dissertations. https://doi.org/10.18297/etd/3233

Golden SD, Earp JA (2012) Social ecological approaches to individuals and their contexts: twenty years of health education & behavior health promotion interventions. Health Educ Behav 39(3): 364–372. https://doi.org/10.1177/1090198111418634

Golden TL, Wendel ML (2020) Public health's next step in advancing equity: re-evaluating epistemological assumptions to move social determinants from theory to practice. Front Public Health 8:131. https://doi.org/10.3389/fpubh.2020.00131

Golden TL, Springs S, Kimmel HJ, Gupta S, Tiedemann A, Sandu CC, Magsamen S (2021) The use of music in the treatment and management of serious mental illness: a global scoping review of the literature. Front Psychol 12:649840. https://doi.org/10.3389/fpsyg.2021.649840

Grinspoon L (1981) LSD reconsidered. Sciences 21:20–23. https://doi.org/10.1002/j.2326-1951.1981.tb01901.x

Haijen E, Kaelen M, Roseman L, Timmermann C, Kettner H, Russ S, Nutt D, Daws RE, Hampshire A, Lorenz R, Carhart-Harris RL (2018) Predicting responses to psychedelics: a prospective study. Front Pharmacol 9:897. https://doi.org/10.3389/fphar.2018.00897

Hartogsohn I (2017) Constructing drug effects: a history of set and setting. Drug Sci Policy Law. https://doi.org/10.1177/2050324516683325

Iyendo TO (2016) Exploring the effect of sound and music on health in hospital settings: a narrative review. Int J Nurs Stud 63:82–100. https://doi.org/10.1016/j.ijnurstu.2016.08.008

Jin Y, Sanger N, Shams I, Luo C, Shahid H, Li G et al (2018) Does the medical literature remain inadequately described despite having reporting guidelines for 21 years? – A systematic review of reviews: an update. J Multidiscip Healthc 11:495–510. https://doi.org/10.2147/JMDH.S155103

Johnson M, Richards W, Griffiths R (2008) Human hallucinogen research: guidelines for safety. Journal of psychopharmacology (Oxford, England) 22(6):603–620. https://doi.org/10.1177/0269881108093587

Kaelen M, Giribaldi B, Raine J, Evans L, Timmerman C, Rodriguez N, Roseman L, Feilding A, Nutt D, Carhart-Harris R (2018) The hidden therapist: evidence for a central role of music in psychedelic therapy. Psychopharmacology (Berl) 235(2):505–519. https://doi.org/10.1007/s00213-017-4820-5

Kaelen M, Roseman L, Kahan J, Santos-Ribeiro A, Orban C, Lorenz R, Barrett FS, Bolstridge M, Williams T, Williams L, Wall MB, Feilding A, Muthukumaraswamy S, Nutt DJ, Carhart-Harris R (2016) LSD modulates music-induced imagery via changes in parahippocampal connectivity. Eur Neuropsychopharmacol 26(7):1099–1109. https://doi.org/10.1016/j.euroneuro.2016.03.018

Kaimal G, Ray K, Muniz J (2016) Reduction of cortisol levels and participants' responses following art making. J Am Art Ther Assoc 33(2):74–80. https://doi.org/10.1080/07421656.2016.1166832

Kettner H, Rosas FE, Timmermann C, Kärtner L, Carhart-Harris RL, Roseman L (2021) Psychedelic communitas: intersubjective experience during psychedelic group sessions predicts enduring changes in psychological wellbeing and social connectedness. Front Pharmacol 12:623985. https://doi.org/10.3389/fphar.2021.623985

Kilic S, Karadag G, Oyucu S, Kale O, Zengin S, Ozdemir E, Korhan EA (2015) Effect of music on pain, anxiety, and patient satisfaction in patients who present to the emergency department in Turkey. Jpn J Nurs Sci 12(1):44–53. https://doi.org/10.1111/jjns.12047

Kondo MC, Fluehr JM, McKeon T, Branas CC (2018) Urban green space and its impact on human health. Int J Environ Res Public Health 15(3):445. https://doi.org/10.3390/ijerph15030445

Kuiper K (2018 Venus of Willendorf. Encyclopedia Britannica. https://www.britannica.com/topic/Venus-of-Willendorf
Lambert P, Betts D, Rollins J, Sonke J, Swanson K (2017) Arts, health, and well-being in America. National Organization for Arts in Health, pp 1–52
Lankston L, Cusack P, Fremantle C, Isles C (2010) Visual art in hospitals: case studies and review of the evidence. J R Soc Med 103(12):490–499. https://doi.org/10.1258/jrsm.2010.100256
Leary T (1961) Drugs, set & suggestibility. Paper presented at the annual meeting of the American Psychological Association, 6 Sept 1961
Leary T, Litwin GH, Metzner R (1963) Reactions to psilocybin administered in a supportive environment. J Nerv Ment Dis 137:561–573. https://doi.org/10.1097/00005053-196312000-00007
Liang Z, Ren D, Choi J, Happ MB, Hravnak M, Hoffman LA (2016) Music intervention during daily weaning trials-A 6 day prospective randomized crossover trial. Complement Ther Med 29: 72–77. https://doi.org/10.1016/j.ctim.2016.09.003
Lyttle T, Montagne M (1992) Drugs, music, and ideology: a social pharmacological interpretation of the acid house movement. Int J Addict 27(10):1159–1177. https://doi.org/10.3109/10826089209047341
Mason AS (2016) Ancient aesthetics, vol 1. Routledge, London, p 8. https://doi.org/10.4324/9781315697444
Mays N, Roberts EPopay J (2001) Synthesising research evidence. In: Fulop N, Allen P, Clarke A, Black N (eds) Methods for studying the delivery and organisation of health services, Routledge
McLeroy KR, Bibeau D, Steckler A, Glanz K (1988) An ecological perspective on health promotion programs. Health Educ Q 15(4):351–377. https://doi.org/10.1177/109019818801500401
Metzner R, Leary T (1967) On programming psychedelic experiences. Psychedelic Rev 9:5–19
Neitzke-Spruill L (2019) Race as a component of set and setting: how experiences of race can influence psychedelic experiences. J Psychedelic Stud 4:1–10. https://doi.org/10.1556/2054.2019.022
Olson JA, Suissa-Rocheleau L, Lifshitz M, Raz A, Veissière S (2020) Tripping on nothing: placebo psychedelics and contextual factors. Psychopharmacology (Berl) 237(5):1371–1382. https://doi.org/10.1007/s00213-020-05464-5
Pallavicini C, Cavanna F, Zamberlan F, de la Fuente LA, Ilksoy Y, Perl YS, Arias M, Romero C, Carhart-Harris R, Timmermann C, Tagliazucchi E (2021) Neural and subjective effects of inhaled N,N-dimethyltryptamine in natural settings. J Psychopharmacol 35(4):406–420. https://doi.org/10.1177/0269881120981384
Perkins D, Schubert V, Simonová H, Tófoli LF, Bouso JC, Horák M, Galvão-Coelho NL, Sarris J (2021) Influence of context and setting on the mental health and wellbeing outcomes of Ayahuasca drinkers: results of a large international survey. Front Pharmacol 12:623979. https://doi.org/10.3389/fphar.2021.623979
Peters MDJ, Godfrey C, McInerney P, Munn Z, Tricco AC, Khalil H (2020) Chapter 11: scoping reviews. In: Aromataris E, Munn Z (eds) JBI manual for evidence synthesis. https://synthesismanual.jbi.global. https://doi.org/10.46658/JBIMES-20-12
Phelps J (2017) Developing guidelines and competencies for the training of psychedelic therapists. J Humanist Psychol 57(5):450–487. https://doi.org/10.1177/0022167817711304
Pinch TJ, Bijker WE (1984) The social construction of facts and artefacts: or how the sociology of science and the sociology of technology might benefit each other. Soc Stud Sci 14(3):399–441
Powell J (2019) Cave paintings that recorded the night sky. In: From cave art to Hubble, Astronomers' Universe. Springer, Cham. https://doi.org/10.1007/978-3-030-31688-4_2
Preller KH, Herdener M, Pokorny T, Planzer A, Kraehenmann R, Stämpfli P, Liechti ME, Seifritz E, Vollenweider FX (2017) The fabric of meaning and subjective effects in LSD-induced states depend on serotonin 2A receptor activation. Curr Biol 27(3):451–457. https://doi.org/10.1016/j.cub.2016.12.030
Raphael K (1987) Recall bias: a proposal for assessment and control. Int J Epidemiol 16(2): 167–170. https://doi.org/10.1093/ije/16.2.167

Roseman L, Ron Y, Saca A, Ginsberg N, Luan L, Karkabi N, Doblin R, Carhart-Harris R (2021) Relational processes in Ayahuasca Groups of Palestinians and Israelis. Front Pharmacol 12: 607529. https://doi.org/10.3389/fphar.2021.607529

Sandison, R. A. (1959). The role of psychotropic drugs in group therapy. Bull World Health Organ, 21(4-5), 505–515.

Sato H, Kawahara JI (2011) Selective bias in retrospective self-reports of negative mood states. Anxiety Stress Coping 24(4):359–367. https://doi.org/10.1080/10615806.2010.543132

Schlicht J, Mitcheson M, Henry M (1972) Medical aspects of large outdoor festivals. Lancet 1(7757):948–952. https://doi.org/10.1016/s0140-6736(72)91508-5

Scott S (2020) Learning spaces: biophilic design in schools. Teacher Mag

Sepeda ND, Clifton JM, Doyle LY, Lancelotta R, Griffiths RR, Davis AK (2020) Inhaled 5-methoxy-N,N-dimethyltryptamine: supportive context associated with positive acute and enduring effects. J Psychedelic Stud 4(2):114–122. Retrieved 11 Sept 2021, from https://akjournals.com/view/journals/2054/4/2/article-p114.xml

Smigielski L, Kometer M, Scheidegger M, Krähenmann R, Huber T, Vollenweider FX (2019) Characterization and prediction of acute and sustained response to psychedelic psilocybin in a mindfulness group retreat. Sci Rep 9(1):14914. https://doi.org/10.1038/s41598-019-50612-3

Staricoff R (2004) Arts in health: a review of the medical literature. Arts Council England, London

Stineman MG, Streim JE (2010) The biopsycho-ecological paradigm: a foundational theory for medicine. PM R 2(11):1035–1045. https://doi.org/10.1016/j.pmrj.2010.06.013

Strickland JC, Garcia-Romeu A, Johnson MW (2020) Set and setting: a randomized study of different musical genres in supporting psychedelic therapy. ACS Pharmacol Transl Sci 4(2): 472–478. https://doi.org/10.1021/acsptsci.0c00187

Stuckey HL, Nobel J (2010) The connection between art, healing, and public health: a review of current literature. Am J Public Health 100(2):254–263. https://doi.org/10.2105/AJPH.2008.156497

Tricco AC, Lillie E, Zarin W, O'Brien KK, Colquhoun H, Levac D, Moher D, Peters MDJ, Horsley T, Weeks L, Hempel S, Akl EA, Chang C, McGowan J, Stewart L, Hartling L, Aldcroft A, Wilson MG, Garritty C, Lewin S, Godfrey CM, Macdonald MT, Langlois EV, Soares-Weiser K, Moriarty J, Clifford T, Tunçalp Ö, Straus SE (2018 Oct 2) PRISMA extension for scoping reviews (PRISMAScR): checklist and explanation. Ann Intern Med. American College of Physicians. https://doi.org/10.7326/M18-0850

Uthaug MV, Mason NL, Toennes SW, Reckweg JT, de Sousa Fernandes Perna EB, Kuypers K, van Oorsouw K, Riba J, Ramaekers JG (2021) A placebo-controlled study of the effects of ayahuasca, set and setting on mental health of participants in ayahuasca group retreats. Psychopharmacology (Berl) 238(7):1899–1910. https://doi.org/10.1007/s00213-021-05817-8

Vitos B (2017) "Go Out on the Weekends to the Forest, Get Trashed and Dance": residual behaviors and reconfigured environments at psychedelic trance festivals. Leis Sci 39:1–18. https://doi.org/10.1080/01490400.2016.1256796

WHO Regional Office for Europe (2021) Green and blue spaces and mental health: new evidence and perspectives for action (2021) Copenhagen. Licence: CC BY-NC-SA 3.0 IGO

Yin J, Yuan J, Arfaei N, Catalano PJ, Allen JG, Spengler JD (2020) Effects of biophilic indoor environment on stress and anxiety recovery: a between-subjects experiment in virtual reality. Environ Int 136:105427. https://doi.org/10.1016/j.envint.2019.105427

Zinberg NE, Jacobson RC, Harding WM (1975) Social sanctions and rituals as a basis for drug abuse prevention. Am J Drug Alcohol Abuse 2(2):165–182. https://doi.org/10.3109/00952997509002732

Psychedelic-Assisted Therapy for Social Adaptability in Autistic Adults

Alicia Danforth

Contents

Abstract Access to the Internet has upended long-standing myths and misconceptions about autism as autistic individuals are enabled through technology increasingly to influence the dialog around neurodiversity, the experience of being autistic, and the effectiveness of mental health interventions for autistic adults. Autistic self-advocates are speaking up in support of including neurodivergent adults as a population that might benefit from the burgeoning psychedelic medicine field, in an absence of many other mental health treatment options that have been researched and shown to be effective for them. Autism is a genetically-determined neurocognitive variant with considerable heterogeneity across the broad autistic phenotype spectrum. Therefore, enthusiasm for investigating psychedelics to cure or alter the course of autism is most likely ill-informed and misdirected; psychiatric and psychopharmacological interventions do not alter the genome. However, autism frequently co-occurs with clinical conditions such as anxiety, depression, obsessive-compulsive disorder, and trauma that have been investigated as indications for clinical trials with classic and atypical psychedelics. The purpose of this chapter will be to inform researchers and clinicians on the history of clinical research with classic psychedelics with autistic minors, recent and current clinical trials of atypical

---

A. Danforth (✉)
The Lundquist Research Institute, Los Angeles, CA, USA
e-mail: adanforth@lundquist.org

Curr Topics Behav Neurosci (2022) 56: 71–92
https://doi.org/10.1007/7854_2021_269
Published Online: 11 November 2021

psychedelics with autistic adults, and considerations for providing psychedelic-assisted psychotherapies that are compatible with autism.

Keywords Asperger's · Autism · MDMA · Psychedelics · Social anxiety

1 Autism and Psychedelics Research

In early studies, more than 100 minors were given psychedelics in approximately a dozen trials of varying sizes and durations. Many of the children treated presented with characteristics and behaviors, such as muteness, self-stimulation, repetitive behaviors, lack of eye-to-eye gazing, and stereotyped behaviors (e.g., flapping, rocking) that were commonly associated with autism, which was also referred to at the time as childhood or juvenile schizophrenia. However, in the 1950s and 1960s, clinicians applied different diagnostic criteria than criteria used today. A retrospective case history review would be required to more accurately discern how many of those children might be likely to receive an autism diagnosis today versus another unrelated serious mental health condition. In recent psychedelics studies with adults, screening for enrollment has required some assessment for autistic traits as an inclusion criterion.

1.1 Early Psychedelics Research with Autistic Minors (1950s–1970s)

From the late 1950s to the early 1970s, researchers in the USA, Europe, and Argentina (Abramson 1967; Bender et al. 1962, 1963; Fisher and Castile 1963; Fisher 1970; Freedman et al. 1962; Rojas-Bermúdez 1960; Simmons et al. 1966; Simmons et al. 1972) conducted research with therapeutic applications of synthetic psychedelic compounds, mainly LSD, UML (a methylated derivative of LSD), and psilocybin, in attempts to treat autistic children and adolescents until clinical interest declined when funding for research and access to psychedelics decreased due to illicit use and political pressure (Sigafoos et al. 2007).

Rationale Researchers shared the common goal of breaking through the "autistic barrier" toward increased empathic affect and attachment (Mogar and Aldrich 1969). The rationales for conducting the research varied. Some studies emphasized a medical approach. For example, the prospect of restoring speech and reducing social isolation inspired some investigators after case reports were published on "improved affective display and contact" in schizophrenic adults who "talked more than usual" and "showed greater interest and emotion" while experiencing the effects of LSD (Freedman et al. 1962, p. 204). Employing a neurobiological approach, Bender et al.

(1963) made early attempts to understand the role of the neurotransmitter serotonin and its anomalies in its utilization in autism by including biochemical assays in LSD study protocols. Other researchers emphasized subjective considerations. Fisher (1970), for example, took a psychoanalytic approach to providing LSD and psilocybin to autistic children and emphasized the potential of psychedelics to augment the psychotherapeutic alliance.

Outcomes Mogar and Aldrich (1969) published a comprehensive review of seven independent studies of LSD for treatment of 91 children. Children as young as 5 years old were given repeated moderate to high doses (100–300 μg) of LSD (e.g., Fisher 1970; Simmons et al. 1972). In retrospect, the methods employed were flawed to varying degrees according to current standards. For example, shortcomings in methodology included: demographic heterogeneity across studies, small sample size (e.g., Rojas-Bermúdez 1960; Rolo et al. 1965; Simmons et al. 1966), ill-defined criteria for drug effects and outcomes (e.g., Bender et al. 1963, Bender et al. 1962; Freedman et al. 1962; Rolo et al. 1965), absent or incomplete baseline data (e.g., Bender et al. 1963; Rojas-Bermúdez 1960), and inadequate follow-up data collection (Mogar and Aldrich 1969).

Even though there were marked inconsistencies across the seven studies, the reviewers concluded that there was general consensus among the research teams that the LSD treatment sessions were "effective in characteristic ways" (Mogar and Aldrich 1969, p. 1):

> The most consistent effects of psychedelic therapy reported in these studies included: (a) improved speech behavior in otherwise mute children; (b) increased emotional responsiveness to other children and adults; (c) an elevation in positive mood including frequent laughter; and (d) decreases in compulsive ritualistic behavior. (p. 13)

In the studies with positive outcomes, increases in behaviors that indicated a greater capacity for comfort with interpersonal connectedness (e.g., eye-to-face contact) were observed more frequently during LSD sessions than in placebo or control sessions. For example, a majority of the studies reported an increase in emotional responsiveness with LSD, such as this one:

> There were definite changes in response to the environment, which was most remarkable in these autistic children. They became gay, happy, laughing frequently . . . Nearly all of them were more alert, aware, and interested in watching other persons. (Bender et al. 1963, p. 87)

Overall, acute negative responses to psychedelic treatments were typical of a "bad trip," characterized as episodes of fear or panic and isolating from others. Some participants had both negative and positive experiences (Mogar and Aldrich 1969). No researchers reported serious side effects, toxicity, or permanent regression following treatment with psychedelics.

A majority of the early researchers were primarily concerned with observing drug effects and proceeded without a clearly stated treatment hypothesis or random assignment to control groups, and reports on the durability of post-session effects were vague and inconsistent.

Only two LSD studies of childhood autism (Simmons et al. 1972; Simmons et al. 1966) applied methods that approached the rigor of current protocol and ethical standards. In one example, the Simmons et al. (1966) twofold experiment employed a single-subject reversal design, in which a pair of identical male twins participated in both arms of the study. In total, both boys participated in six double-blinded, inert placebo observation sessions and nine sessions with a moderate dose of LSD. The LSD sessions were interspersed with placebo and control observations.

One factor that set the Simmons et al. (1966) experiments apart from less methodologically sound studies was that the researchers clearly defined the dependent variables. To enhance the objectivity level of measurements, they also standardized their testing and observation processes, employing state-of-the-art recording equipment, and recorders' reliability was assessed as adequate before, during, and after the study. As a result, the study design allowed investigators to replicate effects and show internal validity. The Simmons et al. (1966) and Simmons et al. (1972) studies are valuable today because they cannot be replicated under current restrictions. Although the findings from the 1972 study with a larger participant group were more variable and less encouraging than the 1966 study, Simmons et al. (1966) concluded that LSD could be an adjunct to psychotherapy for autistic children due to the potential to promote desired behaviors, such as an increase in positive affect and eye-to-face contact with examiners.

Table 1 provides an overview of the scope of the early studies and general outcomes.

Conclusions from Early Studies The general consensus resulting from the early research was that psychedelics were unlikely to affect or alter the course of autism or to significantly restore speech in non-speaking autistic children. Most researchers concluded that psychedelics allowed participants a greater degree of contact with others, regardless of demographics, dose, schedule, or setting (Mogar and Aldrich 1969). Follow-up data gathering ceased by 1974, before firm conclusions were reached about longer-term outcomes.

1.2 Psychedelic Clinical Research with Autistic Adults

In the early studies with presumed-autistic minors, investigators could not collect self-reported baseline or follow-up data or conduct interviews. Fisher and colleagues (Fisher and Castile 1963) reported that participants with verbal skills and less severe disability showed more improvement. However, participants were unable to communicate verbally in most of the early experimental psychedelic treatment sessions. When research with psychedelic-assisted therapy to explore autistic social adaptability resumed after revision and expansion of diagnostic criteria along a broader spectrum, investigators could recruit speaking adults instead of working with non-speaking children with severe disabilities that precluded informed consent and self-reported data.

Table 1 Early psychedelics research with autistic minors: 1950s–1970s

Study	Patients	Age Range in years	Agent	Dose (mcgs/mgs)	Treatments per patient	Sched.	Treatment Effects		
							Excellent	Good	Poor
Abramson (1960)	6	5–14	LSD	40	3–6 (mean 4)	Weekly	5	–	1
Freedman et al. (1962)	12	6–12	LSD	100	1	–	–	5	7
Bender et al. (1962)	14	6–10	LSD	100	45	Daily	7	7	–
Bender et al. (1963)	44	6–15	LSD/ UML	50–150/ 4–12	60	Daily	20	21	3
Fisher and Castile (1963)	12	5–13	LSD/ Psilocybin	50–100/ 10–20	1–11 (mean 4)	Bi-weekly/ Monthly	4	4	4
Rolo et al. (1965)	1	12	LSD	100	28	Daily	–	–	1
Simmons et al. (1966)	2	5	LSD	50	9	Twice weekly	2	–	–
Simmons et al. (1972)	17	5–13	LSD	50	2–3	Weekly	7	3	7
Fisher (1970)	1	12	LSD Psilocybin	100–300	14 3	7–70 days	1	–	–

MDMA An important innovation that supported commencing, research with psychedelics for autistic adults was the reintroduction in the 1970s of the phenethylamine 3, 4-methylenedioxymethamphetamine (MDMA), which became well-known in uncontrolled settings as Ecstasy or Molly. MDMA shares characteristics with classic hallucinogens and stimulants, and research has shown that it can increase affective empathy and prosocial behavior (Hysek et al. 2014). For purposes of this discussion, MDMA will be classified as an atypical psychedelic.

1.2.1 Qualitative Research

Pre-clinical qualitative research that included survey data and Thematic Content Analysis of semi-structured interviews suggested that MDMA use might hold potential for improving social adaptability in autistic adults (Danforth 2013, 2019). In spite of limitations of qualitative research methods, self-selected and self-reported survey responses supported the case that positive effects of MDMA/ecstasy use in non-controlled settings could be both notable and durable and also warranted further exploration in randomized, placebo-controlled, clinical studies of safety, feasibility, and efficacy.

One hundred survey participants who self-reported MDMA/ecstasy use in the range 1–20 times were asked to indicate which items from a list of commonly reported acute subjective MDMA effects, if any, they recalled experiencing when they took MDMA/ecstasy (Table 2). The acute effects selected were consistent with known effects for MDMA, and responses were in alignment with the high level of confidence that participants reported that the substance they consumed contained MDMA. Two notable findings from the survey questions about drug effects were that 91% of respondents reported that they experienced "Increased Feelings of Empathy/Connectedness," and 86% indicated "Ease of Communication" as an effect of their MDMA/ecstasy use.

Respondents rated the intensity of reported effects on a 0–6 Likert-type scale. Positive effects (e.g., joy, openness, enjoying being touched) were reported as more strongly experienced in all examples; whereas, no participants reported strongly experiencing anxiety. Only 2% of participants reported that they did not experience "feeling more emotions than usual," and only 2% indicated that finding it "easier than usual to talk with others" was not a feature of their MDMA/ecstasy experience.

Some participants reported sustained benefits from non-clinical MDMA/ecstasy use. A notable finding was that 72% of MDMA/ecstasy-experienced participants reported "more comfort in social settings," and 12% indicated that the effect lasted for two or more years. Another positive outcome reported was that 78% of the MDMA/ecstasy-experienced group reported "feeling at ease in my own body" as a result of consuming the drug, and 15% indicated that the effect lasted 2 years or longer.

A finding that might have relevance to establishing rapport with therapists in clinical settings was that 77% of the MDMA/ecstasy-experienced group reported that they found it "easier than usual to talk to others" as an effect of taking MDMA/

Table 2 Intensity of effects autistic adults reported experiencing during MDMA/ecstasy

Effect[a]	0—Did not experience (%)	1 (%)	2 (%)	3 (%)	4 (%)	5 (%)	6—Strongly experienced (%)
Increased insight into own thought processes	10	4	6	10	17	16	37
Joy	3	1	1	4	14	14	63
Feeling more emotions than usual	2	2	4	4	14	22	52
Getting a sense of how others feel	8	6	6	7	23	20	30
Anxiousness	39	21	20	12	4	4	0
Understanding why others feel the way they do	16	7	4	18	17	16	22
Openness	3	2	2	0	12	22	59
Feeling at ease in my own body	6	1	0	4	10	19	60
Enjoying being touched	8	3	2	6	9	18	54
Disappointment	63	16	5	4	5	2	5
Increased sense of humor	13	5	4	26	20	18	14
Easier than usual to talk with others	2	2	4	5	8	23	56
More comfort in social settings	3	3	3	6	11	18	56
Better able to discuss emotions	10	0	2	9	14	17	48
Easier to express affection	3	0	5	5	8	21	58
Pleasant body sensations	2	3	1	5	5	20	64
Unpleasant body sensations	41	33	14	6	4	0	2

[a]Self-reported survey responses ($N = 100$)

ecstasy, and 18% indicated that the effect lasted up to 1 year or longer. A final finding about the duration of effects that could have implications for psychotherapy for autistic adults was that 22% of the MDMA/ecstasy-experienced group reported "increased insight into own thought processes" that persisted for two or more years. For more than half of the respondents, the last MDMA/ecstasy use was less than 1 year prior to survey completion. Therefore, the actual durability of all of the reported effects in the survey is unknown, and the duration of reported effects would likely increase over time.

Three overarching metathemes that emerged from the systematic qualitative analysis of interview transcripts were the potential roles of MDMA/ecstasy as a change, transformation, and healing catalyst. Content from multiple interviews described the potential of MDMA to enhance psychotherapy. Examples of constellations of meaning units in interview text clustered around subthemes supporting MDMA's potential to enhance therapeutic rapport, increase affect regulation and coping skills, reduce defenses, increase self-esteem, improve interpersonal skills, enhance psycho-social overall well-being, and minimize resistances to psychotherapeutic processes. In addition, non-clinical MDMA/ecstasy use was self-reported to have decreased symptoms of specific conditions, such as trauma/PTSD, depression, general anxiety, alexithymia, and social anxiety.

Limitations and Delimitations Self-selection bias was likely to have been a notable limitation for this study. The researcher mitigated the risk of overly optimistic responses by encouraging participants to provide accurate descriptions of their experiences and by assuring them that accounts of positive and negative experiences were of equal importance. Other key limitations, such as selective memory, existed with regard to discussing use of an illegal drug from a retrospective point of view. Without robust baseline data or pre- and post-MDMA/ecstasy comparisons, conclusions were limited.

The researcher was not able to confirm purity, dose, extent of reported or non-reported poly drug use, or concomitant use of other substances, including prescription medications. In order to assess the likelihood that the respondent consumed ecstasy that contained MDMA, general data were collected about the source, visual description of substance, and pre-use drug testing measures. All participants who reported ecstasy use were asked to rate their confidence level that the substance they took contained MDMA. Factors such as frequency of use and duration since last use increased variability in respondent profiles and limited generalizing to larger autistic populations.

The Autism-Spectrum Quotient (AQ) screening survey used to determine eligibility for participation in this study is not diagnostic. Latitude in inclusion was afforded because an unknown percentage of adults on the autism spectrum have never been assessed or diagnosed. Considering that the official definition and diagnostic criteria of autism changed and Asperger's disorder was eliminated from a new edition of the *DSM* over the course of the data gathering and analysis, the researcher concluded that being more inclusive would yield more comprehensive, robust, and relevant data.

Despite attempts at global, cross-cultural recruitment, 88% of survey participants and 92% of interview participants reported their ethnicity as "white/non-Hispanic." A recommendation for future studies would be to increase efforts to recruit non-white participants. Efforts to recruit female participants by posting announcements to women's autism groups resulted in better than anticipated female participation. Individuals who did not have access to computers and the Internet would not be able to participate, which could have excluded older persons, less literate persons, and those of lower socioeconomic status.

A lack of time and resources to collect longitudinal data could result in latent or late-emerging positive and adverse outcomes being overlooked. In addition, self-selection bias might have contributed to overly positive findings that are minimally generalizable to the larger autistic population. Furthermore, no data were collected from individuals who engaged in high-risk use, such as taking MDMA/ecstasy in excess of 50 times. Plus, only two survey respondents reported taking higher doses at or above 300 mgs.

To offset personal bias, the researcher wrote down all known opinions and assumptions about MDMA/ecstasy and the autism spectrum, however they came to awareness. Throughout the research process, in addition to logging analysis notes, the researcher kept a personal journal of internal dialog, conscious thoughts, personal assumptions, intuitive inspirations, emotions, opinions, and synchronous life circumstances outside of the research setting that influenced the research. To the fullest extent possible, the researcher's relevant experience was acknowledged and bracketed as present but not intruding on participant experiences or the researcher's interpretation of them.

1.2.2 Recent and Current Clinical Trials with Autistic Adults

Data suggest that a combination of both intrinsic genetic information and multiple extrinsic factors may contribute to the development of autism (Amaral 2017). Even though the ever-increasing body of whole-brain analysis has provided deeper insights into the potential biological bases of social cognition, no single theory dominates regarding an explanation for why autism affects social cognition and behavior (Velikonja et al. 2019). Regardless of ambiguous etiology, there is consensus that autistic adults are at heightened risk of mood and anxiety disorders (Hollocks et al. 2019). Social anxiety disorder (SAD) is far more common in autistic populations than among non-autistic individuals (Spain et al. 2018). Prevalence rates vary widely across studies in autistic adult clinical populations, yet there is general consensus that prevalence rates are much higher than the estimated 7–13% in the general population in the USA (Kessler et al. 2012).

Comparative studies suggest that speaking autistic adults with less readily apparent autistic traits and without significant intellectual disability are faced with considerable pressure to conform to non-autistic social norms. As a result, they are at greater risk for social distress (Tantam 2000). Given the paucity of confirmed efficacy in clinical trials and clinical practice, the search for supportive novel treatments for mental illness in autistic adults is relevant given the heterogeneity of autism and the paucity of effective treatment options (Tantam 2003). Pharmacotherapy generally is indicated for alleviating co-occurring conditions such as depression, anxiety, obsessive-compulsive disorder, and Tourette's syndrome (Tsai 2007). However, responses to psychiatric medications in autistic populations often are atypical, variable, and in some cases, iatrogenic (Coleman et al. 2019).

Phase 2 Pilot Study of MDMA-Assisted Therapy for Social Anxiety The objective of the first controlled study of MDMA-assisted therapy in autistic adults was not to attempt to cure or alter the course of autism (Danforth et al. 2015). Since MDMA-assisted therapy shows promise as a treatment for other anxiety disorders, such as OCD (Moreno et al. 2006) and PTSD (Mithoefer et al. 2019), a blinded, placebo-controlled pilot study was conducted to explore feasibility and safety of MDMA-assisted therapy for reduction of social fear and avoidance (Danforth et al. 2018).

In this pilot study, 12 autistic adults with marked to very severe social anxiety were randomized to receive MDMA (75–125 mg, $n = 8$) or inactive placebo (0 mg, $n = 4$) during two 8-h therapy sessions (experimental sessions) in a controlled clinical setting. Double-blinded experimental sessions were spaced approximately 1 month apart with three preparatory non-drug therapy sessions and three non-drug therapy sessions following each experimental session to support the participants with integrating their experiences. Research findings support mindfulness-based therapies for autistic adults (Poquérusse et al. 2021; Hartmann et al. 2012). Consequently, all participants received standardized mindfulness-based therapy adapted from the Dialectical Behavioral Therapy (DBT) brief Core Mindfulness module during their preparatory therapy sessions (Linehan 2020). A notable advantage of this approach was the introduction of vocabulary and skills, such as somatic grounding through the five senses, intended to support participants with transitioning into MDMA-influenced cognitive and affective states, as well as with communicating with others during novel, often ineffable, altered states of consciousness.

The primary outcome was change in the clinician-administered Liebowitz Social Anxiety Scale (LSAS) (Heimberg et al. 1999) total scores from Baseline to 1 month after the second experimental session. Outcomes were measured again 6 months after the last experimental session. One blinded independent rater administered every LSAS assessment, from Baseline through 6-month follow-up, to mitigate the influence of researcher bias.

This study demonstrated rapid and durable improvement in social anxiety. Improvement in LSAS scores from Baseline to the Primary Endpoint was significantly greater for the MDMA group compared to the placebo group ($P = 0.037$), and placebo-subtracted Cohen's d effect size was very large ($d = 1.4$, CI: −0.074, 2.874). Change in LSAS scores from Baseline to 6-month follow-up showed similar results ($P = 0.036$), with a placebo-subtracted Cohen's d effect size of 1.1 (CI: −0.307, 2.527). The rate of clinically significant changes in SAD symptoms from Baseline was 6/8 (75%) with MDMA versus 2/4 (50%) with placebo. There also was a clear linear relationship between visit and mean LSAS score for the MDMA group, whereas no such relationship existed for the placebo group.

Changes in LSAS scores and subjective observations were consistent with the hypothesis that anxiety interferes with social functioning in autistic adults and can be alleviated with a combination of MDMA and therapy, supportive preparation, and integrative after care. Findings supported more trials of MDMA-assisted therapy in larger samples of adults with social anxiety.

Other Relevant Clinical Research As of this writing, a randomized, triple-blinded cross-over study is in progress to investigate the effects of MDMA on responses to affective touch in individuals with a range of autistic traits. This study is scheduled to conclude in 2022. A related double-blind, within-subject study with healthy normal volunteers concluded that MDMA enhances the pleasantness of affective touch (De Wit and Bershad 2020). Additionally, a growing body of research in animal and human studies into the prosocial, anxiolytic, and other relevant effects of MDMA and potential mechanisms of action will continue to inform future clinical trials that include autistic adults (De Gregorio et al. 2021; Dumont et al. 2009; Edsinger and Dölen 2018; Kamilar-Britt and Bedi 2015; Young et al. 2017).

2 Establishing an Autism-Affirming Set and Setting for Psychedelic-Assisted Therapy

Clinical insight into the importance of set and setting to successful outcomes of psychedelic-assisted therapies is evolving, and reasonable accommodations in these domains are essential for positive outcomes for autistic adults. *Set* refers to the traits, mind state, and expectations of the participant regarding the session (Strassman 1995). The clinician's perception of the nature of the experience, the preparation and pre-session psychotherapy, and the specific technique of guidance employed during the drug experience also contribute to set (Grof 1980). *Setting* refers primarily to the external factors that influence the ambiance of the experience for the participant. An ideal setting includes a safe and pleasant environment with reassuring interpersonal support. Additionally, the set of research team members, including factors such as training, countertransference, empathy, and expectations for drug effects influence the client as elements of the setting.

2.1 Considerations for Establishing a Neurodiversity-Affirming Set

Attending to set for psychedelic-assisted therapy for autistic adults requires fostering a conducive mindset for both the participants and the clinicians well in advance of medication sessions. As a first step, early community engagement can have a positive influence on set for all involved in the treatment process. When possible, research benefits from collaboration with autistic care providers, researchers, and consultants. Their participation has the potential to reduce the risk of both overt and covert ableism (discrimination in favor of able-bodied or neurotypical norms), by reducing the risk of unintentional harms, hurtful language, outdated stereotypes, and general misconceptions about the needs and preferences of autistic adults.

Autism-Informed Practice Therapists preparing to work with autistic adults are advised to complete a thorough examination of their pre-existing biases and understandings of what autism is and what it is not. Myths about autism are still prevalent, even among professionals who provide services specifically for them. Reading information published by autistic writers and researchers is as critical to preparation as reviewing relevant peer-reviewed research literature in preparation for providing treatment. Familiarity with the lived experience of autism as well as well-informed insight into the heterogeneity of the spectrum is essential to maintaining a supportive set. For example, therapists should be thoroughly comfortable with differences such as atypical eye contact, tics, idiosyncrasies in prosody and vocabulary, dyspraxia, aversion to common forms of social touch such as shaking hands, and challenges with both expressive and receptive social communication.

Language Another step that clinicians can take starting at intake to foster a therapeutic set is to demonstrate respect for language and communication preferences. Treatment providers need to feel confident in initiating conversations around terms used to describe autism. For example, a participant might have a strong preference for identity-first language (i.e., "I am autistic.") as opposed to person-first language (i.e., "I am a person with autism."). Terms such as "Aspie" can be considered affirming or demeaning. As the influence of the neurodiversity paradigm becomes more pervasive in research (Chapman 2020), even terms that have been widely accepted in the past, such as "high-functioning," are increasingly viewed as inaccurate, unnecessarily pathologizing, and even offensive. A frank conversation to check-in about language preferences can be an effective early step toward establishing therapeutic rapport and reducing the likelihood of adverse ableist influences (Bottema-Beutel et al. 2021). Additionally, due to notable sex and gender diversity in autistic populations (Strang et al. 2020), using correct pronouns and names is vital to establishing a supportive set.

Alexithymia *Alexithymia* refers broadly to subclinical difficulties with identifying and describing emotions. Given that around 50% of autistic adults are also alexithymic (Poquérusse et al. 2018), assessment with instruments such as the Toronto Alexithymia Scale (TAS-20) or Perth Alexithymia Questionnaire (Preece et al. 2020) prior to psychedelic-therapy is recommended as a factor in establishing rapport and providing appropriate accommodations. In cases where clinically significant alexithymic traits are present, providing alternative means of communicating affective states such as "Emotion Wheel" graphics (e.g., http://feelingswheel.com/) or a deck of emotions flash cards (e.g., https://www.mixed-emotions.com/) can help participants in psychedelic-assisted therapy feel more assured that they will be able to communicate their emotions to others throughout the process.

Communication and Emotion Regulation Normalizing the possibility that someone receiving treatment might find themselves unable to communicate verbally can help reduce anxiety around the process. Researchers should confirm a plan in advance in case speaking becomes inaccessible for participants at any point. Some participants might benefit from having access to assistive technology to type instead of speak, mutually understood sign language/hand signals, or simply an acknowledgement that there might be periods of time when they do not prefer to use spoken language. Similarly, some participants might want to stim (i.e., use repetitive self-soothing movement or gestures), rest on the floor, or adopt an atypical body posture. Knowing in advance that they will not be judged for this type of self-regulation introduces compassion and acceptance into the co-created set and setting.

Structure In general, autistic individuals tend to prefer consistent structures and routines. Therefore, extra effort to make schedules, assessment requirements, study tasks, and steps in the process as explicit and consistent as possible are likely to contribute to a desirable set. Extra time might be required to complete the Informed Consent process, allowing for elaboration and clarification as needed to support a truly informed decision and to manage expectations appropriately. Therapists should be prepared to increase the length or frequency of office visits to accommodate slower communication styles, as needed. Any deviations from the protocol or previously discussed plans should be communicated as far in advance as possible.

Social Challenges Autistic individuals are often uncomfortable meeting strangers. Therapists should prepare participants in advance when they are likely encounter someone new to them during treatment and follow-up, such as phlebotomists, administrative staff, pharmacists, or other team members. Also, group communication can be particularly difficult. Learning more about social situations that might be problematic can be part of the intake process and appropriate accommodations can be implemented, when possible.

Agency An opportunity at the intersection of set and setting to improve both is to invite autistic participants to make adjustments to the treatment room for their comfort. For example, invite them to occupy the seat in the room that is most welcoming to them upon first arriving. Clinicians can solicit input on modifications for comfort and acknowledge openly that they might need guidance to address hypersensitivities that are not overtly apparent to them.

2.2 *Considerations for Establishing Neurodiversity-Affirming Settings*

An important factor to attend to when establishing a supportive setting for psychedelic-assisted therapies for autistic adults is the prevalence and heterogeneous presentation of sensory modulation difficulties in this population (Ben-Sasson et al. 2009). Debilitating consequences of sensory hyper-reactivity and over-arousal as well as sensory hypo-reactivity have the potential to interfere with treatment. Therefore, considerable care is required to create a supportive setting that accommodates sensory atypicalities, reduces the risk of sensory overwhelm, and minimizes sudden, unexpected changes. Careful assessment and remediation of environmental factors that can have an impact on all of the senses is recommended.

2.2.1 Attending to the Senses

Below are examples of the types of considerations clinicians and researchers can address to improve the setting for psychedelic-assisted therapy with autistic adults.

Sight Soothing ambient lighting is important. Autistic individuals occasionally cite strobing fluorescent lighting as a distressing trigger. When possible, treatment facilities should utilize versatile window treatments that allow for an unobstructed pleasant view out a window, filtered daylight, and light blocking. Elements of nature, such as flowers or art that depicts landscapes or natural beauty, provide an option for a pleasant and non-suggestive décor theme that accommodates spiritual experiences without potential intrusion of undesired religious overtones. Items such as battery-operated flickering candles, tapestries, soothing colors, and self-selected personal objects also can work well, when appropriate.

Hearing Research supports the hypothesis that high levels of auditory perceptual capacity in autistic individuals correlate with higher levels of sensory sensitivities (Brinkert and Remington 2020). Therefore, the auditory environment is a critical aspect of setting. Special attention to the comfort and volume control of session music is essential, and therapists need to be prepared for requests for silence, unusually low volume, or ambient music instead of headphones. In addition, accommodations such as white noise machines or sound-cancelling headphones might be necessary to mitigate sound intrusions (e.g., sirens, announcements over PA systems) coming from outside the treatment setting. Each participant should be invited to assess the setting in advance of treatment sessions in order to determine if there are subtle sounds that are distracting or uncomfortable for them that might not be obvious to the treatment team. Furniture and lighting that are as quiet as possible when in use are recommended (e.g., minimal creaking or buzzing). Finally, some autistic individuals have difficulty modulating the volume of their speaking voice or present with conspicuous verbal tics. Therapists are advised to conduct treatment in settings in which sound emitted by participants is unlikely to disturb others nearby.

Smell As a general rule, creating a fragrance-free setting is advisable. Some autistic individuals are extremely sensitive to scents. Perfumes and colognes, body lotions, hair products, undesired intrusive food odors, tobacco smells in clothing/hair, and similar scents should be minimized in treatment settings as much as possible.

Taste Food sensitivity, selectivity, and neophobia have been studied in autistic children and adolescents and can also persist into adulthood (Kuschner et al. 2015). With longer-acting psychedelics, serving a light meal is often an element of the treatment day schedule. In addition to noting food allergies, consultation with participants in advance about food preferences and making a list of items to be avoided is recommended. Charting other dietary restrictions is also important, as many autistic adults adhere to gluten-free or medically restricted diets (Sumathi et al. 2020).

Touch Due to atypical sensory reactivity, extra attention in selecting furniture with comfortable textures, pillows, bolsters, and blankets is advised. Assumptions about touch boundaries and preferences should be made explicit in detail at intake. They can vary from extreme aversion to light, brushing skin contact to preferences for deep pressure or extended contact. Alternatives to wearing eye shades and over-ear headphones can be discussed if they are likely to cause physical or emotional discomfort. Having a weighted blanket available can support sensory regulation (Green et al. 2020) and substitute for high-touch contact, as appropriate; however, care should be taken to avoid overheating. As with any psychedelic-assisted therapy, conversations in advance of treatment about unambiguous boundaries of personal/intimate/sexual touch are a paramount priority. When possible, demonstrate and practice mutually acceptable touch in advance to confirm boundaries and expectations. As a general guideline, supportive touch can be limited to non-sexual/non-overtly sensual contact, upon participant request, from the top of the shoulder to the hand, as appropriate. Therapists can explore the use of signals, such as a no-contact *air high-5* or *thumbs up* or words of encouragement that can serve as substitutes for affirming, reassuring touch. Additionally, washing hands in warm water, offering a warmed moist towelette, or providing a basin to wash the face can sometimes be an effective substitute alternative to touch.

2.2.2 Additional Features of Autism-Friendly Settings

The following list of suggestions is not intended to be exhaustive but rather to encourage more empathic and informed ways of supporting neurodivergent adults in clinical settings during what is often their first experience with psychedelics:

Age-Appropriate Ambiance Autistic adults frequently struggle with feeling as if they lag behind their age-cohort peers in meeting developmental milestones. In spite of a physical transition to full adulthood, autistic individuals commonly report feeling younger than their chronological age. By necessity, many of them have had to seek services in settings with pediatric motifs that can feel insulting and alienating. Accordingly, décor in treatment rooms should be appropriate for an adult setting.

Restroom Facilities Autistic adults who experience anxiety may be in the habit of seeking privacy by retreating for brief respite in a restroom. Optimal treatment locations will include a private, gender-neutral restroom in close proximity that eliminates concerns of encountering strangers unexpectedly. (Note: Seeing one's reflection in a mirror can have the potential to be distressing. The treatment team can agree in advance whether or not to cover mirrors, with access to view them optional upon request.)

Somatic Awareness and Motor Control Professionals working with autistic adults need to be responsive to challenges with dyspraxia, joint hypermobility, poor proprioception, and general problems with motor coordination which are significantly more prevalent than in non-autistic populations (Cassidy et al. 2016). Push-button motorized furniture that reclines from upright to fully reclined allows for an increased measure of agency for participants in positioning the body with enhanced ease in ways that have the potential to improve the setting by increasing self-directed decisions about comfort with minimal disruption to the session flow.

Stim/Fidget Objects Autistic adults are likely to appreciate having a selection of stimming objects readily available, such as fidget spinners, tangles, stress balls, or other items that can be manipulated to relieve excessive motor agitation and for self-soothing (e.g., www.stimtastic.co).

Personal Comfort Objects Bearing in mind that temporary age regression during psychedelic experiences is common and that autistic individuals often already feel younger than their chronological age, an invitation to bring personal objects for grounding and comfort is advised. Participants may opt to bring blankets, pillows, artwork, photos, items of religious significance, or other creature comforts to help them feel more at ease in the treatment room.

Privacy Enhancement Options Autistic adults often present with a history of traumatic experiences of feeling deep shame after episodes of emotion dysregulation that is observed by others and accompanied by bullying. A validating discussion in advance of treatment about protocols to be observed (Lipsky and Richards 2009) if a meltdown or shutdown (i.e., an episode of marked emotion dysregulation) occurs during a session, may enhance both the set and setting by reducing anticipatory anxiety. When possible, designate a safety spot where the participant can temporarily withdraw to *hide* or self-regulate, such as under a blanket or behind a privacy screen, while still feeling supported.

Support Partner In research settings, autistic individuals might benefit from the option to choose a person they know well, such as a family member or friend, for general support related to participation in a clinical trial. This individual would be responsible for driving them to and from drug treatment sessions and providing other assistance, as appropriate, for the duration of the study. They might also be invited to remain available in a designated waiting area on treatment days, in the event the participant requests brief contact with them during a session for ad hoc support.

3 Therapist Role in Psychedelic-Assisted Therapy with Autistic Adults

In comparison with psychedelic-assisted therapies with an emphasis on inner, reflective work, sessions focused on social adaptability can sometimes require a more twofold approach. Although deep insight-oriented process might be a core element of sessions, psychedelic-assisted treatment also can provide opportunities for autistic individuals to explore interpersonal interactions in ways that are often difficult or even inaccessible to them. Therefore, an essential element of the therapist's role during psychedelic-assisted therapy sessions is discernment regarding when to facilitate introspection and when to invite social interaction. For example, in addition to the classic *headphones and eyeshades* method, the therapists might invite or gently encourage:

- experimenting with eye contact and face gazing.
- exploring the participant's affective responses to the music played during the session as a means to practice emotion expression and reciprocal social communication.

- sharing of artwork or writing created during or after the session.
- roleplaying of various social situations, as appropriate.

For autistic individuals, psychedelic-assisted therapy can be analogous to social *training wheels* allowing participants to explore different types of social interactions in a supportive setting.

Regarding autism-specific tips and guidelines for establishing a conducive set in psychedelic-assisted therapies more generally, therapists are encouraged to adapt how they communicate to be sensitive and responsive to obvious or subtle disabilities related to social inference. For example, autistic adults might struggle with open-ended, non-specific questions, figurative language, understated sarcasm, or using language to self-advocate. Therapists will more readily establish effective rapport if they accommodate individuals who have difficulty accurately interpreting facial expressions or body language. Additionally, therapists should be sensitive to the potential need to extend the number, frequency, or duration of visits for autistic adults in treatment (Maddox et al. 2018).

In research settings, all assessments and measures should be evaluated for rightness of fit for autistic individuals. When possible, literature documenting whether or not instruments have been normed in autistic populations with validity shown for autistic adults can be cited in protocols and findings papers. Similarly, when psychotherapy is provided, interventions that have been shown to be effective for adults on the spectrum with substantiating research literature should take priority over other options, when possible.

Substance Choice Autistic research participants occasionally have reported that they perceive the empathic, heart-opening effects of MDMA as somewhat synthetic or *fake*. As a group, autistic adults stereotypically place high value on authenticity, and some individuals prefer the more reliable mystico-spiritual effects of the classic psychedelics for this reason. Anticipating potential future treatment regimens, autistic participants might benefit from a gentler first experience with MDMA to establish familiarity with the coming on, peak, plateau, and return phases of psychedelic altered states of consciousness followed by treatment with more disruptive, unpredictable, and longer-acting classic psychedelics.

4 Risk Mitigation in Psychedelic-Assisted Therapies for Autistic Adults

In addition to the risk of unintentional retraumatizing exposure to ableism due to failing to implement appropriate accommodations, other risks require careful consideration as access to psychedelic therapies for autistic adults expands. At present, MDMA and psilocybin are advancing toward approval as prescription medications at an accelerated pace due to being granted breakthrough therapy designation for other indications based on large effect sizes in Phase 2 clinical trials. Therefore, the current circumstances present a pressing need for timely analysis of potential risks to prevent actual harms. Several potential drawbacks and harms, that have not been explored extensively in research literature yet, merit careful attention here, especially when considering treatment for vulnerable, understudied, and underserved populations.

Age of Consent One of the most significant risks that emerged simultaneously with the announcement of the first clinical trial for MDMA-assisted therapy with autistic adults was the assumption in research communities, media, and the general public that this type of treatment might also be appropriate for children. The well-intentioned, but possibly misguided expectation of regulatory agencies that young children will need to undergo safety testing with psychedelics as part of the standard approval process for new drugs might need to be modified, even bypassed, in order to address safety concerns appropriately for substances that alter consciousness and can affect sense of self and worldview. The effects of psychedelics are so novel and disruptive to ordinary states of consciousness, that making the decision on behalf of someone else to take them for the first time before the age of consent could be considered unethically intrusive.

To elaborate, obtaining truly informed consent or assent is a delicate issue when considering novel psychiatric treatments for children that could have life-long ramifications, including resentment that the decision was made for them and thus undermining their sense of agency. The choice is somewhat analogous to losing a virginity, in that the experience can be a transformative *first* that an individual deserves to choose for oneself when mature enough to consent. The risk to autistic minors is compounded by the likelihood of significant developmental delays, regardless of intellectual ability. Autistic youth often require more time to engage meaningfully with the full sense of agency required to identify, refine, and express a personal intention for treatment. Children and younger adolescents might benefit from training and therapy focused on core mindfulness, emotion regulation, assertive communication, journaling, and other skills that could help to prepare them if they decide at an appropriate age of consent to explore psychedelic-assisted therapy. The minimum age for safety and treatment studies with older adolescents can be reconsidered after more data from adult studies become available.

Furthermore, the autism community has a long and tragic history of quackery and harmful interventions (Herbert et al. 2002). Well-intentioned parents and service providers sometimes take desperate and detrimental measures to *cure* their children in order to remove a perceived *autistic barrier*. In some cases, psychedelic-assisted therapy might be indicated for caregivers to help them reconcile existential difficulties with adapting to parenting a child whose neurotype differs from theirs or from their expectations.

Increased Risk of Suicidality The risk of suicidality and suicidal death is significantly higher in autistic adults than in the general population (Hedley and Uljarević 2018). Risk assessment screening at intake and periodically throughout study participation with measures such as the Columbia-Suicide Severity Rating Scale

(C-SSRS) (Posner et al. 2011) can be included as a risk mitigation and safety strategy.

Non-responders Another risk that is currently in early stages of investigation is the non-responder effect. In the case of MDMA, some individuals in research settings have attenuated physiological effects (e.g., no increase in blood pressure, heart rate, or temperature) as well as atypical or absent subjective effects (Danforth et al. 2018). In some cases, they report no or negligible observable effects of the study drug in comparison with placebo. Some early data analysis suggests that this effect might be due to a prior history of selective serotonin reuptake inhibitor (SSRI) medication, despite tapering off prior to study enrollment (Feduccia et al. 2021). Genetically-determined rapid metabolism is also under investigation as a factor that contributes to atypical attenuated responses (Studerus et al. 2021). The current public enthusiasm for psychedelic-assisted therapies has the potential to contribute to a profound sense of disappointment and even hopelessness if expectations about the full range of known effects, including no discernable effects, are not made explicit as part of the informed consent process prior to medication treatment. Alternative plans for aftercare should be discussed to mitigate potential harms of a non-responder outcome.

Termination and Aftercare An additional risk of short-term treatment, either in research or clinical settings, is the potential failure to adequately prepare participants for what can seem like an abrupt termination process after a profound cathexis with treatment personnel. Psychedelics have been shown to promote hyper-suggestible states (Carhart-Harris et al. 2015). Subjective accounts from autistic adults who have used psychedelics in research and uncontrolled settings often report a profound sense of connection to others that is new for them (Danforth 2013). Anticipating the possibility of strong parental or erotic transference should be included in study design and treatment planning, with preparatory discussions around explicit timelines and boundaries beginning at enrollment/intake. Additionally, clinicians need to be prepared to support clinically significant age regression in participants during psychedelic therapy sessions (Taylor 2014). An intentional and thorough termination process, appropriate transitional objects, and contingency plans for aftercare can be confirmed in advance, as appropriate.

5 Conclusion

Data from early research with presumed-autistic children and teens, academic and community-based qualitative data, and findings from recent basic and applied research all support cautious and committed inclusion of autistic adults in the current psychedelic renaissance. The prevalence of co-occurring mental health conditions among this population, growing interest in peer-support communities (e.g., https://www.autisticpsychedelic.com/), and data from other relevant research provide a solid foundation for embracing neurodiversity in disruptive psychopharmacology that holds promise for new and innovative treatments.

References

Abramson HA (1960) The use of LSD in psychotherapy: transactions of a conference on dlysergic acid diethylamide (LSD-25). Josiah Macy Jr. Foundation, Oxford

Abramson HA (1967) The use of LSD (D-lysergic acid diethylamide) in the therapy of children: a brief review. J Asthma Res 5:139–143

Amaral DG (2017) Examining the causes of autism. Cerebrum: the Dana forum on brain science. Dana Foundation

Bender L, Goldschmidt L, Sankar DS, Freedman AM (1962) Treatment of autistic schizophrenic children with LSD-25 and UML-491. In: Recent advances in biological psychiatry. Springer

Bender L, Faretra G, Cobrinik L (1963) LSD and UML treatment of hospitalized disturbed children. Recent Adv Biol Psychiatry 5:84–92

Ben-Sasson A, Hen L, Fluss R, Cermak SA, Engel-Yeger B, Gal E (2009) A meta-analysis of sensory modulation symptoms in individuals with autism spectrum disorders. J Autism Dev Disord 39:1–11

Bottema-Beutel K, Kapp SK, Lester JN, Sasson NJ, Hand BN (2021) Avoiding ableist language: suggestions for autism researchers. Autism Adulthood 3:18–29

Brinkert J, Remington A (2020) Making sense of the perceptual capacities in autistic and non-autistic adults. Autism 24:1795–1804

Carhart-Harris RL, Kaelen M, Whalley M, Bolstridge M, Feilding A, Nutt D (2015) LSD enhances suggestibility in healthy volunteers. Psychopharmacology 232:785–794

Cassidy S, Hannant P, Tavassoli T, Allison C, Smith P, Baron-Cohen S (2016) Dyspraxia and autistic traits in adults with and without autism spectrum conditions. Mol Autism 7:1–6

Chapman R (2020) Defining neurodiversity for research and practice. In: Neurodiversity studies: a new critical paradigm. Routledge, pp 218–220

Coleman DM, Adams JB, Anderson AL, Frye RE (2019) Rating of the effectiveness of 26 psychiatric and seizure medications for autism spectrum disorder: results of a National Survey. J Child Adolesc Psychopharmacol 29:107–123

Danforth AL (2013) Courage, connection, and clarity: a mixed-methods collective-case study of MDMA (ecstasy) experiences of autistic adults. Institute of Transpersonal Psychology

Danforth AL (2019) Embracing neurodiversity in psychedelic science: a mixed-methods inquiry into the MDMA experiences of autistic adults. J Psychoactive Drugs 51:146–154

Danforth AL, Struble CM, Klosinski B, Grob C (2015) MDMA-assisted therapy: a new treatment paradigm for autistic adults with social anxiety. Prog Neuro-Psychopharmacol Biol Psychiatry 64:237–249

Danforth AL, Grob CS, Struble C, Feduccia AA, Walker N, Jerome L, Yazar-Klosinski B, Emerson A (2018) Reduction in social anxiety after MDMA-assisted psychotherapy with autistic adults: a randomized, double-blind, placebo-controlled pilot study. Psychopharmacology 235:3137–3148

De Gregorio D, Popic J, Enns JP, Inserra A, Skalecka A, Markopoulos A, Posa L, Lopez-Canul M, Qianzi H, Lafferty CK (2021) Lysergic acid diethylamide (LSD) promotes social behavior through mTORC1 in the excitatory neurotransmission. Proc Natl Acad Sci 118:e2020705118

De Wit H, Bershad AK (2020) MDMA enhances pleasantness of affective touch. Neuropsychopharmacology 45:217

Dumont GJ, Sweep F, van der Steen R, Hermsen R, Donders A, Touw DJ, van Gerven J, Buitelaar JK, Verkes R-J (2009) Increased oxytocin concentrations and prosocial feelings in humans after ecstasy (3, 4-methylenedioxymethamphetamine) administration. Soc Neurosci 4:359–366

Edsinger E, Dölen G (2018) A conserved role for serotonergic neurotransmission in mediating social behavior in octopus. Curr Biol 28:3136–3142.e4

Feduccia AA, Jerome L, Mithoefer MC, Holland J (2021) Discontinuation of medications classified as reuptake inhibitors affects treatment response of MDMA-assisted psychotherapy. Psychopharmacology 238:581–588

Fisher G (1970) The psycholytic treatment of a childhood schizophrenic girl. Int J Soc Psychiatry 16:112–130

Fisher G, Castile D (1963) Interim report on research project: an investigation to determine therapeutic effectiveness of LSD-25 and psilocybin on hospitalized severely emotionally disturbed children. Unpublished manuscript

Freedman AM, Ebin EV, Wilson EA (1962) Autistic schizophrenic children: an experiment in the use of D-lysergic acid diethylamide (LSD-25). Arch Gen Psychiatry 6:203–213

Green L, Willis E, Ziev N, Oliveira D, Kornblau B, Robertson S (2020) The impact of weighted blankets on the sleep and sensory experiences of autistic adults. Am J Occup Ther 74:7411515430p1

Grof S (1980) LSD psychotherapy. Hunter House

Hartmann K, Urbano M, Manser K, Okwara L (2012) Modified dialectical behavior therapy to improve emotion regulation in autism spectrum disorders. In: Autism spectrum disorders. Nova Science Publishers, Hauppauge, pp 41–72

Hedley D, Uljarević M (2018) Systematic review of suicide in autism spectrum disorder: current trends and implications. Curr Dev Disord Rep 5:65–76

Heimberg RG, Horner K, Juster H, Safren S, Brown E, Schneier F, Liebowitz M (1999) Psychometric properties of the Liebowitz social anxiety scale. Psychol Med 29:199–212

Herbert JD, Sharp IR, Gaudiano BA (2002) Separating fact from fiction in the etiology and treatment of autism: a scientific review of the evidence. Sci Rev Mental Health Pract 1:23–43

Hollocks MJ, Lerh JW, Magiati I, Meiser-Stedman R, Brugha TS (2019) Anxiety and depression in adults with autism spectrum disorder: a systematic review and meta-analysis. Psychol Med 49:559–572

Hysek CM, Schmid Y, Simmler LD, Domes G, Heinrichs M, Eisenegger C, Preller KH, Quednow BB, Liechti ME (2014) MDMA enhances emotional empathy and prosocial behavior. Soc Cogn Affect Neurosci 9:1645–1652

Kamilar-Britt P, Bedi G (2015) The prosocial effects of 3, 4-methylenedioxymethamphetamine (MDMA): controlled studies in humans and laboratory animals. Neurosci Biobehav Rev 57:433–446

Kessler RC, Petukhova M, Sampson NA, Zaslavsky AM, Wittchen HU (2012) Twelve-month and lifetime prevalence and lifetime morbid risk of anxiety and mood disorders in the United States. Int J Methods Psychiatr Res 21:169–184

Kuschner ES, Eisenberg IW, Orionzi B, Simmons WK, Kenworthy L, Martin A, Wallace GL (2015) A preliminary study of self-reported food selectivity in adolescents and young adults with autism spectrum disorder. Res Autism Spectr Disord 15:53–59

Linehan MM (2020) Dialectical behavior therapy in clinical practice. Guilford Publications

Lipsky D, Richards W (2009) Managing meltdowns: using the SCARED calming technique with children and adults with autism. Jessica Kingsley Publishers

Maddox BB, Kang-Yi CD, Brodkin ES, Mandell DS (2018) Treatment utilization by adults with autism and co-occurring anxiety or depression. Res Autism Spectr Disord 51:32–37

Mithoefer MC, Feduccia AA, Jerome L, Mithoefer A, Wagner M, Walsh Z, Hamilton S, Yazar-Klosinski B, Emerson A, Doblin R (2019) MDMA-assisted psychotherapy for treatment of PTSD: study design and rationale for phase 3 trials based on pooled analysis of six phase 2 randomized controlled trials. Psychopharmacology 236:2735–2745

Mogar RE, Aldrich RW (1969) The use of psychedelic agents with autistic schizophrenic children. Behav Neuropsychiatry 1:44–50

Moreno FA, Wiegand CB, Taitano EK, Delgado PL (2006) Safety, tolerability, and efficacy of psilocybin in 9 patients with obsessive-compulsive disorder. J Clin Psychiatry 67:1735–1740

Poquérusse J, Pastore L, Dellantonio S, Esposito G (2018) Alexithymia and autism spectrum disorder: a complex relationship. Front Psychol 9:1196

Poquérusse J, Pagnini F, Langer EJ (2021) Mindfulness for autism. Adv Neurodev Disord 5:77–84

Posner K, Brown GK, Stanley B, Brent DA, Yershova KV, Oquendo MA, Mann JJ (2011) The Columbia–suicide severity rating scale: initial validity and internal consistency findings from three multisite studies with adolescents and adults. Am J Psychiatr 168(12):1266–1277

Preece DA, Becerra R, Allan A, Robinson K, Chen W, Hasking P, Gross JJ (2020) Assessing alexithymia: psychometric properties of the Perth alexithymia questionnaire and 20-item Toronto alexithymia scale in United States adults. Personal Individ Differ 166:110138

Rojas-Bermúdez J (1960) Tratamiento Combinado deo psicoanalisis y LSD 25 en ninos psicoticos [combined treatment with psychoanalysis and LSD 25 in psychotic children]. Acta Neuropsiquait (Argent) 6:497–500

Rolo A, Krinsky LW, Abramson HA, Goldfarb L (1965) Preliminary method for study of LSD with children. Int J Neuropsychiatr 1(6):552–555

Sigafoos J, Green VA, Edrisinha C, Lancioni GE (2007) Flashback to the 1960s: LSD in the treatment of autism. Dev Neurorehabil 10:75–81

Simmons J 3rd, Leiken S, Lovaas O, Schaeffer B, Perloff B (1966) Modification of autistic behavior with LSD-25. Am J Psychiatry 122:1201–1211

Simmons J, Benor D, Daniel D (1972) The variable effects of LSD-25 on the behavior of a heterogeneous group of childhood schizophrenics. Behav Neuropsychiatry 4:10-6 passim

Spain D, Sin J, Linder KB, Mcmahon J, Happé F (2018) Social anxiety in autism spectrum disorder: a systematic review. Res Autism Spectr Disord 52:51–68

Strang JF, van der Miesen AI, Caplan R, Hughes C, Davanport S, Lai M-C (2020) Both sex-and gender-related factors should be considered in autism research and clinical practice. SAGE, London

Strassman RJ (1995) Hallucinogenic drugs in psychiatric research and treatment: perspectives and prospects. J Nerv Ment Dis 183:127–138

Studerus E, Vizeli P, Harder S, Ley L, Liechti ME (2021) Prediction of MDMA response in healthy humans: a pooled analysis of placebo-controlled studies. J Psychopharmacol 35:556–565

Sumathi T, Manivasagam T, Thenmozhi AJ (2020) The role of gluten in autism. Adv Neurobiol 24:469–479

Tantam D (2000) Psychological disorder in adolescents and adults with Asperger syndrome. Autism 4:47–62

Tantam D (2003) The challenge of adolescents and adults with Asperger syndromes. Child Adolesc Psychiatr Clin N Am 12(1):143–163

Taylor K (2014) Ethical caring in psychedelic work. In: Manifesting minds: a review of psychedelics in science, medicine, sex, and spirituality. Evolver Editions, Berkeley, p 136

Tsai LY (2007) Asperger syndrome and medication treatment. Focus Autism Other Dev Disabil 22:138–148

Velikonja T, Fett A-K, Velthorst E (2019) Patterns of nonsocial and social cognitive functioning in adults with autism spectrum disorder: a systematic review and meta-analysis. JAMA Psychiat 76:135–151

Young MB, Norrholm SD, Khoury LM, Jovanovic T, Rauch SA, Reiff CM, Dunlop BW, Rothbaum BO, Howell LL (2017) Inhibition of serotonin transporters disrupts the enhancement of fear memory extinction by 3, 4-methylenedioxymethamphetamine (MDMA). Psychopharmacology 234:2883–2895

Foundations for Training Psychedelic Therapists

Janis Phelps and James Henry

Contents

Abstract This chapter presents an integrative model for training psychedelic therapists. It begins with an overview of the psychedelic healing process – remembering the disconnected parts of ourselves and reestablishing fundamental trust. The authors then walk through six therapist competencies that describe the root principles of practice, breadth of knowledge, transpersonal perspective, and therapeutic stance of a psychedelic practitioner. This model includes 12 curricular areas and didactic methods that support the therapist's process of growth. Finally, the authors link their training philosophy to the current needs of society, offering a vision of how to express psychedelic integration through the honoring of diversity and collaboration.

J. Phelps
Center for Psychedelic Therapies and Research, California Institute of Integral Studies, San Francisco, CA, USA

J. Henry (✉)
Karunava.org, Berkeley, CA, USA

Curr Topics Behav Neurosci (2022) 56: 93–110
https://doi.org/10.1007/7854_2021_266
Published Online: 4 November 2021

Keywords Diversity · Empathetic abiding presence · Ethics · Integrative curriculum · Psychedelic assisted therapy · Therapist competencies · Training program

Contemporary psychedelic therapy is a convergence of healing practices. Cultures around the world have developed methods for working with psychedelic plants and fungi; now, in light of promising research, medical providers and psychotherapists are preparing to enhance their treatments with psychedelics. How can psychedelic training programs weave together these modalities, so that practitioners can offer care across the full spectrum of body, mind, and spirit? Is it possible to retain the benefits of diverse traditions while defining consistent, measurable standards? In this chapter we describe an integrative approach, using as guideposts six essential competencies distilled from clinical research with psychedelics. Together, these competencies provide an adaptive inner foundation for therapists – the breadth of knowledge and principles of relationship that support transformation.

We present here a navigational map for future psychedelic therapy training programs. These are the methods, curricular areas of study, and orientation that we find most valuable as curators of a pluralistic certificate program for therapists, clergy, and healthcare providers. We hope to convey perennial therapeutic qualities, while also pointing toward strategies that uniquely serve this moment of evolution. Our approach is open to elaboration, and we share it with the intention of supporting optimal outcomes for therapy and collective healing.

1 Overview of the Psychedelic Healing Process

What defines a successful treatment with psychedelics? To begin exploring this question, we offer here one theoretical framework of the psychedelic healing process. Our model strives to give each person the freedom to choose their own treatment goals, while facilitating discussion among trainees about the progress of therapeutic work. It is derived from accounts of research participants and psychedelic practitioners, integrating concepts from humanistic, transpersonal, and depth psychologies.

The condition that psychedelics treat is disconnection. Whether it is a trauma that separates one from others, a loss that makes one withdraw from life, or an environment that constricts the expression of one's spirit, all are circumstances of disconnection from one's whole being.

The work of psychedelic therapy, then, is to help people remember who they are in their fullness (Metzner 2015). Psychedelics open a door to the feelings, relations, and parts of oneself that were placed outside the sphere of awareness. The task of the therapist is to accompany with acceptance, so there is freedom for the process to flow with honesty and without impingement. This environment of safety and support becomes a resting place in the center of our being, from which we can see the full circle of who we are.

Experiencing the natural design and intelligence within a healing process can reestablish a fundamental trust in the unfolding of life (Richards 2015). With this trust it is easier to dive into the stream of life and let go of trying to control it. The result is an enhanced openness, empathy, love, and capacity to make meaning (MacLean et al. 2011; Bossis 2014; Griffiths et al. 2011; Grob and Dobkin de Rios 2013; Jesse and Griffiths 2014; Richards 2014).

In our training model, we seek to highlight the therapeutic principles that help a person return to this seat of acceptance, trust, and interconnection.

2 The Competencies of the Psychedelic Therapist

Every therapist has a unique path toward their authentic way of practice. The job of a psychedelic therapy training program is to prepare guiding principles, collected knowledge, and evidence based practices which can serve as footholds on the arc of development. The following educational framework offers a place of entry for people from a variety of training backgrounds. This expands on the descriptions that were originally put forth in Phelps (2017), incorporating new wisdom from the evolution of the field.

Competencies 1 and 2 – "Ethical Integrity and Self-Reflection" and "Trust Enhancement" – are the foundational roots from which the therapist grows. Competencies 3 and 4 – "Knowledge of the Physical and Psychological Effects of Psychedelics" and "Proficiency in Complementary Techniques" – comprise the body of knowledge and practices necessary to facilitate the full range of psychedelic healing sequences. Competency 5 – "Transpersonal Awareness" – is the spiritual perspective at the heart of our psychedelic training philosophy. Competency 6 – "Empathetic Abiding Presence" – describes the fruit of the training process, an integrated stance of dedication, acceptance, and understanding.

2.1 Competency 1: Ethical Integrity and Self-Reflection

What are the principles that should guide a therapist's decisions? This competency includes six components of the therapist's acumen related to ethics and self-awareness: understanding one's personal motives for this work; integrity in protecting boundaries with participants; well-developed capacities for building therapeutic alliances; skills in attachment theories and transference-countertransference analysis; understanding the influence of cultural prejudice and the value of reciprocity; and personal self-care, which protects both the therapist and the people receiving treatment. Therapists who embody this competency have concurrent awareness of self and other; they wisely reflect on their motives for conducting psychedelic therapy while simultaneously attending to participants' emotional and relational processes (Fisher and Martin 1969; Grof 1968; Penn et al. 2021; Sherwood et al. 1962; Strassman 2010). The impediments to this competency are personal needs, cultural conditioning, and fatigue.

A therapist can find grounding in ethical principles from their professional associations, established psychedelic therapy guidelines, and wisdom traditions. The Counsel on Spiritual Practices (CSP), a collaboration of psychedelic elders and scholars, created a code of ethics specific to guiding people on transformational paths (Jesse 1995). The code has nine foci: intention, serving society, serving individuals, competence, integrity, quiet presence, not for profit, tolerance, and peer review. Other pioneering professional codes for psychedelic work include the *Usona Code of Ethics for Entheogen Guides in a Research Setting,* the *Ethics for Holotropic Breathwork Practitioners,* and the *North Star Ethics Pledge* (Chang et al. 2020; Cooper 2014; Grof Transpersonal Training 2016). A theme common to therapeutic and religious codes of ethics is non-harming. The Hippocratic oath, a foundational document of western medicine, invites clinicians to "abstain from all intentional wrong-doing and harm" (Jones 1868). From our personal experience, discovering what one reacts to with destructive emotions is a key to becoming a healer.

Every therapist has areas of personal imbalance or need which can obscure their therapeutic motivation. The work of self-reflection is to make these influences conscious so that they do not determine one's actions. Metzner (2015) cautions therapists to be ever aware of the potential for their own grandiosity and the over-idealizing of their perceptions of what is meaningful. Grof (1980) also recommends the cultivation of thorough self-reflection to prevent acting impressively or demonstrating authority over participants. An antidote to such dynamics can be found in humility and guardianship – remembering that one is not in control of the healing process, and committing to hold feelings of powerlessness or loneliness within oneself. A therapist must also stay mindful of the interaction between their personal history and their recruitment into transference. According to psychoanalytic theory transference is an inevitable and helpful process in therapeutic relationship, but only if the therapist remains aware of what is happening. This is particularly important for the vulnerability of psychedelic work, during which a person can return to original states where they were wounded and now rely on the therapist for a corrective experience. Only with that protection can a person learn to rely on something deep within (Fisher and Martin 1969).

Special training is necessary to become mindful of cultural prejudices which distort the therapist's lens of perception and can lead to reenactments of exploitation or neglect. This is particularly important during therapy with people who have been underrepresented and discriminated against (Naranjo 2013). A guiding principle when working with people from different life experiences is to *ask* what they need and to *listen* to what they request. Taking time to explore one's own blind spots, with the help of others, is foundational work. It allows therapists to illuminate their shadow, or unconscious aspects of their personality, and to assist participants in doing the same. This practice extends to blind spots, injustices, and gaps in access that exist in systems of care delivery (Suite et al. 2007). Therapists should understand how indigenous communities are impacted by the rapidly growing global interest in psychedelic medicine, and consider ways to honor and support these communities as an expression of reciprocity (Chacruna Institute 2021). In training to notice and correct exploitation at the systemic level, providers are practicing healthy and reparative dynamics that will serve their work with individuals.

Finally, therapists are most likely to become disoriented when they are under-resourced and disconnected from their own place of refuge. Therefore, self-care is essential for being able to remain attuned to each unique person's journey. Working together with a co-therapist or team opens opportunities for processing, receiving feedback, and mutual support (Cooper 2014; Eisner and Cohen 1958; Mithoefer 2016; Smith 1988). Imitation is common in human development – children watch how parents treat themselves, which becomes a template for their own inner life. Similarly, the sensitivity of people in psychedelic states makes the therapist's relationship with themselves an important factor in this work (Buckman 1967; Fischer 2015; Grof 1980; Johnson et al. 2008).

Therapists who achieve this competency live with self-awareness, knowing what guides their choices and what nurtures their lives. This understanding is the foundation for supporting the growth of their patients.

2.2 Competency 2: Trust Enhancement

Trust is the foundation for psychedelic therapy and also what arises from its success. The therapist seeks to enhance the participant's trust in three arenas: trust in the process of psychedelic sessions, trust in the relationship with the guide, and trust in one's own inner healing capacity. These conditions of safety bring a person into alignment with their healing process, supporting what Passie (2012) calls "a unity with oneself."

A therapist should be knowledgeable about the types of subjective states that participants may experience and be able to normalize these states during preparation. Unexpected or paradoxical experiences are welcomed, with trust that confusion can give way to meaning as sessions unfold. Difficult feelings including "grief, rage and fear or panic" are introduced as natural, intrinsic to the process, and safe to be embraced (Mithoefer 2009, p. 32). This expansion of one's window of tolerance for emotions is critical to self-integration. It allows access to previously inaccessible feelings, so that a person can dream "undreamt dreams" and complete "interrupted cries" (Ogden 2005). Trust in mystical states, which may also include unfamiliar and initially overwhelming emotions, can be a gateway to healing through awe and gratitude.

Trust in the therapist allows a person to let go of vigilance, surrender within, and bring forward sensitive material. A therapist becomes trustworthy by developing ethical integrity, as detailed in the previous section, and by demonstrating "congruence" between their inner life and outer presentation (Rogers and Stevens 1967). This honesty and consistency allows a participant to trust their therapist's signal that what is happening is safe and just as it needs to be. Maté (2010) observes that, as an addiction therapist, trustworthiness entails the practitioner's full acceptance and honoring of addicts. This quality of non-judgment gives people the opportunity to bring forward whatever it is that they have held in shame. At the beginning of the therapeutic relationship, before any psychedelic work proceeds, trust is established through mutual agreement and consent about rules for the session. From the perspective of trauma-informed care, this safety is the medicine itself (Badenoch 2018).

What allows participants to trust their own healing capacity? It begins with the therapist's knowing of this innate capacity, conveyed to the participant through confidence in the treatment process (Cosimano 2021). During psychedelic therapy, the person's trust in their own nature grows step by step with each experience of letting go that leads to progress or relief. Trust in this inner healing intelligence can ultimately give way to a trust in the process of life. To quote Richards (2015, p. 211), in his epilogue to *Sacred Knowledge*, "When we trust and act in the world, a meaningful process unfolds within us."

2.3 Competency 3: Knowledge of the Physical and Psychological Effects of Psychedelics

With this competency we transition into the body of knowledge about psychedelics that is necessary to facilitate therapy. The components of this expertise range from: knowledge of the anthropology of shamanism, neurobiology, neuropharmacology, and drug disposition; skills in the creation of safe and artful sets and settings; and optimally, knowledge from subjective, phenomenological experience of personal psychedelic-assisted therapy.

Psychedelic therapy and research has benefited from knowledge of ancient and contemporary indigenous practices of ceremonial use of plant medicines, which have time-honored sets and settings (Bourzat 2019; Labate 2014). Psychedelic scholars have noted the need for guides in psychedelic-assisted therapy to have in-depth, theoretical, and experiential knowledge of the cross-cultural roots of plant medicine use (Bravo and Grob 1989; Grob and Dobkin de Rios 2013; Metzner 1998; Narby 1998; Smith et al. 2004). These traditions hold invaluable wisdom about preparation through diet and behavioral restrictions, forming relationships with psychedelic plants, choosing fungal species for different purposes, and the transpersonal and socio-integrative applications of psychedelic medicines (Loizaga-Velder 2021).

Therapists must be competent in their knowledge of the physiological and psychological effects of psychedelics. This includes knowledge of pharmacology, drug disposition and interactions, anatomy and physiology, neurobiology, the neuropharmacology of psychedelic drugs, and the normative effects of different psychedelic drugs at varying dosages in a variety of sets and settings (Pahnke and Richards 1966; Nichols 2004; Walsh and Grob 2006). To best accompany this broad range of effects, it would serve a therapist to receive special training in the following areas: grief and existential processing; the safe management of acute distress; child and adult development; unusual somatic drug reactions; and psychedelic-induced psychotic episodes or "spiritual emergencies" (Grof and Grof 1989). Practitioners are encouraged to stay up to date on strategies for integrating psychotic states while respecting patient autonomy (Fletcher et al. 2021). A comprehensive summary of safety practices for psilocybin-assisted therapy and MDMA-assisted therapy can be found in Johnson et al. (2008) and Mithoefer (2016). In addition to these areas, therapists must learn to assess participants for eligibility, be familiar with contraindications, and create physical environments that support psychedelic work (Bonny and Pahnke 1972; Metzner 2015).

The most direct way for a therapist to understand the effects of psychedelics is to receive a training experience with a psychedelic medicine (Hoffer and Osmond 1967; Mithoefer 2009; Nielson and Guss 2018; Phelps 2017). Researchers attest that therapists with such experiences are more likely to have "empathy and effectiveness" when working with patients and "sensitivity to the potency of factors such as trust, honesty, courage, and openness" (Richards 2015, p. 148). These benefits must be balanced with the potential negative impact, as Forstmann and Sagioglou (2021) found that researchers who self-report substance use may be perceived as less objective. When direct psychedelic experience is not available to a therapist due to legal or health reasons, alternative practices such as Holotropic Breathwork, guided imagery, legally accessible cannabis-assisted psychotherapy, or contemplative retreats are recommended, along with access to videos of treatment sessions. Ketamine-assisted therapy can be a valuable method for accessing non-ordinary states of consciousness; further research is necessary to clarify the unique benefits, risks, and qualities of ketamine compared to classic psychedelics (Bennett 2019; Wolfson and Hartelius 2016).

2.4 Competency 4: Proficiency in Complementary Techniques

Complementary therapeutic methods are useful across the phrases of preparation, medicine sessions, and integration (Grob 2002; Phelps 2019). These techniques include somatic, psychotherapeutic, shamanic, and contemplative practices.

Somatic methods consist of therapeutic bodywork, stress inoculation, and touch (Cooper 2014; Eisner and Cohen 1958); felt sensing and focusing (Danforth 2009); techniques of eye-gazing, whether with a mirror or with the therapist (Grof 1968); somatic experiencing and sensorimotor therapies (Metzner 2013); and integrated somatic techniques such as Hakomi and Holotropic Breathwork (Taylor 2007). Dance is an expression of and catalyst for non-ordinary states of consciousness across history; it warrants further research in conjunction with psychedelic therapy (Ferrucci 1990; Renye 2012).

Complementary psychotherapeutic methods include logotherapy, existential and narrative therapy (Bossis 2014; Frankl 1984); psychoanalysis (Naranjo 2013); Acceptance and Commitment Therapy (Sloshower et al. 2020); post-hypnotic suggestion (Hastings 2006); family-oriented techniques with photos or analytic inquiry, such as Internal Family Systems work (Mithoefer 2016; Stolaroff 1997); shadow work (Shulgin 2001); group therapy (Anderson et al. 2020; Trope et al. 2019); and Gestalt (Dass et al. 2010). Further methods are expressive arts therapy and the Bonny Method of Guided Imagery and Music (GIM) (Bonny and Savary 1973); guided Affective Imagery (Leuner 1969); creativity exercises (Savage et al. 1964); music therapy (Kaelen 2021); and the creation of mandalas and art (Grof and Grof 2010).

With proper training, a therapist might use strategies drawn from shamanism, ethnobotany, or ceremonial aspects of healing within community (Davis 1996; Furst 1976; Labate 2014; McKenna 2007; Schultes 1979; Smith 1988; Wasson et al. 2008). These methods may include the creation of a personal altar for the participant, healing songs, music, and prayer (Bourzat 2019; Cooper 2014; Metzner 2015). Indigenous community advocates within the psychedelic movement are in ongoing conversation about the responsible and non-appropriated incorporation of cross-cultural practices, and how to respectfully request guidance and training within a particular tradition.

Patients who adhere to a major world spiritual tradition may call upon contemplative practices during the therapeutic process. In the preparation phase, for example, people may use strategies of mindfulness or prayer to ground themselves and open to new experiences; researchers have also incorporated meditation into their integration sessions (Griffiths et al. 2011; Watts 1970). Throughout treatment, therapists must allow patients the freedom to make meaning of psychedelic experiences within their chosen spiritual paradigm. Griffiths et al. (2006) established that psilocybin can occasion mystical states that are among the most spiritually significant events of a person's life – it is therefore natural that some patients turn toward their religious communities for guidance in deepening and integrating these experiences. This may include time-honored contemplative techniques such as retreat and engaging in service; conversations across the disciplines of psychiatry and spirituality will help to characterize which practices best sustain the insights from psychedelic sessions (Henry 2019). The development of training institutions that bring together therapists and clergy from various faiths is an excellent first step, and will broaden access to the full range of techniques supporting transformation that are held within spiritual traditions (Richards 2009; Roberts and Hruby 2002; Vaughan 2001).

A time-limited training program can only expose therapists to the complementary methods such as those above. True proficiency comes through practice, as therapists discover which methods enhance the psychedelic experience. A skilled psychedelic therapist should apply complementary methods as needed and permitted, depending on the therapeutic issues of the client.

2.5 Competency 5: Transpersonal Awareness

Transpersonal awareness is the competency that brings psychedelic therapists through a gateway, moving beyond psycho-somatic healing and into the realm of spiritual awakening. It implies "awareness of our relationship to the transcendent, to each other, to the earth and all beings" (Vaughan 2002, p. 18). This intelligence includes a deep understanding of entelechy, the inherent movement of consciousness toward a meaningful wholeness. When therapists come to see that psychological and spiritual content emerges in an orderly way, they can relinquish control to a process "infinitely wiser than the egos of the patient and therapist" (Richards 2003, p. 33).

Both MDMA and psilocybin-assisted therapy can bring about transpersonal experiences and insights. Passie (2012) describes six characteristics of spiritual intelligence that can arise with MDMA: (a) the ability to open up toward inner awareness, (b) feeling free to accept oneself, (c) feelings of relaxed detachment to an extent that everything appears safe, (d) a felt contact with an inner core, (e) a feeling for what love is, and (f) awareness of non-material aspects of being and happiness (p. 38). Bossis (2014) derives four themes of experience from psilocybin in end-of-life anxiety research: (a) transcendence into the nature of self, (b) cultivation of equanimity, (c) acceptance of change and impermanence, and (d) the experience of love (p. 267). Self-compassion is emphasized as a means and an end in mindfully cultivating a larger sense of self (Germer 2009; Kamjol et al. 2015). The lists for MDMA and psilocybin overlap, suggesting arrival at a similar destination of love and acceptance. Further research is encouraged to characterize the transpersonal experience from each psychedelic medicine, and how these states uniquely offer resolution to relational or existential issues.

Psychedelic therapists develop familiarity with the themes of transpersonal awakening so that they can recognize transformational processes and reflect them back to participants. A common and central revelation, based on participant accounts, is one's "interrelatedness within the great unity of all human beings and perhaps all life forms" (Richards 2014, p. 653). Feeling the full capacity of human consciousness brings awe, and is intrinsically healing – having recognized ourselves, our mistakes are put in perspective, giving way to a profound love. The smaller concerns of life, in juxtaposition to this brilliance, are held more lightly. While the content of a psychedelic experience may fade from memory, a deep gratitude and humility remain. Held within the greater web of life, there is a renewed capacity to bear suffering and be present with the reality of one's life (Cosimano 2021).

A line of future inquiry will be to investigate the correlations between this embodied spiritual perspective in therapists and the clinical outcomes of psychedelic-assisted psychotherapy. In the section that follows, we will explore the relational expression of this transpersonal awareness.

2.6 Competency 6: Empathetic Abiding Presence

A psychedelic therapist established in trust and acceptance can remain steadily attuned to the participant's full expression. Components of empathetic abiding presence include composure, evenly suspended attention, mindfulness, empathetic listening, responding to distress with calmness, and equanimity (Phelps 2017).

Naranjo (2006) conceives of the therapist's non-attachment as a form of love. Rather than trying to shape the participant's healing in a particular way, a therapist provides the conditions that nourish growth. Presence and empathy, like sunlight and water, support life without determining its course. To offer this quality of care, a therapist must have done "sufficient inner work themselves" to refrain from imposing an outcome and accept whatever course emotions need to flow (Mithoefer 2016, p. 11). This includes a recognition of when prior training "needs to be set aside" to allow for authentic and spontaneous human connection (Fadiman 2011, p. 77). Mindfulness, self-compassion, and meditation practice are essentials for cultivating this presence (Germer et al. 2013). By listening with curiosity and evenly suspended attention, therapists make space for something new and undefined by their preferences (Epstein 1984; Naranjo 2013; Richards 2009; Stolaroff 1997). Another way to say this, in the parallel modality of Holotropic Breathwork, is "doing not doing" (Sparks 1987). The therapist abides in patience, becoming ground to the figure of the participant's own process.

For patients to feel supported, the therapist's non-directive presence must be coupled with empathetic responsivity. This combined presence and responsivity helps patients to feel accompanied wherever their healing process leads. Mithoefer (2016) describes this attuned, abiding companionship during MDMA-assisted therapy:

> The therapists act as empathic listeners, trustworthy guides, facilitators of deep emotional expression and catharsis, and supporters of the participant's own inner healing intelligence. (p. 29)

With encouragement and a hand to hold, patients can access the difficult emotions they have been unable to feel on their own. This "balance of protection, permission and connection" helps patients to weave together their life experiences into an integrated self (Taylor 2007, p. 133).

The therapist's equanimity allows them to move gracefully with the peaks and valleys of each unique person's session. The term "abiding" is used to convey a witnessing of the mystery of life in action during psychedelic-assisted therapy. It points to the wider perspective that is possible when there is a continuity of presence – the view from which a therapist can see the whole humanity of a person. This inclusive vision, when mirrored back to the patient, can be profoundly healing.

Drawing upon their knowledge of psychedelic effects and trust in the process, the therapist now has confidence to accompany the patient in whatever way is needed. In so doing, the therapist embodies an integration of the skills and wisdom from all the previous competencies.

3 Twelve Curricular Areas of Study for Psychedelic Therapy Training

A training program's emphasis on each of these 12 areas will uniquely depend upon the intended scope of practice of their graduates. The first seven domains are primarily didactic, focusing on the development of knowledge and dispositions necessary for the six competencies of the psychedelic therapist. The last five areas of study use skills-based learning to strengthen and integrate all of the competencies:

1. The history of clinical research and current legal status of psychedelic-assisted therapy
2. Models of consciousness, transpersonal awareness, and mystical experiences
3. Building connection: Use of psychedelics in religious and community settings
4. Best practices in sets and settings: Preparation, psychedelic session, and integration
5. Variations in therapeutic models: Client-centered, psycholytic, and emerging psychedelic therapies
6. Neurobiology, neuropharmacology, drug disposition, and drug interactions
7. Psychedelics and therapeutic relationships: Transference, boundaries, ethics, and self-care
8. Supervised role-plays, experiential practices, and observation of psychedelic session videos
9. Somatic and complementary therapeutic techniques in psychedelic-assisted therapy
10. Individual and group clinical supervision during and after training as a psychedelic therapist in approved clinical trials or expanded access clinical research programs
11. Personal experience of being guided as a research participant in an approved study
12. Co-therapy methods and interprofessional skills for working on multidisciplinary teams

4 Didactic Methods that Support Therapist Development

Psychedelic therapy training is unique for the depth of personal engagement it requires. The therapist's discovery of their own wounds, gifts, and patterns is central to understanding the psychedelic journey toward an inclusive whole self. To facilitate this process, the training program structure and staff can demonstrate the supportive, attuned accompaniment which helps a person to explore and grow. This provides the environment for trainees to remember the unintegrated parts of their lives, along with a personal understanding of what impedes or supports this path. By the completion of training, a therapist arrives at a more developed *state* of integration while also being able to impart the *process* of integration.

The training competencies are recognizable within each didactic method facilitating this process. Mentoring and apprenticeship can offer an experience of trust enhancement and supportive relationship. Lectures by prominent scholars, therapists, and elders provide a reliable base of knowledge about psychedelics and complementary techniques. Self-reflection papers and mentor assessments guide therapists to face their own growth areas with honesty and acceptance. Role plays and discussion groups offer opportunities to work as part of a team. Non-ordinary state experiences, through breathwork, drumming circles, guided imagery, or legal psychedelic therapy, provide exposure to the transpersonal. Eliciting feedback from trainees, with real time adaptations of programing, demonstrates how to listen and respond to participants. When feedback reveals shortcomings, there is an opportunity to role model how to take responsibility and make amends. Finally, volunteer work brings trainees back in touch with their guiding intention and emphasizes an orientation toward service.

Further details about the first university-accredited training program for psychedelic therapists at the California Institute of Integral Studies (CIIS) are accessible in a previous publication (Phelps 2019). In the next section, we will describe how the training principles described thus far can be expressed at the level of community.

5 Flourishing Through Diversity: The Evolving Landscape of Training

We are beginning to see a great variety of psychedelic training programs emerge. In addition to foundational programs like that found at CIIS, there are training programs that focus on specific psychedelics (MDMA through the Multidisciplinary Association of Psychedelic studies; psilocybin through the Heffter Research Institute and Usona Institute) and areas of interest such as trauma and social justice (Sage Institute 2021). Formats range from university settings to field work with indigenous carriers of medicine. Each of these programs offers an invaluable perspective. Like different trees in a garden, they provide unique fruits that nourish and heal the community.

As a result of our cultural structures, there is a risk of imbalance in the delivery of psychedelic medicines and training. Commercial interests, which reward rapid growth, provide an incentive for programs to compete. Under the influence of this power dynamic, institutions and companies may expand without awareness, crowding out marginalized voices and forgetting the importance of fair access to treatment. In light of this tendency, prominent psychedelic advisors have produced agreements of collaboration among practitioners and research organizations, such as the Statement on Open Science and Open Praxis (Jesse 2017). Training programs must also weave the practice of collaboration into the fabric of their structures. At the organizational level, this can take the form of diverse advisory councils, centering black, indigenous, and people of color (BIPOC) presenters, and regularly surveying the needs of community stakeholders. A diverse cohort of trainees will help bring psychedelic therapy to a larger distribution of communities. In addition to supporting the movement toward equitable access, programs with diverse trainees provide a higher quality of education due to the wider range of voices in conversation (Hurtado and Guillermo-Wann 2013).

To truly evolve as a community, we propose that each person involved in psychedelic training must cross through a new threshold of integration.

Transpersonal experiences allow us only a moment of unified togetherness. The integration of this awareness cannot be done as an individual; it is only available when people widen the circles of their identity to include one another. This is possible when we recognize ourselves as belonging to the same body of nature. Each of us has an essential function that allows the wondrous complexity to unfold. As conscious parts of the whole, we each deserve to experience life and express our blossom within it. With this understanding comes a natural devotion to diversity and concomitant activism for equity and inclusion. When we see our lives as intertwined, the ego's focus gives way to a deeper current moving within us – the instinct to safeguard and honor all our relations.

We have now been given the circumstances to realize this stage of development. The task of this process is reconciliation, and the feeling of its achievement is peace. When this understanding and commitment informs our training programs, only then can therapists accompany people along the full path of their journey toward wholeness.

6 Conclusion

In this chapter we have presented an integrative model for training psychedelic therapists, synthesizing perspectives from various psychological schools, cultures, and spiritual traditions. We began with an overview of the psychedelic healing process – remembering the disconnected parts of ourselves and reestablishing fundamental trust. We then walked through the six therapist competencies that describe the root principles of practice, breadth of knowledge, transpersonal perspective, and therapeutic stance of a psychedelic practitioner. We presented the 12 curricular areas and didactic methods that support the therapist's process of growth. Finally, we linked our training philosophy to the current needs of our society, offering a vision of how to express psychedelic integration through the honoring of diversity and collaboration.

The theme of this collection – Disruptive Psychopharmacology – invites us to consider what patterns are interrupted by psychedelic medicines. At the level of the individual, psychedelics seem to disrupt functions of our ego that have adapted to conditions of fear. We hope this chapter has demonstrated why a safe, supportive therapeutic environment helps a person to resume natural development toward their full self. At the level of community, psychedelics can disrupt the structures of competition and injustice that arise from a perspective of separation. In the clear seeing of our interconnection, there is an invitation to surrender our defensive strategies and join together in alignment with our common nature.

In a field as vast and complex as psychedelic therapy, it serves to remember a guiding simplicity: to help others on the path, we must follow the course of our own healing. Only through that journey can we learn how to love ourselves and build the vehicle of a loving relationship. Acceptance makes room for every part of us, and trust opens the sail. Then we need only let go, and let the winds carry us back to our center. Safeguarded here in our hearts is the template for an integrated life, and the seed of a harmonious world. The therapist is a fellow traveler on this difficult and worthwhile passage, lending support as we remember our way home.

References

Anderson BT, Danforth A, Daroff R, Stauffer C, Ekman E, Agin-Liebes G, Trope A, Boden MT, Dilley J, Mitchell J, Woolley J (2020) Psilocybin-assisted group therapy for demoralized older long-term AIDS survivor men: an open-label safety and feasibility pilot study. EClinicalMedicine 27:100538. https://doi.org/10.1016/j.eclinm.2020.100538

Badenoch B (2018) The heart of trauma. W.W. Norton & Company, New York

Bennett R (2019) Paradigms of ketamine treatment. MAPS Bull 29(1):48–49

Bonny HL, Pahnke WN (1972) The use of music in psychedelic (LSD) psychotherapy. J Music Ther 9(2):64–87. https://doi.org/10.1093/jmt/9.2.64

Bonny HL, Savary LM (1973) Music and your mind: listening with a new consciousness. Harper & Row, Oxford

Bossis A (2014) Psilocybin and mystical experience: implications for the alleviation of existential and psycho-spiritual distress at end of life. In: Ellens JH (ed) Seeking the sacred with psychoactive substances, vol 2. Praeger, Santa Barbara, pp 251–284

Bourzat F (2019) Consciousness medicine. North Atlantic Books, Berkeley

Bravo G, Grob C (1989) Shamans, sacraments and psychiatrists. J Psychoactive Drugs 21 (1):123–128

Buckman J (1967) Theoretical aspects of LSD therapy. In: Abramson H (ed) The use of LSD in psychotherapy and alcoholism. Bobbs-Merrill Company, Indianapolis, pp 83–100

Chacruna Institute (2021) Indigenous reciprocity initiative of the Americas [Internet]. https://chacruna.net/indigenous-reciprocity-initiative/. Accessed 26 Mar 2021

Chang T, Conour K, Gillooly L, McGaughey D, Noble S (2020) North Star ethics pledge. https://northstar.guide/ethicspledge/. Accessed 25 Apr 2020

Cooper K (2014) Guide manual for pharmacokinetics of psilocybin in healthy adult volunteers study [unpublished manuscript]. University of Wisconsin, Madison

Cosimano M (2021) The role of the guide in psychedelic-assisted treatment. In: Grob C, Grigsby J (eds) Handbook of medical hallucinogens. Guilford Press, New York, pp 377–394

Danforth A (2009) Focusing-oriented psychotherapy as a supplement to preparation for psychedelic therapy. J Transpers Psychol 41(2):151–160

Dass R, Metzner R, Bravo G (2010) The birth of a psychedelic culture. Synergetic Press, Santa Fe

Davis W (1996) One river: explorations and discoveries in the Amazon rain forest. Simon and Schuster, New York

Eisner B, Cohen S (1958) Psychotherapy with lysergic acid diethylamide. J Nerv Ment Dis 127 (6):528–539

Epstein M (1984) On the neglect of evenly suspended attention. J Transpers Psychol 16(2):193–205

Fadiman J (2011) The psychedelic explorer's guide: safe, therapeutic and sacred journeys. Park Street Press, Rochester

Ferrucci P (1990) Inevitable grace. Penguin Books, London

Fischer F (2015) Therapy with substance: psycholytic psychotherapy in the twenty-first century. Muswell Hill Press, Dorset

Fisher G, Martin J (1969) The psychotherapeutic use of psychodysleptic drugs. Voices Art Sci Psychother 5(4):67–72

Fletcher EH, Anderson M, Henry J, Moore V, Sullivan P, Unger R (2021) Panel 2 Q&A: breakdown or breakthrough? Understanding experiences of madness, awakening and altered states beyond pathology. psychedelics, madness, & awakening: harm reduction and future visions 2021. In: Digital conference, held January-April, 2021. Video of panel. https://youtu.be/LRmvcO8V_Hk

Forstmann M, Sagioglou C (2021) How psychedelic researchers' self-admitted substance use and their association with psychedelic culture affect people's perceptions of their scientific integrity and the quality of their research. Public Underst Sci 30(3):302–318. https://doi.org/10.1177/0963662520981728

Frankl V (1984) Man's search for meaning: an introduction to logotherapy. Simon & Schuster, New York

Furst P (1976) Hallucinogens and culture. Chandler and Sharp Publishers, Novato

Germer C (2009) The mindful path to self-compassion: freeing yourself from destructive thoughts and emotions. Guilford Press, New York

Germer C, Siegel R, Fulton P (2013) Mindfulness and psychotherapy, 2nd edn. Guilford Press, New York

Griffiths RR, Richards WA, McCann U, Jesse R (2006) Psilocybin can occasion mystical-type experiences having substantial and sustained personal meaning and spiritual significance. Psychopharmacology 187(3):268–283. https://doi.org/10.1007/s00213-006-0457-5

Griffiths RR, Johnson MW, Richards WA, Richards BD, McCann U, Jesse R (2011) Psilocybin occasioned mystical-type experiences: immediate and persisting dose-related effects. Psychopharmacology 218(4):649–665. https://doi.org/10.1007/s00213-011-2358-5

Grob C (2002) Psychiatric research with hallucinogens: what have we learned? In: Grob C (ed) Hallucinogens: a reader. Jeremy P. Tarcher/Putnam, New York, pp 263–291

Grob CS, Dobkin de Rios M (2013) Hallucinogens and related compounds. In: Rosner R (ed) Clinical handbook of adolescent addiction. John Wiley and Sons, New York, pp 213–222

Grof S (1968) Tentative theoretical framework for understanding dynamics of LSD psychotherapy. In: Shlien JM (ed) Research in psychotherapy. American Psychological Association, Washington, pp 449–465. https://doi.org/10.1037/10546-021

Grof S (1980) LSD psychotherapy. Hunter House, Alameda

Grof S, Grof C (eds) (1989) Spiritual emergency. TarcherPerigee, New York

Grof S, Grof C (2010) Holotropic breathwork: a new approach to self-exploration and therapy. SUNY Press, Albany

Grof Transpersonal Training (2016) Ethical agreements of holotropic breathwork practitioners. Grof Transpersonal Training application pack. Author, Mill Valley, p 19. http://www.holotropic.com/pdf_docs/GTT%20Training%20Application%20Pack%202015.pdf. Accessed 4 Mar 2016

Hastings A (2006) An extended nondrug MDMA-like experience evoked through post-hypnotic suggestion. J Psychoactive Drugs 38(3):273–283. https://doi.org/10.1080/02791072.2006.10399853

Henry J (2019) A call for spirit: understanding the reemergence of psychedelics in psychiatry. Paper presented at International Congress in Spirituality and Psychiatry 4th Global Meeting in Spirituality and Mental Health, Jerusalem, Israel; 2019 Dec 1–4

Hoffer A, Osmond H (1967) The hallucinogens. Academic Press, New York

Hurtado S, Guillermo-Wann C (2013) Diverse learning environments: assessing and creating conditions for student success – final report to the Ford Foundation. University of California, Higher Education Research Institute, Los Angeles

Jesse R, Griffiths RR (2014) Psilocybin research at Johns Hopkins: a 2014 report. In: Ellens JH (ed) Seeking the sacred with psychoactive substances: chemical paths to spirituality and to god, vol 2. Praeger, Santa Barbara, CA, pp 29–43

Jesse R (1995) Code of ethics for spiritual guides. Council on Spiritual Practices. https://csp.org/docs/code-of-ethics-for-spiritual-guides. Accessed 26 Nov 2015

Jesse R (2017) Statement on open science and open praxis with psilocybin, MDMA, and similar substances. Council on Spiritual Practices. https://csp.org/open. Accessed 12 Aug 2018

Johnson MW, Richards WA, Griffiths RR (2008) Human hallucinogen research: guidelines for safety. J Psychopharmacol 22(6):603–620

Jones W (1868) Hippocrates collected works. Harvard University Press, Cambridge

Kaelen M (2021) The use of music in psychedelic therapy. In: Grob C, Grigsby J (eds) Handbook of medical hallucinogens. Guilford Press, New York, pp 363–376

Kamjol S, Kilford E, Minchin S, Moss A, Lawn W, Das R et al (2015) Recreational 3,4-methylenedioxy-N-methylamphetamine (MDMA) or 'ecstasy' and self-focused compassion: preliminary steps in the development of a therapeutic psychopharmacology of contemplative practices. J Psychopharmacol 29(9):961–970. https://doi.org/10.1177/0269881115587143

Labate B (2014) Ayahuasca shamanism in the Amazon and beyond. Oxford University Press, Oxford

Leuner H (1969) Guided affective imagery (GAI). Am J Psychother 23(1):4–10

Loizaga-Velder A (2021) The therapeutic potentials of psilocybin containing mushrooms from a transdisciplinary perspective. Paper presented at sacred plants in the Americas II; virtual; 2021 Apr 23–25. https://www.chacruna.net/sacred-plants-americas-2-conference/. Accessed 24 Apr 2021

Maté G (2010) In the realm of the hungry ghosts. North Atlantic Books, Berkeley

McKenna DJ (2007) The healing vine: ayahuasca as medicine in the 21st century. In: Winkelman MJ, Roberts TB (eds) Psychedelic medicine: new evidence for hallucinogenic substances as treatments, vol 1. Praeger, Westport, pp 21–44

Metzner R (1998) Hallucinogenic drugs and plants in psychotherapy and shamanism. J Psychoactive Drugs 30(4):333–341. https://doi.org/10.1080/02791072.1998.10399709

Metzner R (2013) The toad and the jaguar. Regent Press, Berkeley

Metzner R (2015) Allies for awakening: guidelines for productive and safe experiences with entheogens. Regent Press, Berkeley

Mithoefer M (2009) A phase 1 placebo-controlled, double-blind crossover study to assess psychological effects of MDMA when administered to healthy volunteers. Multidisciplinary Association for Psychedelic Studies. http://www.maps.org/researcharchive/mdma/mt1_docs/mt1_amend2_protocol_dec_2_09_final.pdf. Accessed 19 May 2017

Mithoefer M (2016) A manual for MDMA-assisted psychotherapy in the treatment of posttraumatic stress disorder. Multidisciplinary Association for Psychedelic Studies. https://s3-us-west-1.amazonaws.com/mapscontent/research-archive/mdma/MDMAAssistedPsychotherapyTreatmentManualVersion+8_25May16_Formatted.pdf. Accessed 8 Sep 2016

Naranjo C (2006) The way of silence and the talking cure. Blue Dolphin Publishing, Nevada City

Naranjo C (2013) The healing journey. MAPS, Santa Cruz

Narby J (1998) The cosmic serpent. Jeremy Tarcher/Putnam, New York

Nichols D (2004) Hallucinogens. Pharmacological Therapy 101:131–181

Nielson EM, Guss J (2018) The influence of therapists' first-hand experience with psychedelics on psychedelic-assisted psychotherapy research and therapist training. J Psychedelic Stud 2 (2):64–73

MacLean KA, Johnson MW, Griffiths RR (2011) Mystical experiences occasioned by the hallucinogen psilocybin lead to increases in the personality domain of openness. J Psychopharmacol 25:1453–1461. https://doi.org/10.1177/0269881111420188

Ogden T (2005) This art of psychoanalysis: dreaming undreamt dreams and interrupted cries. Routledge, London

Pahnke WN, Richards WA (1966) Implications of LSD and experimental mysticism. J Relig Health 5(3):175–208. https://doi.org/10.1080/02791072.1970.10471366

Passie T (2012) Healing with entactogens. MAPS, Santa Cruz

Penn A, Phelps J, Rosa W, Watson J (2021) Psychedelic-assisted psychotherapy practices and human caring science: toward a care-informed model of treatment. J Humanist Psychol. https://doi.org/10.1177/00221678211011013

Phelps J (2017) Developing guidelines and competencies for the training of psychedelic therapists. J Humanist Psychol 57(5):450–487

Phelps J (2019) Training psychedelic therapists. In: Winkelman M, Sessa B (eds) Advances in psychedelic medicine. Praeger, Santa Barbara, pp 274–294

Renye D (2012) Spontaneous embodied spiritual experience of movement and being moved: a qualitative analysis. PhD dissertation, California Institute of Integral Studies, San Francisco. Available from: ProQuest Dissertations Publishing; publication no. AAT 3539734
Richards WA (2003) Navigation within consciousness: insights from four decades of psychotherapy research with imagery, music, and entheogens. J Assoc Music Imag 9:27–39
Richards WA (2009) The rebirth of research with entheogens: lessons from the past and hypotheses for the future. J Transpers Psychol 41(2):139–150
Richards WA (2014) Here and now: discovering the sacred with entheogens. Zygon J Relig Sci 49 (3):652–665. https://doi.org/10.1111/zygo.12108
Richards WA (2015) Sacred knowledge: psychedelics and religious experience. Columbia University Press, New York
Roberts T, Hruby P (2002) Toward an entheogen research agenda. J Humanist Psychol 42(1):71–89
Rogers C, Stevens B (1967) Person to person: the problem of being human. Real People Press, Lafayette
Sage Institute (2021) Sage Institute internship training. Berkeley, CA. https://sageinst.org/internshiptraining
Savage C, Savage E, Fadiman J, Harmon W (1964) LSD: therapeutic effects of the psychedelic experience. Psychol Rep 14:111–120
Schultes RE (1979) The plants of the gods: origins of hallucinogenic use. McGraw-Hill, New York
Sherwood JN, Stolaroff MJ, Harmon WW (1962) The psychedelic experience: a new concept in psychotherapy. J Neuropsychiatr 4(2):69–80
Shulgin A (2001) The new psychotherapy: MDMA and the shadow. In: Roberts T (ed) Psychoactive sacramentals: essays on entheogens and religion. Council on Spiritual Practices, San Francisco, pp 196–204
Sloshower J, Guss J, Krause R, Wallace RM, Williams MT, Reed S, Skinta MD (2020) Psilocybin-assisted therapy of major depressive disorder using acceptance and commitment therapy as a therapeutic frame. J Contextual Behav Sci 15:12–19. https://doi.org/10.1016/j.jcbs.2019.11.002
Smith ED (1988) Evolving ethics in psychedelic drug taking. J Drug Issues 18(2):201–214
Smith H, Grob C, Jesse R, Bravo G, Agar A, Walsh R (2004) Do drugs have religious import?: a 40-year retrospective. J Humanist Psychol 44(2):120–140. https://doi.org/10.1177/0022167804263209
Sparks T (1987) Doing not doing. Holotropic and the Grof Transpersonal Training, Mill Valley
Stolaroff M (1997) The secret chief. MAPS, Charlotte
Strassman R (2010) Hallucinogenic drugs in psychiatric research and treatment: perspectives and prospects. J Nerv Ment Dis 183(3):127–138. https://doi.org/10.1097/00005053-199503000-00002
Suite D, La Bril R, Primm A, Harrison-Ross P (2007) Beyond misdiagnosis, misunderstanding and mistrust: relevance of the historical perspective in the medical and mental health treatment of people of color. J Natl Med Assoc 99(8):879–885
Taylor K (2007) Considering holotropic breathwork. Hanford Mead Publishers, Santa Cruz
Trope A, Anderson B, Hooker A, Glick G, Stauffer C, Woolley J (2019) Psychedelic-assisted group therapy: a systematic review. J Psychoactive Drugs 51(2):174–188
Vaughan F (2001) Transpersonal counseling: some observations regarding entheogens. In: Roberts T (ed) Psychoactive sacramentals: essays on entheogens and religion. Council on Spiritual Practices, San Francisco, pp 191–195
Vaughan F (2002) What is spiritual intelligence? J Humanist Psychol 42(2):16–33. https://doi.org/10.1177/0022167802422003
Walsh R, Grob CS (2006) Early psychedelic investigators reflect on the psychological and social implications of their research. J Humanist Psychol 46(4):432–448. https://doi.org/10.1177/0022167806286745
Wasson G, Hofmann A, Ruck C (2008) The road to Eleusis: unveiling the secrets of the mysteries. North Atlantic Books, Berkeley
Watts A (1970) The joyous cosmology. Random House, New York
Wolfson P, Hartelius G (eds) (2016) The ketamine papers. MAPS, Santa Cruz

Part II
Disruptive Psychopharmacology for Mood Disorders

Ayahuasca for the Treatment of Depression

Fernanda Palhano-Fontes, Bruno Lobão Soares, Nicole Leite Galvão-Coelho, Emerson Arcoverde, and Draulio B. Araujo

Contents

Abstract Ayahuasca, the vine of the souls in Quechua, is a psychedelic brew with a few formulations that most often include the bark of a liana in the *Malpighiaceae* family (*Banisteriopsis caapi*), with leaves from a shrub in the coffee family *Rubiaceae* (*Psychotria viridis*). Mixed with water and boiled for hours or days, it produces a brownish-colored liquid with a strong and characteristic taste. Ayahuasca contains the psychedelic tryptamine N,N-Dimethyltryptamine (DMT), and Monoamine Oxidase Inhibitors (MAOi), and in the past few years, it has been tested. In

F. Palhano-Fontes and D. B. Araujo (✉)
Brain Institute, Federal University of Rio Grande do Norte (UFRN), Natal, Brazil

Hospital Universitário Onofre Lopes (UFRN), Natal, Brazil
e-mail: draulio@neuro.ufrn.br

B. L. Soares
National Science and Technology Institute for Translational Medicine, Natal, Brazil

Department of Biophysics and Pharmacology (UFRN), Natal, Brazil

N. L. Galvão-Coelho
National Science and Technology Institute for Translational Medicine, Natal, Brazil

Department of Physiology and Behavior, Laboratory of Hormone Measurement (UFRN), Natal, Brazil

E. Arcoverde
Hospital Universitário Onofre Lopes (UFRN), Natal, Brazil

Curr Topics Behav Neurosci (2022) 56: 113–124
https://doi.org/10.1007/7854_2021_277
Published Online: 11 November 2021

recent years its antidepressant properties have been put to the test. Evidence from open and randomized placebo-controlled clinical trials has shown encouraging results, indicating significant and rapid antidepressant effects, starting as early as 1 day after the ayahuasca intervention. In addition, we have explored the nature of these effects using multivariate measures. In this article, we will review the history, pharmacology, clinical trials, and clinical and behavioral markers associated with the antidepressant effects of ayahuasca.

Keywords Ayahuasca · DMT · MAOi · Monoamine Oxidase Inhibitors · N,N-Dimethyltryptamine · Treatment-resistant depression

1 Ayahuasca

Ayahuasca, the *vine of the souls* in Quechua, is a psychedelic brew with a few formulations that most often include the bark of a liana in the *Malpighiaceae* family (*Banisteriopsis caapi)*, with leaves from a shrub in the coffee family *Rubiaceae* (*Psychotria viridis*) (Pinkley 1969). Mixed with water and boiled for hours or days, it produces a brownish-colored liquid with a strong and characteristic taste.

At the end of the nineteenth century, the Amazon rubber boom favored an intense social exchange between Amerindian populations who used ayahuasca and rubber tappers, mostly Christians. This contact expanded the traditional use of ayahuasca by indigenous groups into a sacrament of different syncretic religions, such as the *Santo Daime* founded in 1930, the *Barquinha* in 1945, and the *União do Vegetal* (UDV) in 1961 (Labate and MacRae 2016). In 1987, the Brazilian government provisionally authorized the ceremonial use of ayahuasca. In 2006 this decision was made permanent by the Brazilian National Council on Drug Policies, and in 2010 the use of ayahuasca was regulated for religious purposes. In 2006 the Supreme Court of the USA allowed the religious use of ayahuasca by a UDV center in New Mexico, and since then more and more churches have been allowed to have ayahuasca as a sacrament (Labate and Jungaberle 2011).

2 The Pharmacology of Ayahuasca

The leaves of *P. viridis* are rich in psychedelic tryptamine N,N-dimethyltryptamine (N,N-DMT), while *B. caapi* is rich in reversible Monoamine Oxidase inhibitors (MAOi), such as harmine and harmaline (Callaway et al. 1999; Riba et al. 2003). N, N-DMT when administered orally is degraded by monoamine oxidase (MAO) enzymes present in the gastrointestinal tract. However, the presence of MAOi in the brew prevents N,N-DMT deamination, allowing it to reach the bloodstream, cross the blood-brain barrier, and lead to effects on the central nervous system

(Callaway et al. 1999). From a pharmacological point of view, ayahuasca resembles other classic psychedelics. N,N-dimethyltryptamine (N,N-DMT) is a hallucinogenic tryptamine that acts mainly as a serotonergic agonist, particularly to the 5HT-2A receptor but also acts on glutamate, dopamine, acetylcholine, TAAR, and sigma-1 receptors (Carbonaro and Gatch 2016).

3 The Phenomenology of Ayahuasca

The effects of ayahuasca are heterogeneous and depend on the synergistic interaction between its components (Schenberg et al. 2015), in addition to the individuals' personality, expectations, and intentions (set), and the social and environmental context (setting) (Pontual et al. 2021). The acute effects begin around 20–40 min after ingestion, peaking around 60–120 min, and last approximately 4 h (Riba et al. 2001). During this period, for typical doses, most individuals remain quiet and immobile, communicate coherently, remain aware of the environment, what they are doing, and understand that they are under the influence of ayahuasca (Shanon 2002).

The effects include changes in the perception of the body, such as temperature, numbness, weight, and yawning. Nausea and vomiting are common and, less often, diarrhea, all considered part of a purging process. Visual phenomena are also frequent and occur mainly with the eyes closed – visions vary from colorful geometric patterns to complex scenes as in a dream (Shanon 2002). Other subjective effects are usually described, such as increased introspection, transient anxiety, dissociation, depersonalization, distortions in the sense of time and space, and mystical experiences, even in laboratory settings (Callaway et al. 1999; Riba et al. 2001; Shanon 2002; de Araujo et al. 2012; Palhano-Fontes et al. 2019; Riba 2003).

Ayahuasca induces slight increases in blood pressure (systolic and diastolic), respiratory and heart rates, body temperature, and pupil diameter (Callaway et al. 1999; Dos Santos et al. 2011). To date, all studies have demonstrated the safety of ayahuasca, with reports from individuals who have used it for more than 30 years with no evidence of health damage (Callaway et al. 1999; Grob et al. 1996; Riba et al. 2001; Jiménez-Garrido et al. 2020). In addition, it has been shown that the traditional use of ayahuasca is not associated with the psychosocial problems that are common with other drugs (Fábregas et al. 2010).

4 Ayahuasca and Depression

4.1 Evidence from Naturalistic Settings

Initial evidence of the therapeutic potential of ayahuasca came from early studies conducted in regular members of syncretic churches. These studies reported that the frequent use of the brew in religious settings did not appear to be associated with psychopathological, personality, or cognitive deterioration. Instead, they suggested a benefit for mental health and mild psychiatric symptoms (Grob et al. 1996; Barbosa et al. 2005, 2009, 2012; Santos et al. 2007).

A double-blind study conducted in a naturalistic setting evaluated the use of ayahuasca as a modulator of anxiety, panic, and hopelessness and observed a reduction in the scores of panic and hopelessness in healthy subjects who were under the effect of the brew compared to the control group (Santos et al. 2007). In another study, individuals were followed for 6 months from their first experience with ayahuasca, and a significant reduction of minor psychiatric symptoms improved, increased confidence and optimism (Barbosa et al. 2009). Long-lasting antidepressant effects of ayahuasca (1–12 months) were also observed after a single ayahuasca dose in a naturalistic context, which was followed by decreases in stress and anxiety levels and rises in well-being and life satisfaction (Uthaug et al. 2018; González et al. 2020). Another placebo-controlled study in a naturalistic setting found that the emotional empathy to negative stimuli increased after a single ayahuasca session, but not a session with placebo (Uthaug et al. 2021).

A large online survey made with ayahuasca users (12,000 respondents) in naturalistic settings also has supported its antidepressant effects, 32% of them related that the depression was "completely resolved," and 46% "very much" improved (Sarris et al. 2021), which seem to be related to traditional ceremonies practices (Perkins et al. 2021). These findings, although limited by the selective sample problems of both the naturalistic and online survey studies, associated with anecdotal reports of the benefits of ayahuasca use encouraged further investigation of its therapeutic effects in a clinical context.

4.2 *Recent Evidence from the Laboratory*

To date, three clinical trials have been performed to evaluate the antidepressant effects of ayahuasca. In a first open-label trial, six patients with refractory depression participated in a single dosing session with ayahuasca (Osório et al. 2015). Patients were evaluated with clinical scales for depression, the Hamilton Depression Rating Scale (HAM-D), and the Montgomery-Åsberg Rating Scale (MADRS). Both scales were applied before, during, and after the acute effects of a single session with ayahuasca. A significant reduction in symptoms was observed 24 h after the session, which remained reduced for 21 days (Osório et al. 2015). In a subsequent study, also using an open-label design, 17 patients participated in a single session with ayahuasca, and subacute changes in cerebral blood flow were assessed 8 h after the dosing session (Sanches et al. 2016). A significant decrease in the symptoms of depression was observed already during the experimental session (40 min after ingestion), remaining significantly reduced for 21 days. In addition, increased blood flow was observed in the nucleus accumbens, right insula and subgenual anterior cingulate gyrus, brain areas frequently involved in the pathophysiology of depression (Otte et al. 2016; Sanches et al. 2016). Although promising, these results have some limitations as the small number of patients and the lack of a placebo group.

A follow-up randomized placebo-controlled trial (RCT) with ayahuasca for treatment-resistant depression was conducted (Palhano-Fontes et al. 2019). We assessed 218 patients for eligibility, and 35 met clinical criteria for the trial. All patients had used at least two different antidepressant medications unsuccessfully, and were in a current depressive episode, in need of medication change. Following recruitment, patients underwent a washout period of about 2 weeks prior to the dosing session to reduce interaction between their medications and ayahuasca. During this period, patients were followed by our psychiatrists for medication weaning. Seven days after the dosing session (D7) patients were prescribed a new medication, different than the one the patient was using before the trial.

The trial was carried out between January/2014 and June/2016 in 29 patients with treatment-resistant depression, i.e., all had tried at least two different antidepressant medications unsuccessfully. It occurred at the Onofre Lopes University Hospital (HUOL), in Natal, Brazil. The University Hospital Research Ethics Committee approved the study, and all patients provided written informed consent before participation. They were all naïve to ayahuasca.

The trial had a parallel two-arm design, in which 14 patients participated in a single session with ayahuasca, and 15 in a session with the placebo (Palhano-Fontes et al. 2019). The placebo was a brownish liquid prepared with water, beer yeast, citric acid, zinc sulfate, and caramel colorant. It was designed to simulate organoleptic properties of ayahuasca, such as its sour and bitter flavor, and could provoke modest gastrointestinal distress due to the presence of zinc sulfate. We used a single batch of ayahuasca, produced specially for the study by a Barquinha church, in Ji-Paraná, Brazil. Patients received a single dose of 1 mL/kg of placebo or ayahuasca at an adjusted dose of 0.36 mg/kg of N,N-DMT (Palhano-Fontes et al. 2019). In addition, the mean concentrations of the beta-carbolines were 1.86 mg/kg of harmine, 0.24 mg/kg of harmaline, and 1.20 mg/kg of tetrahydroharmine.

Depression severity was evaluated with the MADRS and the HAM-D scales at baseline and followed up at 1, 2, 7, 14 days, and monthly for the next 6 months after the dosing session. Our trial was also intended to explore the acute and subacute effects of ayahuasca on a number of markers that might help to explain the antidepressant effects observed. For that, we made several assessments 1 day before, during, and 1 day after the session with ayahuasca or placebo. The whole experimental protocol lasted 4 days. Baseline assessments, which occurred 1 day before the session, included psychiatric and neuropsychological evaluations, functional magnetic resonance imaging (fMRI), sleep electroencephalography (sEEG), saliva and blood samples. Patients were reassessed 1 day after the session (D1), when psychiatric and neuropsychological, fMRI, sEEG, saliva, and blood samples were recollected.

Each patient participated in a single session, ingesting either a single dose of ayahuasca or placebo. The session lasted approximately 9 h and took place in a bedroom-like environment specifically designed for the study, equipped with a recliner, a bed, a desk, controlled temperature, and natural and dimmed light

Fig. 1 Setting where dosing sessions took place. The room was a bedroom-like hospital site equipped with a recliner, a bed, and a desk, controlled temperature, natural and dimmed light. Patients stayed in a recliner most of the time and were continuously monitored with EEG/ECG/EMG

(Fig. 1). Throughout the dosing session, at least two members of our team accompanied the participant, assisting them when necessary.

During the session, patients were monitored with EEG, ECG, and EMG. Saliva samples were collected at three time points (1 h40, 2 h40, and 4 h after ingestion), together with psychiatric evaluations of depressive, dissociative, psychotic, and mania-like symptoms.

When the acute effects had subsided, patients freely described their experiences and responded to three questionnaires: the Hallucinogenic Rating Scale (HRS) (Mizumoto et al. 2011), the Mystical Experience Questionnaire (MEQ) (Maclean et al. 2012), and the Amsterdam Resting-State Questionnaire (ARSQ) (Diaz et al. 2013).

Psychiatric evaluations during the dosing session showed that patients had slight transient changes in dissociative and psychotomimetic symptoms, and no significant increases in mania-like symptoms, as measured by the Brief Psychiatric Rating Scale (BPRS), Young Mania Rating Scale (YMRS), and Clinician-Administered Dissociative States Scales (CADSS). Transient nausea (ayahuasca = 71%, placebo = 26%), vomiting (ayahuasca = 57%, placebo = 0%), transient anxiety (ayahuasca = 50%, placebo = 73%), and transient headache (ayahuasca = 42%, placebo = 53%) were also commonly described.

Patients were asked to return for follow-up assessments at 7 days (D7), 14 days (D14), 1 month (M1) and from then on, every month, for 6 months (M2, M3, M4, M5, M6). In the return at D7, a new antidepressant medication was prescribed,

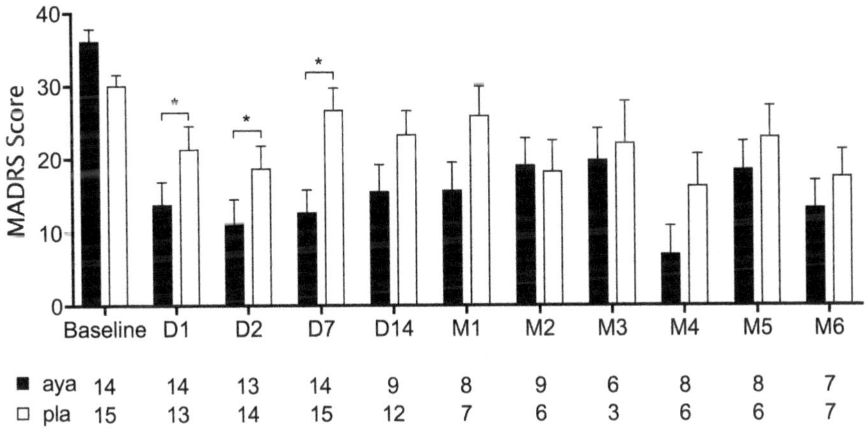

Fig. 2 Depression severity at baseline and follow-up assessments. We observed significant differences between ayahuasca and placebo at D1 ($p = 0.04$), D2 ($p = 0.04$), and D7 ($p < 0.0001$). No differences were observed after D14. The numbers of patients in the ayahuasca and placebo groups who attended psychiatric evaluation at those time points are indicated below the graph. Values are (mean ± SEM). MADRS scores: mild depression (11–19), moderate (20–34), severe (≥35) $*p < 0.05$

which was chosen based on the clinical history of each patient. For more details about the whole protocol, please see Palhano-Fontes et al. (2019).

Figure 2 shows changes in depression severity over time, at baseline up to 6 months after dosing. Unfortunately, the number of patients who returned to the follow-up assessments significantly decreased after D14 (inset Fig. 2). We observed significantly decreased depression severity 1 day (D1) after the session with ayahuasca, which remained reduced for at least 7 days (D7). We observed a significant placebo effect, but the antidepressant effects of ayahuasca were more robust and lasted longer (Palhano-Fontes et al. 2019). Due to the small sample size, no statistical tests were performed for the follow-up assessments and the results presented are merely exploratory. No statically significant changes were observed in the follow-up assessment, likely due to the small sample size after D14 (Fig. 2).

We also observed significant between-groups differences in response and remission rates at D7, with 64% of responders and 36% of remitters in the ayahuasca group versus 27% and 7% in the placebo group (Palhano-Fontes et al. 2019). Furthermore, our results suggest that the ayahuasca session significantly reduced suicidality (i.e., suicide attempts, suicide planning, and suicidal ideation) from baseline to D1, D2, and D7 after the dosing session (Zeifman et al. 2019).

Our results are comparable with trials that used ketamine, although they differed in their temporal dynamics. Whereas the effect sizes observed for ayahuasca are smallest at D1 and most prominent at D7, trials with ketamine report the largest effect sizes 1 day after the session, which is reduced 1 week after the session (Palhano-Fontes et al. 2019). A recent meta-analysis suggested that the largest antidepressant effects of classic psychedelics, compared to placebo, are observed between 2 and 7 days after dosing (SMD = −0.841, CIs −1.359 to −0.323, $p = 0.001$; I2 = 7.0%) (Galvão-Coelho et al. 2021).

5 Potential Mechanisms of Action

The mechanism of action of psychedelic substances has been twofold. Some authors argue that the observed therapeutic effects of these substances are due to changes in broad biological mechanisms, such as neuroplasticity, and that the acute subjective experience is not necessary for the long-lasting positive outcome (Olson 2020; Cameron et al. 2021). Others support the idea that the subjective experience, particularly the intensity of the mystical experience, mediates the beneficial outcomes (Bogenschutz et al. 2015; Garcia-Romeu et al. 2015; Majić et al. 2015; Griffiths et al. 2016; Ross et al. 2016; Yaden and Griffiths 2020). Are they mutually exclusive, or do the effects arise from the combination of both? Our trial helps to explore this question a little bit further, as both aspects have been assessed.

In our RCT, we used two questionnaires to assess the subjective effects felt during the session. Twenty-seven patients responded to the HRS, and 15 responded to the MEQ. We found significantly higher scores in ayahuasca than placebo in almost all subscales of the HRS and MEQ, except for affect in the HRS and positive mood in the MEQ. The HRS subscale *perception* positively correlated with the observed antidepressant effects of ayahuasca (Palhano-Fontes et al. 2019). Visions are a well-known effect of ayahuasca, and we have previously observed that ayahuasca has an important impact on the activity of our visual system (de Araujo et al. 2012). Furthermore, it has been argued that they can serve as a mechanism to initiate reflections on personal relationships and traumatic events, which could impact the therapeutic benefits (Frecska et al. 2016).

We found a negative correlation between changes in depression severity and the transcendence of time and space. Previous studies with psilocybin have found that the characteristic of the patients' subjective experience as measured in the mystical experience questionnaire has a specific role apart from the overall intensity of the psychedelic effects measured by the HRS (Griffiths et al. 2016; Ross et al. 2016). In our study, we found that changes in visual perception were strongly correlated with the positive clinical outcome, suggesting that other dimensions, and not only the mystical experience, may play a role in the therapeutic response of psychedelics.

Neuroimaging evidence also suggests that functional brain changes influence the therapeutic response (Sanches et al. 2016; Carhart-Harris et al. 2017; Roseman et al. 2018). Recently, we have suggested that ayahuasca may exert its long-lasting effects on mood by modulating neural networks supporting interoceptive, affective, and self-referential processes, such as the Default Mode Network (DMN) and the Salience Network (SN) (Pasquini et al. 2020). As depicted in Fig. 3, we found decreased functional connectivity (fc) in the DMN and increase in the SN 1 day after a single dose of ayahuasca or placebo in healthy individuals (Pasquini et al. 2020). These results are in line with our previous observation that the activity and

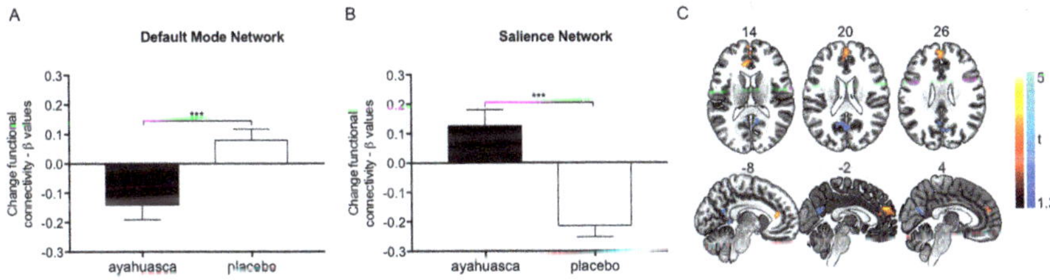

Fig. 3 Subacute changes in brain functional connectivity (fc) after ayahuasca. (**a**) default mode network fc decreases for the ayahuasca compared to the placebo group. (**b**) salience network fc increases for the ayahuasca compared to the placebo group. Values are (mean ± SEM). (**c**) Brain slices showing DMN fc decreases within the posterior cingulate cortex (cold colors) and SN fc increases within the anterior cingulate cortex (warm colors). The left side is on the left, bar reflects t-values. $^{***}p < 0.0001$

connectivity within the SN are increased during the acute effects of ayahuasca while are decreased in the DMN (Palhano-Fontes et al. 2015).

Although subjective effects of the experience seem to have influenced the antidepressant effects of ayahuasca, we also found significant changes in three different physiological systems known to be related to depression. First, the stress system, which we assessed through cortisol, an important hormone related to stress responses. Our patients presented significantly low levels of cortisol at baseline (hypocortisolemia) and blunted salivary cortisol awakening response. After the session, patients who received ayahuasca, but not a placebo, changed their salivary cortisol to normal levels (Galvão et al. 2018). Second, changes in neuroplasticity are associated with the Brain-Derived Neurotrophic Factor (BDNF) (Ly et al. 2018). We found a significant increase in serum BDNF levels, which were significantly correlated with decreased depressive symptoms (MADRS scores) 2 days (D2) after dosing (de Almeida et al. 2019). Third, changes in the inflammatory system, by means of serum Interleukin-6 (IL-6) and the C-Reactive Protein (CRP). Patients presented significantly increased CRP levels with respect to controls. No differences were found in IL-6 levels. Two days after the session (D2), CRP levels decreased significantly only in the group of individuals who had an ayahuasca session. We also found a positive correlation between decreased CRP levels and depressive symptoms (Galvão-Coelho et al. 2020).

6 Final Remarks

Early evidence of the antidepressant effects of ayahuasca has been explored both in ayahuasca communities, particularly in Brazil, and later in laboratory settings. When used in appropriate settings, both lines of evidence point to a safe and beneficial profile of ayahuasca to alleviate the symptoms of depression.

Our recent study was the first RCT to test a classic psychedelic in treatment-resistant depression (Palhano-Fontes et al. 2019). Ayahuasca showed a rapid antidepressant effect starting already 1 day after dosing, and lasting for at least 7 days, a promising observation taking into consideration that currently available antidepressants take around 2 weeks for the onset of the therapeutic response (Cai et al. 2015; Otte et al. 2016).

As the number of studies suggesting the antidepressant effects of psychedelics increase (Osório et al. 2015; Carhart-Harris et al. 2016; Sanches et al. 2016; Griffiths et al. 2016; Ross et al. 2016; Palhano-Fontes et al. 2019; Davis et al. 2021), research is starting to unravel the origin of these effects. Our evidence suggests that these are the result of a combination of the subjective changes experienced during the acute effects of the substance, and the modulation of neuroimmune, neuroendocrine, and neuroplasticity pathways important for homeostatic regulation. Such observations indicate that the underlying effects of ayahuasca are indeed multifactorial ranging from biochemical changes to subtle cognitive processes.

References

Barbosa PCR, Giglio JS, Dalgalarrondo P (2005) Altered states of consciousness and short-term psychological after-effects induced by the first time ritual use of ayahuasca in an urban context in Brazil. J Psychoactive Drugs 37:193–201

Barbosa PCR, Cazorla IM, Giglio JS, Strassman R (2009) A six-month prospective evaluation of personality traits, psychiatric symptoms and quality of life in Ayahuasca-naive subjects. J Psychoactive Drugs 41:205–212

Barbosa PCR, Mizumoto S, Bogenschutz MP, Strassman RJ (2012) Health status of ayahuasca users. Drug Test Anal 4:601–609

Bogenschutz MP, Forcehimes AA, Pommy JA et al (2015) Psilocybin-assisted treatment for alcohol dependence: a proof-of-concept study. J Psychopharmacol 29:289–299

Cai S, Huang S, Hao W (2015) New hypothesis and treatment targets of depression: an integrated view of key findings. Neurosci Bull 31:61–74

Callaway JC, McKenna DJ, Grob CS et al (1999) Pharmacokinetics of Hoasca alkaloids in healthy humans. J Ethnopharmacol 65:243–256

Cameron LP, Tombari RJ, Lu J et al (2021) A non-hallucinogenic psychedelic analogue with therapeutic potential. Nature 589:474–479

Carbonaro TM, Gatch MB (2016) Neuropharmacology of N,N-dimethyltryptamine. Brain Res Bull 126:74–88

Carhart-Harris RL, Bolstridge M, Rucker J et al (2016) Psilocybin with psychological support for treatment-resistant depression: an open-label feasibility study. Lancet Psychiatry 3:619–627

Carhart-Harris RL, Roseman L, Bolstridge M et al (2017) Psilocybin for treatment-resistant depression: fMRI-measured brain mechanisms. Sci Rep 7:13187

Davis AK, Barrett FS, May DG (2021) Effects of psilocybin-assisted therapy on major depressive disorder: a randomized clinical trial. JAMA Psychiat 78(5):481–489

de Almeida RN, Galvão ACM, da Silva FS et al (2019) Modulation of serum brain-derived neurotrophic factor by a single dose of Ayahuasca: observation from a randomized controlled trial. Front Psychol 10:1234

de Araujo DB, Ribeiro S, Cecchi GA et al (2012) Seeing with the eyes shut: neural basis of enhanced imagery following Ayahuasca ingestion. Hum Brain Mapp 33:2550–2560

Diaz BA, Van Der Sluis S, Moens S et al (2013) The Amsterdam resting-state questionnaire reveals multiple phenotypes of resting-state cognition. Front Hum Neurosci 7:446

Dos Santos RG, Valle M, Bouso JC et al (2011) Autonomic, neuroendocrine, and immunological effects of ayahuasca: a comparative study with d-amphetamine. J Clin Psychopharmacol 31: 717–726

Fábregas JM, González D, Fondevila S et al (2010) Assessment of addiction severity among ritual users of Ayahuasca. Drug Alcohol Depend 111:257–261

Frecska E, Bokor P, Winkelman M (2016) The therapeutic potentials of Ayahuasca: possible effects against various diseases of civilization. Front Pharmacol 7:35

Galvão ACM, de Almeida RN, Silva EADS et al (2018) Cortisol modulation by Ayahuasca in patients with treatment resistant depression and healthy controls. Front Psych 9:185

Galvão-Coelho NL, de Menezes Galvão AC, de Almeida RN et al (2020) Changes in inflammatory biomarkers are related to the antidepressant effects of Ayahuasca. J Psychopharmacol 34:1125–1133

Galvão-Coelho NL, Marx W, Gonzalez M et al (2021) Classic serotonergic psychedelics for mood and depressive symptoms: a meta-analysis of mood disorder patients and healthy participants. Psychopharmacology 238:341–354

Garcia-Romeu A, Griffiths R, Johnson M (2015) Psilocybin-occasioned mystical experiences in the treatment of tobacco addiction. Curr Drug Abuse Rev 7:157–164

González D, Cantillo J, Pérez I et al (2020) Therapeutic potential of ayahuasca in grief: a prospective, observational study. Psychopharmacology 237:1171–1182

Griffiths RR, Johnson MW, Carducci MA et al (2016) Psilocybin produces substantial and sustained decreases in depression and anxiety in patients with life-threatening cancer: a randomized double-blind trial. J Psychopharmacol 30:1181–1197

Grob CS, McKenna DJ, Callaway JC et al (1996) Human psychopharmacology of hoasca, a plant hallucinogen used in ritual context in Brazil. J Nerv Ment Dis 184:86–94

Jiménez-Garrido DF, Gómez-Sousa M, Ona G et al (2020) Effects of ayahuasca on mental health and quality of life in naïve users: a longitudinal and cross-sectional study combination. Sci Rep 10:1–12

Labate BC, Jungaberle H (2011) The internationalization of Ayahuasca. LIT Verlag Münster

Labate BC, MacRae E (2016) Ayahuasca, ritual and religion in Brazil. Routledge

Ly C, Greb AC, Cameron LP et al (2018) Psychedelics promote structural and functional neural plasticity. Cell Rep 23:3170–3182

Maclean KA, Leoutsakos J-MS, Johnson MW, Griffiths RR (2012) Factor analysis of the mystical experience questionnaire: a study of experiences occasioned by the hallucinogen psilocybin. J Sci Study Relig 51:721–737

Majić T, Schmidt TT, Gallinat J (2015) Peak experiences and the afterglow phenomenon: when and how do therapeutic effects of hallucinogens depend on psychedelic experiences? J Psychopharmacol 29:241–253

Mizumoto S, da Silveira DX, Barbosa PCR, Strassman RJ (2011) Hallucinogen Rating Scale (HRS) – Versão brasileira: tradução e adaptação transcultural Hallucinogen Rating Scale (HRS) – A Brazilian version: translation and cross-cultural adaptation. Rev Psiq Clín 38:231–237

Olson DE (2020) The subjective effects of psychedelics may not be necessary for their enduring therapeutic effects. ACS Pharmacol Transl Sci. https://doi.org/10.1021/acsptsci.0c00192

Osório FL, Sanches RF, Macedo LR et al (2015) Antidepressant effects of a single dose of ayahuasca in patients with recurrent depression: a preliminary report. Rev Bras Psiquiatr 37: 13–20

Otte C, Gold SM, Penninx BW et al (2016) Major depressive disorder. Nat Rev Dis Primers 2: 16065

Palhano-Fontes F, Andrade KC, Tofoli LF, Santos AC, Crippa JA, Hallak JE, Ribeiro S, de Araujo DB (2015) The psychedelic state induced by ayahuasca modulates the activity and connectivity of the default mode network. PLoS One 10(2):e0118143

Palhano-Fontes F, Barreto D, Onias H et al (2019) Rapid antidepressant effects of the psychedelic ayahuasca in treatment-resistant depression: a randomized placebo-controlled trial. Psychol Med 49:655–663

Pasquini L, Palhano-Fontes F, Araujo DB (2020) Subacute effects of the psychedelic ayahuasca on the salience and default mode networks. J Psychopharmacol 34(6):623–635

Perkins D, Schubert V, Simonová H et al (2021) Influence of context and setting on the mental health and wellbeing outcomes of ayahuasca drinkers: results of a large international survey. Front Pharmacol 12:469

Pinkley HV (1969) Plant admixtures to Ayahuasca, the South American hallucinogenic drink. Lloydia 32:305–314

Pontual AADD, Tófoli LF, Collares CF et al (2021) The setting questionnaire for the Ayahuasca experience: questionnaire development and internal structure. Front Psychol 12:2347

Riba J (2003) Human pharmacology of Ayahuasca. Universitat Autònoma de Barcelona

Riba J, Rodríguez-Fornells A, Urbano G et al (2001) Subjective effects and tolerability of the South American psychoactive beverage Ayahuasca in healthy volunteers. Psychopharmacology 154: 85–95

Riba J, Valle M, Urbano G et al (2003) Human pharmacology of ayahuasca: subjective and cardiovascular effects, monoamine metabolite excretion, and pharmacokinetics. J Pharmacol Exp Ther 306:73–83

Roseman L, Demetriou L, Wall MB et al (2018) Increased amygdala responses to emotional faces after psilocybin for treatment-resistant depression. Neuropharmacology 142:263–269

Ross S, Bossis A, Guss J et al (2016) Rapid and sustained symptom reduction following psilocybin treatment for anxiety and depression in patients with life-threatening cancer: a randomized controlled trial. J Psychopharmacol 30:1165–1180

Sanches RF, de Lima OF, Dos Santos RG et al (2016) Antidepressant effects of a single dose of Ayahuasca in patients with recurrent depression: a SPECT study. J Clin Psychopharmacol 36: 77–81

Santos RG, Landeira-Fernandez J, Strassman RJ et al (2007) Effects of ayahuasca on psychometric measures of anxiety, panic-like and hopelessness in Santo Daime members. J Ethnopharmacol 112:507–513

Sarris J, Perkins D, Cribb L et al (2021) Ayahuasca use and reported effects on depression and anxiety symptoms: an international cross-sectional study of 11,912 consumers. J Affect Disord Rep 4:100098

Schenberg EE, Alexandre JF, Filev R et al (2015) Acute biphasic effects of Ayahuasca. PLoS One 10:e0137202

Shanon B (2002) The antipodes of the mind: charting the phenomenology of the Ayahuasca experience, 1st edn. Oxford University Press, Oxford

Uthaug MV, van Oorsouw K, Kuypers KPC et al (2018) Sub-acute and long-term effects of ayahuasca on affect and cognitive thinking style and their association with ego dissolution. Psychopharmacology 235:2979–2989

Uthaug MV, Mason NL, Toennes SW et al (2021) A placebo-controlled study of the effects of ayahuasca, set and setting on mental health of participants in ayahuasca group retreats. Psychopharmacology. https://doi.org/10.1007/s00213-021-05817-8

Yaden DB, Griffiths RR (2020) The subjective effects of psychedelics are necessary for their enduring therapeutic effects. ACS Pharmacol Transl Sci. https://doi.org/10.1021/acsptsci.0c00194

Zeifman RJ, Palhano-Fontes F, Hallak J et al (2019) The impact of Ayahuasca on suicidality: results from a randomized controlled trial. Front Pharmacol 10:1325

Psilocybin for the Treatment of Depression: A Promising New Pharmacotherapy Approach

Gabrielle Agin-Liebes and Alan K. Davis

Contents

Abstract Depression is highly prevalent and represents the leading cause of global disability and primary contributor to overall global burden of disease. Several lines of evidence from early-phase experimental trials suggest that serotonergic psychedelics, particularly psilocybin, with therapeutic support show great promise in the treatment of depression with large effect sizes. Neuroimaging data have also revealed the dynamic effects of psilocybin on functional activity within and between neural regions. This chapter reviews the methods and findings from three small human laboratory clinical trials examining the effects of psilocybin therapy for patients with major depressive disorder and treatment-resistant depression. Insights from functional magnetic resonance imaging and qualitative analyses are also

---

G. Agin-Liebes (✉)
Department of Psychiatry, University of California, San Francisco, San Francisco, CA, USA

Neuroscape, Sandler Neurosciences Center, San Francisco, CA, USA
e-mail: Gabrielle.Agin-Liebes@ucsf.edu

A. K. Davis
Department of Psychiatry, College of Social Work, The Ohio State University, Columbus, OH, USA

Center for Psychedelic and Consciousness Research, Johns Hopkins University, Baltimore, MD, USA

Curr Topics Behav Neurosci (2022) 56: 125–140
https://doi.org/10.1007/7854_2021_282
Published Online: 23 November 2021

presented, as well as a discussion of study limitations and future directions for the research.

Keywords Depression · Psilocybin · Psychedelics

1 An Essential Need to Help Ease Mental Suffering

Over 300 million people are affected by depression worldwide, rendering it a massive global public health concern (World Health Organization 2019). Depression represents the number one cause of disability (World Health Organization 2019), and the relative risk of all-cause mortality among depressed people is 1.7 times greater than the non-depressed population (Walker et al. 2015). In terms of prevalence, approximately 10% of the United States (U.S.) adult population has been diagnosed with Major Depressive Disorder (MDD) in the past year (Hasin et al. 2018). In addition to the personal burden, the annual economic burden is estimated to be a staggering $210 billion (Greenberg et al. 2015).

A variety of pharmaco- and psychotherapies have been developed to address the public health and personal burden of this condition (Cuijpers et al. 2014; Otto and Hearon 2016). Despite the evidence supporting the relative effectiveness of existing interventions, many people lack access to these treatments and many who are treated experience depression relapse (Collimore and Rector 2014; Coull and Morris 2011; Hofmann et al. 2012; Sinyor et al. 2010; Koen and Stein 2011). For example, although many patients will experience a reduction or remission in depression symptoms after treatment with existing pharmacotherapies (Morilak and Frazer 2004), up to 50% will not respond fully. Additionally, as many as 30% are considered to have treatment-resistant depression (TRD), resulting in effects that are only marginally superior to placebo (Gaynes et al. 2009; Nemeroff 2007). The disparity between the need for effective treatments and an absence of available effective options highlights the need to explore novel ways to treat this debilitating condition.

Within psychiatry, there is an urgent need for improved mental healthcare and novel pharmacological approaches. Many leaders in the field have urged the adoption of a "disruptive pharmacology" approach (Heifets and Malenka 2019) to investigate novel interventions with historically restricted substances, such as classic psychedelics, for treating disabling conditions such as depression (Carhart-Harris et al. 2016, 2018, 2021; Ross et al. 2016; Griffiths et al. 2016; Agin-Liebes et al. 2020; Davis et al. 2021a). Psilocybin (4-phosphoryloxy-N,N-dimethyltryptamine) belongs to a chemical group of naturally occurring serotonergic tryptamines. Psilocybin therapy combines supportive psychotherapy with one or two doses of the classic psychedelic psilocybin, which acts primarily as an agonist of the serotonin-2A (5HT-2A) receptor and has the capacity to occasion profound changes in sensory perceptions, cognition, mood, and alterations in self-related processing mediated by increased neural plasticity (Araújo et al. 2015). Recent studies examining this treatment approach have suggested that it is potentially efficacious in decreasing symptoms of depression among a variety of patient populations (Carhart-Harris et al. 2016, 2018, 2021; Ross et al. 2016; Griffiths et al. 2016; Agin-Liebes et al. 2020; Davis et al. 2021a). For example, two placebo-controlled studies published in 2016 showed that psilocybin therapy continued to decrease depressive symptoms among patients with life-threatening illness up to 6 months after the treatment was completed, and up to 4.5 years in one long-term follow-up study (Ross et al. 2016; Griffiths et al. 2016; Agin-Liebes et al. 2020). In 2017, an open-label study was published assessing the effect of psilocybin therapy for patients with TRD, demonstrating that this treatment decreased depression symptoms lasting up to 3 months (Carhart-Harris et al. 2018). In 2020, the first randomized controlled trial (RCT) was published that examined psilocybin therapy among the general depression population, showing large decreases in depression with about one-half (54%) in remission 1 month after treatment (Agin-Liebes et al. 2020). Furthermore, findings from another recent RCT suggested that psilocybin therapy may be as or more effective than selective serotonin reuptake inhibitors (SSRIs), which are the most commonly prescribed pharmacological treatment for MDD (Davis et al. 2021a).

Given the positive outcomes from several trials, the U.S. Food and Drug Administration (FDA) has granted psilocybin therapy "Breakthrough" status, on the premise that there is meaningful evidence to support rapid and prioritized examination of this treatment approach in large multisite trials. This chapter will review published studies from the extant literature that have explored psilocybin therapy as a treatment for depression. This chapter will limit its review to the trials that have focused exclusively on MDD as a primary clinical indication (Carhart-Harris et al. 2016, 2018; Agin-Liebes et al. 2020; Davis et al. 2021a). We will also include data from neuroimaging and qualitative studies reports and a discussion of possible psychological mechanisms that may explain its antidepressant effects. Future directions for this line of research will also be discussed.

2 Psilocybin Therapy for Depression

All three of the following trials of psilocybin therapy for a primary indication of depression shared a common treatment model. Prior to the psilocybin sessions, participants received preparatory sessions with psychoeducation about the upcoming medication sessions as well as explanation of common reactions and challenges. During the psilocybin experiences, participants were provided eye masks and headphones and were encouraged to listen to a pre-recorded playlist. Participants were encouraged to direct their attention inward during these 8-h psilocybin sessions. The following day participants met with their session facilitators to process insights and emotions that emerged during their psilocybin sessions (commonly referred to as integration).

3 Psilocybin Therapy for People Diagnosed with Treatment-Resistant Depression

In 2016, Carhart-Harris and colleagues at Imperial College, London published findings from a small open-label feasibility study assessing the effects of psilocybin on TRD (Carhart-Harris et al. 2016, 2018). The investigators attempted to increase the sample size from 12 to 20 midway through the trial reportedly to achieve adequate statistical power for brain imaging analyses; the investigators conducted a 6-month follow-up analysis with the larger ($N = 20$) sample (Carhart-Harris et al. 2017). All participants met criteria for MDD and endorsed a score of 16 or greater on the Hamilton Depression Rating Scale (HAM-D) (Iannuzzo et al. 2006). They also were required to have failed two previous courses of antidepressant medication treatment (from separate pharmacological classes) with 6 weeks of duration during the depressive episode.

The study involved two psilocybin administration sessions separated by 1 week. In the first session, participants received 10 mg of psilocybin, followed by 25 mg in the second session. In addition to feasibility and safety, the investigators assessed the preliminary efficacy of psilocybin in reducing self-reported depression scores with the Quick Inventory of Depressive Symptomatology (QIDS-SR16) (Rush et al. 2003). Secondary outcomes included trait anxiety (STAI-T) (Spielberger 2010) and anhedonia (Snaith-Hamilton Pleasure Scale; SHAPS) (Snaith et al. 1995). Participants also underwent a functional MRI (fMRI) scanning session at baseline and 1 day after the high-dose psilocybin session (see further details below) (Araújo et al. 2015).

Results revealed significant within-group reductions on the QIDS-SR16 relative to baseline at the 1-week time point, and these reductions persisted until the 6-month follow-up. These findings were corroborated by clinician-administered ratings. Within-subject effect sizes were very large (Cohen's d 1.4–2.3). In the original analysis ($N = 12$)' 67% of participants met clinical criteria for remission (defined as a score of ≤ 9 on the Beck Depression Inventory; BDI (Beck et al. 1996)) from depression at the 1-week time point, and 42% continued to meet remission criteria at the 3-month follow-up. There were also significant reductions, relative to baseline, on suicidality at the 1-week and 2-week follow-ups and on trait anxiety and anhedonia measures at the 1-week and 3-month follow-ups. None of the participants resumed antidepressant medications within 5 weeks after their psilocybin sessions. Significant reductions on the QIDS-SR16 remained significant at the 6-month follow-up (Cohen's $d = 1.4$). Reductions in depressive symptoms at the 5-week follow-up were significantly associated with ratings of acute psychedelic experience [on the 11-Dimension altered states of consciousness (11D-ASC)] (Studerus et al. 2010).

There were no serious adverse events (SAEs) reported in the trial. The most common AEs reported were transient anxiety (100%), confusion (75%), and nausea (33%) during the psilocybin session and headache 1 day after the session. All AEs resolved by the end of the session, with the exception of headache, which tended to resolve 1–2 days following psilocybin treatment. This pilot study provided important scientific insights informing future research but was limited by the absence of a control group, possible favorable expectancy effects and participant self-selection biases. Additionally, for the majority of participants, positive outcomes did not persist beyond the 3-week follow-up time point. Lastly, the majority of participants (75%) were non-Hispanic White, which limits the generalizability to more diverse groups of individuals.

The investigators also conducted a series of fMRI analyses with the same TRD study sample at baseline and 1 day after participants' high-dose psilocybin session. They assessed changes (relative to baseline) in cerebral blood flow, functional connectivity, and amygdala responsiveness during an emotional face paradigm (Araújo et al. 2015; Roseman et al. 2018; Mertens et al. 2020). fMRI whole-brain analysis data revealed significant reductions in cerebral blood flow in the temporal cortex (including the amygdala) that correlated with reductions in depressed mood. They also found increased resting state functional connectivity in the following two brain regions: (1) the default mode network (DMN), a network involved in self-related processing and mind wandering and has been found to be hyper-engaged in a variety of psychiatric disorders, including MDD (Hamilton et al. 2015) and (2) the ventromedial prefrontal cortex (vmPFC)-bilateral inferior-lateral parietal cortex (ilPC), a network involved in regulating and inhibiting emotional responses. These findings contrast with previous findings of acute decreases in DMN activity captured *during* psilocybin sessions (Carhart-Harris et al. 2012). The investigators speculate that these acute and post-acute (1 day after psilocybin) dynamics may reflect an initial *acute* disintegration of DMN connectivity during psilocybin followed by reintegration of normalized functioning, which may serve a psychological "reset" function (Araújo et al. 2015). Additionally, exploratory analyses showed that acute, subjective mystical-type experiences during the psilocybin session predicted changes in these DMN and limbic (emotion network) regions (Araújo et al. 2015).

With regard to amygdala responsiveness in the same group of participants, the investigators found that participants showed increased amygdala responses to both fearful and joyful faces during an emotional face-processing paradigm administered 1 day after the high-dose psilocybin session (Studerus et al. 2010; Roseman et al. 2018). The post-acute increases in amygdala activity predicted reductions in depression scores (on the QIDS-SR16 and BDI) at the 1-day, 1-week, and 3-week follow-ups. Additionally, the investigators found reductions in connectivity between the vmPFC and right amygdala. Although these alterations did not predict depression changes, they did predict rumination levels (a strong maintaining factor of depression) at the 1-week follow-up. Taken together, these findings could suggest that psilocybin may heighten depressed individuals' sensitivity to experiencing emotions while reducing the top-down inhibitory control of prefrontal control regions on limbic regions. The investigators propose that these mechanisms might explain phenomenological insights regarding participants' increased access to a full range of previously avoided emotions and subsequent emotional catharsis, processing, and release (Studerus et al. 2010; Roseman et al. 2018).

4 Psilocybin Therapy for the Treatment of Major Depressive Disorder

In 2021, Davis and colleagues reported on findings from a randomized controlled trial (RCT) conducted at Johns Hopkins University. The investigators randomized 27 participants (24 completers) with moderate-to-severe MDD to either an immediate intervention group or delayed (waitlist) group. The study involved two psilocybin medication sessions separated by 1 week (i.e., 20 mg/70 kg in session 1 and 30 mg/70 kg in session 2). The intervention period was 8 weeks in duration. Participants in the waitlist group were monitored on a weekly basis and completed assessments at weeks 5 and 8 (corresponding to the 1-week and 1-month time points in the immediate intervention group). The primary study outcome was the GRID-HAM-D, which was assessed by blinded clinician raters. The QIDS-Self Report (QIDS-SR) (Iannuzzo et al. 2006) was also administered as a self-report measure of the rapid (1-day) response to the treatment. The study also included secondary measures of suicidality, anxiety, trait anxiety, and persisting positive effects.

There were significant between-group effects favoring the immediate intervention group at the 1-week follow-up, which remained significant at the 4-week follow-up, compared to the delayed intervention group. Between-group effect sizes were very large at the 1-week and 4-week follow-ups in the immediate treatment group compared to weeks 5 and 8 in the waitlist group (Cohen's d range = 2.5–2.6). Across the entire sample, reductions in depression on the QIDS-SR revealed immediate reductions in depression between baseline and 1 day after the psilocybin session, which remained significant at the 4-week follow-up (following session 2). At 1 month after treatment, 71% of participants demonstrated a clinically significant response (defined as greater than 50% decrease in GRID-HAM-D scores), and 54% were in complete remission from depression (a score of ≤ 7 on the GRID-HAM-D). All secondary depression, suicidality, and anxiety measures showed significant between-group differences and within-subjects reductions at all time points. Participants overwhelmingly (85–90%) rated their psilocybin experiences to be one of the top five most personally meaningful and psychologically insightful experiences of their lives. Decreases in depression scores at 1 month were significantly correlated with peak ratings of the degree to which the psilocybin sessions were personally meaningful ($r = -0.70, p < 0.01$), psychological insightful ($r = -0.60, p < 0.01$), and spiritually significant ($r = -0.57, p < 0.01$). There were no SAEs reported in the trial. The most frequent physiological AE was transient headache (29–33%). Greater than half the participants endorsed experiencing a variety of challenging emotions and physical sensations during psilocybin sessions that resolved before the end of the session. As in other contemporary clinical trials (Carhart-Harris et al. 2016, 2018; Griffiths et al. 2016; Ross et al. 2016) there were no reports of persistent visual perceptual changes, psychosis, or psilocybin use/misuse behaviors as none of the participants reported consuming psilocybin outside of the trial during the 4-week follow-up period.

Despite some notable strengths including the randomized controlled design and the use of a clinician-administered rating scale with blinded raters, the study carried a few limitations. Although the delayed waitlist design controlled for the passage of time, it did not control for other non-pharmacological therapeutic elements such as the preparatory and post-psilocybin therapy sessions and expectancy effects. As in the Carhart-Harris and colleagues' study (Koen and Stein 2011), the majority of participants were non-Hispanic White (92%), and presented with low risk of suicide, which may not generalize to other individuals with moderate-to-severe depression. The study sample was also relatively small, and although the sample size was determined to be well-powered enough to detect significant effects, trials with larger sample sizes are needed.

5 Direct Comparison of Psilocybin Therapy and Antidepressant Medication Therapy

More recently, Carhart-Harris et al. (2021) conducted a double-blind RCT at Imperial College London with 59 participants with moderate-to-severe MDD (Davis et al. 2021a). Participants were randomized to one of two groups. One group received psilocybin (25 mg) followed by daily capsule doses of placebo (microcrystalline cellulose) for the subsequent 3 weeks. The second group received low-dose psilocybin (1 mg) followed by daily doses of escitalopram (10 mg). At the 3-week time point, participants in the former group received a second dose of high-dose psilocybin followed by daily capsule doses of placebo for the subsequent 3 weeks. The latter group received a second dose of low-dose psilocybin followed by additional daily doses of escitalopram (20 mg) for the subsequent 3 weeks. Both groups were provided one preparatory therapeutic session prior to the first dosing day and psychological debriefing/integration on the days after each dosing day. At the 6-week time point (3 weeks after the second dosing day), participants completed final study assessments. The participants in the escitalopram/active comparator group collaborated with their general physicians to discontinue the escitalopram medication. The primary study outcome was self-reported depression change-score (relative to baseline) on the QIDS-SR16 (Iannuzzo et al. 2006). Secondary outcomes included depression response (defined as a decrease in score of $\geq$50% from baseline) and remission rate at 6 weeks according to the QIDS-SR-16 (defined as a score of 0–1).

The investigators found larger decreases in depression scores on the QIDS-SR16 at the 6-week time point (-8.0 ± 1.0 from baseline) in the psilocybin group compared with the escitalopram group (-6.0 ± 1.0). However, this difference did not reach statistical significance in the 95% confidence interval [CI], -5.0 to 0.9; $p = 0.17$. Although there was no significant difference in response rate at the 6-week point, there were significant differences in depression remission favoring the psilocybin group (57%) over the escitalopram group (28%; CI, 2.3–53.8). There were also significant between-group differences in secondary measures of change-scores on measures of depression, anxiety, emotional avoidance, work and social functioning, anhedonia, well-being, and suicidality, all favoring the psilocybin therapy group.

There were no SAEs reported during the 6 weeks of the trial among either group, with equivalent total rates of AEs reported in the psilocybin (87%) and escitalopram groups (83%). As in other contemporary clinical trials (Agin-Liebes et al. 2020; Carhart-Harris et al. 2016, 2018; Griffiths et al. 2016; Ross et al. 2016) there were no reports of persistent visual perceptual changes, psychosis, or use/misuse behaviors during the 6-week period of the study. There were, however, significantly greater reports of anxiety (14%) and dry mouth (14%), sexual dysfunction (3%) and flat affect (7%) in the escitalopram group compared to the psilocybin group (in which all of these symptoms were absent). Further, 17% of the participants in the escitalopram group either prematurely terminated or halved their daily dose of escitalopram due to perceive adverse effects of this medication. Among the psilocybin group, AEs generally occurred during the 24 h after the psilocybin dosing day, with the most common AE being transient headache (67%) (compared with 52% in the escitalopram group), which is consistent with previous studies (Agin-Liebes et al. 2020; Carhart-Harris et al. 2016, 2018; Griffiths et al. 2016; Ross et al. 2016). Additionally, participants in the psilocybin group reported greater access to emotions and catharsis.

Although this study provides important scientific insights and improves upon previous research designs in rigor, it is limited by several important methodological issues. First, there was no inactive placebo group, which limits the ability to determine the treatment effect of either the psilocybin or escitalopram intervention alone. This also rendered the weak antidepressant effect in the escitalopram group less apparent compared to the effects seen in escitalopram versus placebo trials (Gründer and Mertens 2021). Although the investigators pre-registered their hypotheses, they did not adjust for multiple comparisons, and therefore no definitive clinical or statistical conclusions can be drawn about the data. Further, the duration of escitalopram treatment in the trial was only 6 weeks, which is much briefer than typically used in clinical practice, as the drug is known to have a delayed antidepressant therapeutic effect (Trivedi et al. 2006). As in the open-label trial (Carhart-Harris et al. 2016), the majority of participants were non-Hispanic White (83–93%). Additionally, the selection of study volunteers was likely biased toward selection of individuals favorably disposed toward psilocybin, which further limits generalizability of the findings. The titration of participants off of antidepressant medication might have also weakened the antidepressant effects in the escitalopram group. Lastly, expectancy effects were not controlled for, and the integrity of the blind was not assessed. Participants may have been able to identify to which study medication group they were assigned, which could have introduced additional bias in favor of the psilocybin effect.

6 Qualitative Reports from Depression Study Volunteers

Despite a resurgence in research involving psilocybin therapy as a treatment for depression, its therapeutic mechanism of action remains poorly understood. Accumulating evidence suggests that psilocybin stimulates a host of neural changes including modulations in brain connectivity, cerebral blood flow, and transient amygdala activation in response to emotional cues (Araújo et al. 2015; De Gregorio et al. 2018; Studerus et al. 2010; Yaden and Griffiths 2020). The functional and structural neural plasticity effects of psilocybin have been linked to emotion regulation and learning processes, however, the phenomenological effects can vary widely (Aday et al. 2021). In this section we present qualitative narratives highlighting the salient subjective features of the psilocybin therapy experience, which suggest key psychological processes.

As a follow-up to the Imperial open-label trial, Watts and colleagues (2017) conducted a thematic analysis of participant reports ($N = 20$) and identified two primary psychological change processes (Watts et al. 2017). Participants reported enhanced feelings of connectedness with self, others, and the world and a sense of interconnectedness with all of humanity, which helped to counteract long-standing feelings of disconnection that characterized their depressive states. Participants also reported an enhanced ability to confront, process, and accept difficult emotions that had been suppressed or avoided. Many participants described a deep catharsis and release, which culminated in resolution of long-held grief, as the following reported:

> *There was a lot of sadness, really, really deep sadness: the loss the grief, it was love and sadness together, and letting go, I could feel the grief and then let it go because holding onto it was hurting me, holding me back. It was a process of unblocking. (P2).*
>
> *Excursions into grief, loneliness and rage, abandonment. Once I went into the anger it went 'pouf' and evaporated. I got the lesson that you need to go into the scary basement, once you get into it, there is no scary basement to go into [anymore]. (P3).*
>
> *There was this huge terrifying creature with a rifle, and instead of running away, I looked at it, and it wasn't as scary as it had seemed. [My] fear subsided, it suddenly seemed ridiculous, I started laughing. If I had avoided it, it would have got more terrifying. (P4).*

Participants in the RCT at Johns Hopkins University described similar therapeutic processes of emotional release and resolution (Davis et al. 2021a). The following description is from one young adult participant who had been depressed and anxious for many years. (The information has been augmented to protect their confidentiality; the singular use of the pronoun "they" is used in order to mask the gender of the participant.) This participant presented to the trial with moderate levels of depression and anxiety. During the preparation therapy in the weeks leading up to the psilocybin session they reported feeling guarded and nervous but hoped to discover relief from the depression that had kept him feeling emotionally stuck.

During the first psilocybin session (20 mg/70 kg) the participant experienced the following (condensed from qualitative report):

> *Shivering cold. The concept of time is lost. Body is rocking, rolling, and writhing with the music. On inhales I'm soaring up a mountain in the clouds in unison with the crescendo. On exhales my body and mind are slipping into a darkness and quiet that has no end to its*

depths. Other times I feel the loud, turbulent, diffuse mix of thoughts, feelings, memories, anticipations, and the occasional moment of stillness.

I felt an infinite series of expansions and contractions. I wanted to see and have ultimate realities revealed to me. I kept asking for answers but I saw nothing. What I did receive was an internal brief feeling of peace, a knowing, and an understanding without really knowing what I was understanding. Acceptance.

I thought the experience would have a deep long-lasting impact on me, but I do not feel a strong emotional attachment to what happened. Like it wasn't powerful enough to create a profound noticeable change in me. I don't feel like I learned something clear and concise for me to take away from it. Like it didn't make a powerful or lasting impression. I'm disappointed in myself. I felt like I was on the edge of something deeper. I felt like I could have seen or understood things more clearly if I was able to go deeper.

During the second psilocybin session (30 mg/70 kg) the participant experienced the following (condensed from qualitative report):

The main theme was healing and letting go. Healing came in many forms and feelings. I felt as if pain and sadness and trauma and guilt were draining from my body. As if it were being washed away and cleansed. I felt guided and compelled to let go of hurtful memories and recognized it was time to let go and okay to move on from the past. I felt the music was playing an integral role in guiding where the healing needed to take place in my body, mind, and heart. I felt it flowing through my body as a bright silver light traveling where it needed to go and guiding my thoughts to tranquility and acceptance.

I truly felt embraced and taken care of completely by some un-named force. That force allowed me to be open and willing to dissolve and dissipate the depression that was stuck in my mind. It allowed me to trust in myself and to be comfortable opening my once closed and guarded heart. I had a profoundly clear understanding that everything was happening exactly as it needed to be and everything will be okay no matter what.

The first psilocybin session dramatically broke up and cracked and loosened all the stuck energy and rigid patterns of my mind. Once it was loosened and broken up, in the second session it was able to clear all that once hardened debris free. Step 1. Jackhammer my mind and soul and show me I can survive that pain. Step 2. Gently and lovingly cleanse everything out and allow myself to let go and heal. I felt such a strong sense of soft, glowing, healing light assuring me that there was nothing to fear or feel sorrow for. I felt I was in the presence of a pure and perfect energy that was healing me.

I never once felt fear during the session. I knew I was going to be okay. I am okay. And for the first time in my life, I discovered the possibility of what it feels like to love myself. I have never felt that way before. And I truly value the psilocybin journey for showing me that I can honor myself and others as equals on a deep and soulful level.

This participant's mental health outcomes are displayed in Fig. 1. As the figure shows, this participant experienced a large reduction in depression and anxiety symptoms throughout the study, which was persistent until the 12-month follow-up.

How psychedelics promote psychological processes in ways conducive to resolving depression is currently a fervent area of theoretical and empirical inquiry. As the narrative account and study data suggest, this participant reported enormous psychological benefit from psilocybin therapy, including notable improvements in depression, anxiety, and quality-of-life. Although not every participant in this study improved, or improved this dramatically, for some participants the psilocybin sessions provided a powerful opportunity to reevaluate or reappraise their past behaviors and life circumstances in a therapeutically productive manner. This participant reported acute psilocybin effects that ranged from physically challenging

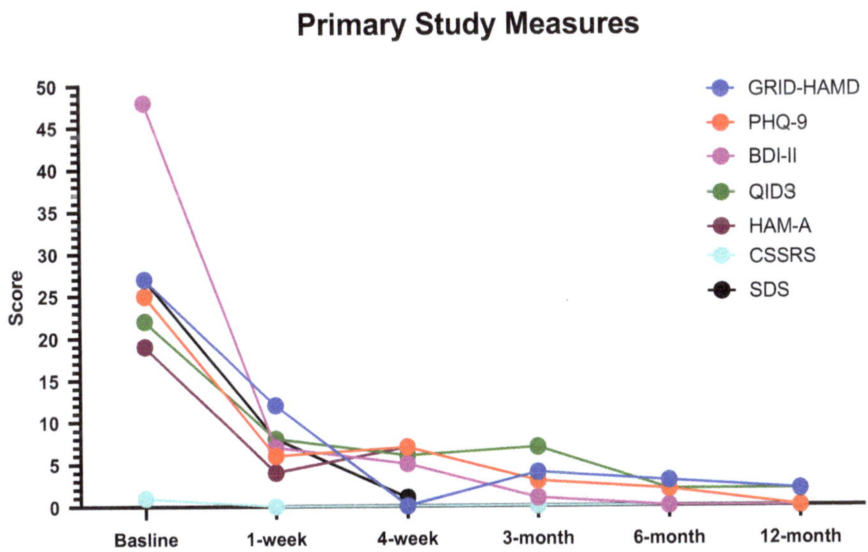

Fig. 1 Primary outcome measures for participant from the John's Hopkins Depression Trial. Measures include: Grid Hamilton Depression Rating Scale (GRID-HAMD), Nine item Patient Health Questionnaire (PHQ-9), Beck Depression Inventory – 2nd version (BDI-II), Quick Inventory of Depressive Symptoms (QIDS), Hamilton Anxiety Scale (HAM-A), Columbia Suicide Severity Rating Scale (CSSRS), Sheehan Disability Scale (SDS)

to deeply meaningful with psychological insights. While some reports documented in the literature feature elements of spiritual or mystical-type experiences (Ross et al. 2016; Griffiths et al. 2016), other experiences, such as the one featured here, involve more psychological and relational feelings of self- and other- forgiveness, compassion, and love, as well as emotional catharsis and acceptance. The psilocybin sessions tend to evoke psychological content and processes that are personally meaningful and relevant to each individual. These qualitative descriptions are supported by Carhart-Harris and Friston's relaxed beliefs under psychedelics (REBUS) model (Carhart-Harris and Friston 2019), which proposes that psychedelics promote conditions conducive to relaxing and revising high-level, maladaptive beliefs (e.g., self-protective but maladaptive self-beliefs, "I am a failure, I will no longer venture anything new"). However, it is unclear to what extent the biological mechanisms account for therapeutic benefits and to what extent they explain the subjective psychological phenomena. It is also unclear to what extent the treatment depends on "set and setting" variables including the psychological support, preparation, and post-psilocybin therapy sessions. These questions deserve rigorous scientific investigation and may eventually reveal important insights regarding key change mechanisms.

7 Future Directions

Although the preliminary quantitative and qualitative data are promising, it is premature to conclude the treatment is efficacious in the much larger population of depressed individuals. The studies described above carry a number of limitations including the absence of control conditions in some studies and small sample sizes (Carhart-Harris et al. 2016), which increase the possibility that statistical effects will be detected when in fact there is indeed no treatment effect. The study samples in psychedelic research are also characterized by higher rates of self-selection biases and positive expectancies, which have not been adequately controlled for. Additionally, most trial participants have primarily been White and college educated, which greatly limits the generalizability of findings to other ethnic and socioeconomic groups, of which there are many. Indeed, there is a substantial lack of diversity within psychedelic research and a gross underrepresentation of Black, Indigenous and People of Color (BIPOC). As a developing field, the psychedelic research community faces an ethical imperative to conduct culturally inclusive and equitable trials through proactive community outreach recruitment efforts, anti-racism trainings for research staff, and the provision of meaningful compensation for participants' time for those with financial barriers. Additionally, the focus on depression as a clinical outcome also neglects the potential role that psilocybin therapy holds in the treatment of conditions that are prevalent in BIPOC communities such as anxiety and depression that emerges as a result of ethnic/racial trauma (Davis et al. 2021b; Williams et al. 2021), which may be diagnosed as post-traumatic stress disorder. In fact, preliminary self-report evidence suggests that the use of psychedelics is increasing among BIPOC communities in non-clinical settings use as a means to address racial trauma, demonstrating an important need to focus on this topic in new clinical trials.

Despite substantial limitations, there is mounting evidence to suggest that psilocybin may carry rapid and sustained antidepressant properties, which have meaningful implications for the clinical management of depression. The findings have informed the basis of phase II/III trials that are currently underway for the treatment of MDD and TRD. The U.S. FDA and European Medicines Agency has granted two organizations fast-track status to conduct phase II multisite trials. The pharmaceutical biotechnology company COMPASS Pathways PLC is conducting a psilocybin therapy RCT for the treatment of TRD (NCT03775200), and the non-profit organization Usona Institute is conducting a psilocybin therapy RCT for the treatment of MDD (NCT03866174). Additional psilocybin studies for depression are taking place at various other sites including Johns Hopkins (depression associated with mild cognitive impairment; NCT04123314), Imperial College (MDD; NCT03429075), Yale (MDD; NCT03554174), University of Zurich (MDD; NCT03715127), and Central Institute of Mental Health/Charité Berlin (TRD; NCT04670081).

An important area of concern is identifying for whom psilocybin therapy is most clinically appropriate. In depressed patients, it will be important to consider various psychological, social, and genetic variables that may increase risk or predispose an individual to respond to psilocybin in a manner that is unsafe or ineffective. For instance, individuals presenting with acute suicidality might not be appropriate candidates for this type of intervention as these experiences can be acutely psychologically challenging and destabilizing. An additional consideration and area of research is which individual and situational features predict acute and longer-term responses to psilocybin. Baseline trait variables such as absorption, openness, and acceptance have been found to strongly moderate the effects of psychedelics (Hartogsohn 2016; Haijen et al. 2018). These and other factors should be carefully considered to help inform the development of optimal clinical and empirical guidelines, which will serve to mitigate adverse reactions and maximize the likelihood of positive therapeutic effects.

An additional challenge that clinician-researchers will face in the coming years is the development of rigorous, effective, and scalable clinical training methodologies. It will be important to establish training programs that ensure good clinical practice and that are structured, manualized, and closely monitored for quality assurance. Thoughtfully designed educational programs have been developed such as the training program at the California Institute of Integral Studies, which spearheaded the first academically accredited certificate program in Psychedelic-Assisted Therapy and Research in 2014 (Phelps 2017). Other programs have emerged, and many more are likely on the horizon. An additional key challenge facing clinicians and researchers are the barriers to dissemination and delivering the treatment at scale. Although only a couple of medication doses are typically required, current models are expensive as they involve intensive clinician time and high costs associated with regulatory challenges.

To overcome these limitations, additional research is needed. However, funding for psychedelic research in the U.S. is mostly restricted to private, philanthropic sources. It will be a critical landmark if/when the National Institutes of Health grants funding for advanced phase research into psilocybin for depression. Given the growing clinical evidence base, regulatory support, and cultural momentum surrounding this movement, public demand for further scientific inquiry may encourage such funding. This could enable a pathway for psilocybin to be rescheduled as a medication, making it available for therapeutic use in the treatment of depression in clinical populations.

8 Conclusion

The urgent need to address the mental health crisis associated with depression demands a dedicated approach to supporting quality research as well as the fortitude to pave a new road with innovative treatments that address the complex nature of such suffering. Psilocybin therapy may provide an opportunity to address the impact of debilitating depression as many people fail to respond to traditional behavioral, cognitive, and pharmacological evidence-based interventions. Although current research indicates that psilocybin therapy could improve depression, further research is needed with larger samples, more diverse participants, and across geographic settings. Furthermore, psilocybin therapy is not a panacea, human suffering is not a simple problem, and depression is a dynamic, heterogenous process. Even if psilocybin therapy garners regulatory approval in the coming years, it will not dissolve the structures and determinants that create an environment wherein a diathesis for depression is fostered. Whether the effects of psilocybin therapy will meet the impossibly high expectations it is garnering in the public is also yet to be determined. Nevertheless, because psilocybin therapy offers a disruptive pharmacological approach that threatens our current understanding of the chronic and debilitating lifetime course of depression, there is hope that it may offer an alternative effective treatment option for those in desperate need of relief.

References

Aday JS, Davis AK, Mitzkovitz CM, Bloesch EK, Davoli CC (2021) Predicting reactions to psychedelic drugs: a systematic review of states and traits related to acute drug effects. ACS Pharmacol Transl Sci 4(2):424–435

Agin-Liebes GI, Malone T, Yalch MM, Mennenga SE, Ponté KL, Guss J, Bossis AP, Grigsby J, Fischer S, Ross S (2020) Long-term follow-up of psilocybin-assisted psychotherapy for psychiatric and existential distress in patients with life-threatening cancer. J Psychopharmacol 34(2):155–166

Araújo AM, Carvalho F, Bastos MD, Pinho PG, Carvalho M (2015) The hallucinogenic world of tryptamines: an updated review. Arch Toxicol 89(8):1151–1173

Beck AT, Steer RA, Brown GK (1996) Beck depression inventory (BDI-II). Pearson

Carhart-Harris RL, Friston KJ (2019) REBUS and the anarchic brain: toward a unified model of the brain action of psychedelics. Pharmacol Rev 71(3):316–344

Carhart-Harris RL, Erritzoe D, Williams T, Stone JM, Reed LJ, Colasanti A, Tyacke RJ, Leech R, Malizia AL, Murphy K, Hobden P (2012) Neural correlates of the psychedelic state as determined by fMRI studies with psilocybin. Proc Natl Acad Sci U S A 109(6):2138–2143

Carhart-Harris RL, Bolstridge M, Rucker J, Day CMJ, Erritzoe D, Kaelen M, Bloomfield M, Rickard J, Forbes B, Feilding A et al (2016) Psilocybin with psychological support for treatment-resistant depression: an open label feasibility study. Lancet Psychiatry 3(7):619–627

Carhart-Harris RL, Roseman L, Bolstridge M, Demetriou L, Pannekoek JN, Wall MB, Tanner M, Kaelen M, Mcgonigle J, Nutt DJ (2017) Psilocybin for treatment-resistant depression: fMRI-measured brain mechanisms. Sci Rep 7(1)

Carhart-Harris RL, Bolstridge M, Day CM, Rucker J, Watts R, Erritzoe DE, Kaelen M, Giribaldi B, Bloomfield M, Pilling S, Rickard JA (2018) Psilocybin with psychological support for treatment-resistant depression: six-month follow-up. Psychopharmacology (Berl) 235(2):399–408

Carhart-Harris R, Giribaldi B, Watts R, Baker-Jones M, Murphy-Beiner A, Murphy R, Martell J, Blemings A, Erritzoe D, Nutt DJ (2021) Trial of psilocybin versus escitalopram for depression. New Engl J Med 384(15):1402–1411

Collimore KC, Rector NA (2014) Treatment of anxiety disorders with comorbid depression: a survey of expert CBT clinicians. Cogn Behav Pract 21(4):485–493

Coull G, Morris PG (2011) The clinical effectiveness of CBT-based guided self-help interventions for anxiety and depressive disorders: a systematic review. Psychol Med 41(11):2239–2252

Cuijpers P, Sijbrandij M, Koole SL, Andersson G, Beekman AT, Reynolds CF (2014) Adding psychotherapy to antidepressant medication in depression and anxiety disorders: a meta-analysis. World Psychiatry 13(1):56–67

Davis AK, Barrett FS, May DG et al (2021a) Effects of psilocybin-assisted therapy on major depressive disorder: a randomized clinical trial. JAMA Psychiatry 78(5):481–489

Davis AK, Xin Y, Sepeda ND, Garcia-Romeu A, Williams MT (2021b) Increases in psychological flexibility mediate the relationship between acute psychedelic drug effects and decreases in racial trauma symptoms among people of color. Chron Stress:1–10

De Gregorio D, Enns JP, Nuñez NA, Posa L, Gobbi G (2018) d-lysergic acid diethylamide, psilocybin, and other classic hallucinogens: mechanism of action and potential therapeutic applications in mood disorders. Prog Brain Res 242:69–96

Gaynes BN, Warden D, Trivedi MH, Wisniewski SR, Fava M, Rush AJ (2009) What did STAR*D teach us? Results from a large-scale, practical, clinical trial for patients with depression. Psychiatr Serv 60(11):1439–1445. https://doi.org/10.1176/ps.2009.60.11.1439

Greenberg PE, Fournier A-A, Sisitsky T, Pike CT, Kessler RC (2015) The economic burden of adults with major depressive disorder in the United States (2005 and 2010). J Clin Psychiatry 76(02):155–162. https://doi.org/10.4088/JCP.14m09298

Griffiths RR, Johnson MW, Carducci MA, Umbricht A, Richards WA, Richards BD, Cosimano MP, Klinedinst MA (2016) Psilocybin produces substantial and sustained decreases in depression and anxiety in patients with life-threatening cancer: a randomized double-blind trial. J Psychopharmacol 30(12):1181–1197

Gründer G, Mertens L (2021) Psilocybin for depression. N Engl J Med 385(9):863

Haijen EC, Kaelen M, Roseman L, Timmermann C, Kettner H, Russ S, Nutt D, Daws RE, Hampshire AD, Lorenz R, Carhart-Harris RL (2018) Predicting responses to psychedelics: a prospective study. Front Pharmacol 9:897

Hamilton JP, Farmer M, Fogelman P, Gotlib IH (2015) Depressive rumination, the default-mode network, and the dark matter of clinical neuroscience. Biol Psychiatry 78(4):224–230

Hartogsohn I (2016) Set and setting, psychedelics and the placebo response: an extra-pharmacological perspective on psychopharmacology. J Psychopharmacol 30(12):1259–1267

Hasin DS, Sarvet AL, Meyers JL et al (2018) Epidemiology of adult *DSM-5* major depressive disorder and its specifiers in the United States. JAMA Psychiatry 75(4):336. https://doi.org/10.1001/jamapsychiatry.2017.4602

Heifets BD, Malenka RC (2019) Disruptive psychopharmacology. JAMA Psychiatry 76(8):775–776

Hofmann SG, Asnaani A, Vonk IJ, Sawyer AT, Fang A (2012) The efficacy of cognitive behavioral therapy: a review of meta-analyses. Cognit Ther Res 36(5):427–440

Iannuzzo RW, Jaeger J, Goldberg JF, Kafantaris V, Sublette ME (2006) Development and reliability of the HAM-D/MADRS interview: an integrated depression symptom rating scale. Psychiatry Res 145(1):21–37

Koen N, Stein DJ (2011) Pharmacotherapy of anxiety disorders: a critical review. Dialogues Clin Neurosci 13(4):423–437

Mertens LJ, Wall MB, Roseman L, Demetriou L, Nutt DJ, Carhart-Harris RL (2020) Therapeutic mechanisms of psilocybin: changes in amygdala and prefrontal functional connectivity during emotional processing after psilocybin for treatment-resistant depression. J Psychopharmacol 34(2):167–180

Morilak DA, Frazer A (2004) Antidepressants and brain monoaminergic systems: a dimensional approach to understanding their behavioural effects in depression and anxiety disorders. Int J Neuropsychopharmacol 7(2):193–218. https://doi.org/10.1017/S1461145704004080

Nemeroff CB (2007) Prevalence and management of treatment-resistant depression. J Clin Psychiatry 68(Suppl 8):17–25

Otto MW, Hearon BA (2016) Combination treatment for anxiety and mood disorders: benefits and issues for the combination of cognitive-behavioral therapy and pharmacotherapy. In: The

Oxford handbook of cognitive and behavioral therapies. Oxford University Press, New York, pp 482–493
Phelps J (2017 Sep) Developing guidelines and competencies for the training of psychedelic therapists. J Humanist Psychol 57(5):450–487
Roseman L, Nutt DJ, Carhart-Harris RL (2018) Quality of acute psychedelic experience predicts therapeutic efficacy of psilocybin for treatment-resistant depression. Front Pharmacol 17(8):974
Ross S, Bossis T, Guss J, Agin-Liebes G, Malone T, Cohen B, Mennenga S, Belser A, Kalliontzi K, Babb J et al (2016) Rapid and sustained symptom reduction following psilocybin treatment for anxiety and depression in patients with life-threatening cancer: a randomized controlled trial. J Psychopharmacol 30(12):1165–1180
Rush AJ, Trivedi MH, Ibrahim HM, Carmody TJ, Arnow B, Klein DN, Markowitz JC, Ninan PT, Kornstein S, Manber R, Thase ME (2003) The 16-item quick inventory of depressive symptomatology (QIDS), clinician rating (QIDS-C), and self-report (QIDS-SR): a psychometric evaluation in patients with chronic major depression. Biol Psychiatry 54(5):573–583
Sinyor M, Schaffer A, Levitt A (2010) The sequenced treatment alternatives to relieve depression (star*d) trial: a review. Can J Psychaitry 55(3):126–135
Snaith RP, Hamilton M, Morley S, Humayan A, Hargreaves D, Trigwell P (1995) A scale for the assessment of hedonic tone the Snaith–Hamilton Pleasure Scale. Br J Psychiatry 167(1):99–103
Spielberger CD (2010) State-trait anxiety inventory. In: The Corsini encyclopedia of psychology, vol 30, p 1
Studerus E, Gamma A, Vollenweider FX (2010) Psychometric evaluation of the altered states of consciousness rating scale (OAV). PLoS One 5:e12412
Trivedi MH, Rush AJ, Wisniewski SR, Nierenberg AA, Warden D, Ritz L, Norquist G, Howland RH, Lebowitz B, McGrath PJ, Shores-Wilson K (2006) Evaluation of outcomes with citalopram for depression using measurement-based care in STAR* D: implications for clinical practice. Am J Psychiatry 163(1):28–40
Walker ER, McGee RE, Druss BG (2015) Mortality in mental disorders and global disease burden implications: a systematic review and meta-analysis. JAMA Psychiatry 72(4):334. https://doi.org/10.1001/jamapsychiatry.2014.2502
Watts R, Day C, Krzanowski J, Nutt D, Carhart-Harris R (2017) Patients' accounts of increased "connectedness" and "acceptance" after psilocybin for treatment-resistant depression. J Humanist Psychol 57(5):520–564
Williams MT, Davis AK, Xin Y, Sepeda ND, Grigas PC, Sinnott S, Haeny AM (2021) People of color in North America report improvements in racial trauma and mental health symptoms following psychedelic experiences. Drugs (Abingdon Engl) 28(3):215–226
World Health Organization. Depression fact sheet. World Health Organization. Published December 2019. Accessed 11 Jan 2020. http://www.who.int/mediacentre/factsheets/fs369/en/
Yaden DB, Griffiths RR (2020) The subjective effects of psychedelics are necessary for their enduring therapeutic effects. ACS Pharmacol Transl Sci 4(2):568–572

Ketamine for Depression: Advances in Clinical Treatment, Rapid Antidepressant Mechanisms of Action, and a Contrast with Serotonergic Psychedelics

Marina Kojic, Johan Saelens, Bashkim Kadriu, Carlos A. Zarate Jr, and Christoph Kraus

Contents

Marina Kojic and Johan Saelens contributed equally to this work.

M. Kojic
Department of Pediatrics and Adolescent Medicine, Medical University of Vienna, Vienna, Austria

J. Saelens
Department of Psychiatry and Psychotherapy, Medical University of Vienna, Vienna, Austria

B. Kadriu
Section on the Neurobiology and Treatment of Mood Disorders, National Institute of Mental Health, National Institutes of Health, Bethesda, MD, USA

Department of Neuroscience, Janssen Research & Development, LLC, San Diego, CA, USA

C. A. Zarate Jr
Section on the Neurobiology and Treatment of Mood Disorders, National Institute of Mental Health, National Institutes of Health, Bethesda, MD, USA

C. Kraus (✉)
Department of Psychiatry and Psychotherapy, Medical University of Vienna, Vienna, Austria

Section on the Neurobiology and Treatment of Mood Disorders, National Institute of Mental Health, National Institutes of Health, Bethesda, MD, USA
e-mail: christoph.kraus@muv.ac.at

Curr Topics Behav Neurosci (2022) 56: 141–168
https://doi.org/10.1007/7854_2022_313
Published Online: 22 March 2022

Abstract The approval of ketamine for treatment-resistant depression has created a model for a novel class of rapid-acting glutamatergic antidepressants. Recent research into other novel rapid-acting antidepressants – most notably serotonergic psychedelics (SPs) – has also proven promising. Presently, the mechanisms of action of these substances are under investigation to improve these novel treatments, which also exhibit considerable side effects such as dissociation. This chapter lays out the historical development of ketamine as an antidepressant, outlines its efficacy and safety profile, reviews the evidence for ketamine's molecular mechanism of action, and compares it to the proposed mechanism of SPs. The evidence suggests that although ketamine and SPs act on distinct primary targets, both may lead to rapid restoration of synaptic deficits and downstream network reconfiguration. In both classes of drugs, a glutamate surge activates α-amino-3-hydroxy-5-methyl-4-isoxazolepropionic acid receptor (AMPAR) throughput and increases in brain-derived neurotrophic factor (BDNF) levels. Taken together, these novel antidepressant mechanisms may serve as a framework to explain the rapid and sustained antidepressant effects of ketamine and may be crucial for developing new rapid-acting antidepressants with an improved side effect profile.

Keywords Antidepressants · Brain-derived neurotrophic factor · Glutamate · Synaptic plasticity · Treatment resistant depression

1 Introduction

1.1 Clinical Studies Leading to the Approval of Esketamine as an Antidepressant: A Historical Overview

Since it was first approved by the US Food and Drug Administration (FDA) in 1970, ketamine[1] has become a crucial and versatile drug and is listed as an essential medicine by the World Health Organization (World Health Organization 2019). Originally developed as a derivative of phencyclidine (PCP) by the pharmaceutical company Parke Davis, ketamine has been used for decades as an anesthetic and analgesic agent; more recently, esketamine (racemic ketamine's (*S*)-stereoisomer) received FDA approval for treatment-resistant depression (TRD) for adults with

[1] Unless otherwise specified, the term ketamine refers to racemic (*R*,*S*)-ketamine.

major depressive disorder (MDD) as well as with acute suicidal ideation or behavior. TRD is commonly defined by nonresponse to at least two adequate antidepressant trials.

PCP was first trialed in humans as an anesthetic, but researchers soon realized that its unpleasant side effect profile – which included loss of feeling in limbs and prolonged sensory deprivation post-treatment – precluded its clinical use. Short-acting PCP derivatives were subsequently synthesized. After pharmacological testing, ketamine – which has similar anesthetic potential to PCP but a more favorable side-effect profile – was selected for human trials. Though ketamine and PCP are both noncompetitive N-methyl-D-aspartate receptor (NMDAR) inhibitors, notable pharmacological differences exist. For example, PCP has pro-convulsive effects while ketamine does not, and ketamine has a faster induction rate of anesthesia than PCP but a shorter duration of action (McCarthy et al. 1965). At the time of its development as an anesthetic, PCP's sensory deprivation effects led researchers to investigate it in a model of schizophrenia (Cohen et al. 1959; Rosenbaum 1959); indeed, a study of nine schizophrenic and nine non-schizophrenic patients showed that PCP had psychotomimetic effects in both groups (Rosenbaum 1959). As a result, such effects were assessed during ketamine's development. Though both drugs exerted psychotomimetic effects, ketamine's were less intense, and of shorter duration, than PCP's (Domino et al. 1965). Nonetheless, Parke Davis worried that ketamine's classification as a psychotomimetic drug would hinder its development and thus employed an internal psychiatrist to observe patients after application of ketamine anesthesia.

In recent decades, our understanding of mood disorders and depression has evolved, and researchers realized that the pathophysiology of mood disorders extends beyond monoaminergic neurotransmission. By the 1990s, emerging preclinical work supported the hypothesis that glutamate and its ionotropic NMDAR might be involved in the pathophysiology of affective disorders and, commensurately, that its study might lead to the development of new classes of antidepressants (Skolnick et al. 1996). Within the context of this preclinical evidence, as well as a broad understanding of ketamine's mechanism of action, Berman and colleagues were the first to administer ketamine to individuals with depression in a placebo-controlled, double-blind trial (Berman et al. 2000). The trial included seven participants with major depression who were treated intravenously with a subanesthetic dose of 0.5 mg/kg ketamine hydrochloride or a saline solution on 2 days 1 week apart. The results showed that ketamine infusion had rapid-acting antidepressant effects, as assessed by its ability to significantly improve depression rating scale scores within hours. Moreover, these antidepressant effects continued for at least 3 days after ketamine's acute dissociative effects had faded.

In retrospect, these results marked a milestone. However, they did not attract lasting interest after publication (Gallagher et al. 2021). Explanations included skepticism that the robust and rapid antidepressant effects observed were due to the drug rather than its intravenous application, as previous work had similarly demonstrated rapid response to an intravenously administered tricyclic antidepressant (Berman et al. 2000; Malhotra and Santosh 1996; Sallee et al. 1997). Another factor was the low number of participants enrolled in the trial. Finally, Berman and colleagues had mentioned that a potential limitation associated with clinical applications for ketamine would be its abuse potential and its well-established psychotomimetic profile, suggesting that these properties would need to be eliminated before further testing could be conducted. Nevertheless, due to persistent interest in ketamine's potential antidepressant effects, in 2006 Zarate and colleagues replicated the findings of the original trial by Berman and colleagues in a randomized, double-blind, placebo-controlled study of 18 patients with TRD (Zarate et al. 2006).

Since then, research into ketamine's mechanism of action has sought to understand the results of these initial clinical trials and the possible connection between the glutamatergic system and the underlying pathophysiology of depression and other stress-related disorders. Notably, the surge of interest in ketamine's antidepressant effects did not occur until independent, placebo-controlled studies substantiated the results of these initial trials underscoring the importance of funding replication studies. Indeed, a recent study that compared publications that were initially highly cited with their replication rate found that nearly half of these publications were not replicated in the following 10 years (Aarts et al. 2015).

Despite ketamine's potential as a rapid-acting antidepressant, the impracticalities of intravenous application in outpatient clinics and private settings led investigators to research alternative routes of administration. Lapidus and colleagues conducted the first trial of intranasally-delivered ketamine and found that this application led to sufficiently high plasma concentrations of the agent to induce antidepressant effects (Lapidus et al. 2014). In combination with positive clinical trials, these findings led to a patent for the intranasal administration of esketamine and culminated in the FDA approval of esketamine in 2019 for adults with TRD and in 2020 for adults with acute suicidal ideation and behavior. Esketamine showed roughly four-fold higher NMDAR binding affinity than the *R*(+)enantiomer but retained ketamine's anesthetic and dissociative properties (Fukumoto et al. 2017; Zhang et al. 2014).

Results of the initial clinical trials with ketamine also led to a surge in research to uncover or develop agents with similar rapid-acting antidepressant properties but fewer psychotomimetic side effects. This effort has, to date, been unsuccessful, suggesting that ketamine might have unique pharmacological characteristics. Nonetheless, the search for new and similar substances continues, as do efforts to improve ketamine's own side effect profile (Kadriu et al. 2020).

1.2 The Need for Rapid-Acting Antidepressants

First-line antidepressant medications such as selective serotonin reuptake inhibitors (SSRIs) target the monoaminergic system. Despite their positive safety profile, these agents have some key limitations. First, SSRIs exhibit response rates in first-line trials of about 30–60%, underscoring that a significant number of patients with major depression do not respond to these agents. The large NIMH-funded Sequenced Treatment Alternatives to Relieve Depression (STAR*D) study found that, in a group of approximately 4,000 patients with MDD who received as many as four different psychotropic treatment combinations, about 33% did not respond to any of the standard medications (Gaynes et al. 2009). Such treatment-resistant patients are vulnerable to disease-associated debilitating impairments in psychosocial functioning, personal relationships, working capacity, and general well-being, and are at particularly high risk for suicidal behavior (Whiteford et al. 2013). Second, even when they will ultimately prove successful, standard antidepressant treatments have a lag time of several weeks before symptom alleviation; this long delay to reach full therapeutic potential makes them relatively ineffective for immediate treatment of emergencies, including suicidal behavior and acute relief of depressive symptoms. Third, side effects, genetic polymorphisms, and other related factors may all critically affect drug efficacy and general drug adherence; interactions with other drugs also remain a key consideration (Goethe et al. 2007). As an example, side effects such as nausea might occur before the onset of antidepressant response given the significant lag time associated with standard antidepressants.

In this context, the discovery of a rapid-acting antidepressant like ketamine was a major paradigm shift in the development of novel antidepressants and the treatment of patients with depression. The potential to treat previous non-responders, including individuals with TRD who experience unremitting depressive episodes is a major advantage, especially considering the high prevalence of depression in community settings. However, it should be noted that, compared with standard antidepressant agents, ketamine's long-term side effects require further research because most of this research has been conducted only in short-term ketamine trials.

1.3 Safety of Ketamine as an Antidepressant

Given that ketamine is one of the most commonly used anesthetics, safety data on single use applications of anesthetic doses were more readily available than for newly developed drugs. Nevertheless, given that anesthetic and antidepressant ketamine doses are quite different, ketamine's antidepressant side effect profile needed to be considered separately.

A systematic assessment of ketamine's side effects across five separate studies (n = 188) found that single-dose IV ketamine (0.5 mg/kg) was associated with several, mostly transient, side effects compared to placebo (Acevedo-Diaz et al. 2020). Dissociative side effects were most commonly reported (e.g., feeling strange (80%), a feeling of floating (>50%), and visual distortion (>50%)). Less common side effects included difficulty speaking (>50%), numbness (>40%), confusion (>40%), euphoria (>20%), and blurred vision (>20%) (Acevedo-Diaz et al. 2020). Side effects that occurred in fewer than 20% of patients included hypertension, dry mouth, tingling, difficulty concentrating, changes in body temperature, hallucinations, headaches, and gastrointestinal problems. As noted above, most of these side effects were transient, resolving after 2 h and peaking within an hour of ketamine administration. No significant cognitive or memory deficits or long-lasting side effects were observed in association with a single-dose infusion (Acevedo-Diaz et al. 2020).

As mentioned previously, most randomized clinical trials have monitored ketamine's short-term side effects, with relatively fewer data available regarding the safety of frequent or long-term use (Short et al. 2018). However, clinical trials surged (Bahr et al. 2019) with the development and FDA approval of the esketamine nasal spray, raising potential safety concerns about repeat- dose applications. Few studies have systematically assessed the side effects of long-term, repeat-dose ketamine and, to date, no clinical trials have monitored patients receiving ketamine for longer than one year. The extant data, however, suggest that ketamine's dissociative side effects remain transient and seem to decrease with repeated dosing; in addition, side effects such as dizziness and nausea vary depending on dose (Bahr et al. 2019). Interestingly, some data have come from substance abuse disorder studies of recreational ketamine users; while such data are not directly comparable to intravenously administered, antidepressant-dose ketamine administered in controlled settings, they do provide information regarding the risks and side effects associated with long-term ketamine use. Most notably, these data suggest a risk for ulcerative cystitis in chronic ketamine users (Jhang et al. 2015), though a recent study found that neither single nor repeat-dose esketamine led to urothelial toxicity (Findeis et al. 2020). Long-term recreational ketamine use has also been associated with impaired cognitive functioning, including episodic and semantic memory (Curran et al. 2001) as well as cognitive processing speed and verbal learning (Chan et al. 2013); these deficits persisted after matching with polydrug controls. Researchers have suggested that these effects may be at least partially explained by the observed upregulation of D1 receptors in the dorsolateral prefrontal cortex of long-term ketamine users (Narendran et al. 2005). Studies in rodents have also suggested that repeated subanesthetic administration of ketamine may have neurotoxic effects (Li et al. 2017; McIntyre et al. 2021).

With regard to esketamine in particular, one Phase 3 long-term study of over 800 enrolled patients assessed the long-term safety and efficacy of esketamine nasal spray plus a novel oral antidepressant for a year. The most common side effects were dizziness (32,9%), dissociation (27,6%), nausea (25,1%), and headache (24,9%). Cognitive function increased or remained stable at baseline throughout the study. Most side effects were mild or moderate, occurred on dosing day, and resolved on the same day; dissociative symptoms typically resolved 1.5 h after dosing (Wajs et al. 2020). Although some patients reported urinary tract side effects (17%), symptoms were mild to moderate and resolved after two weeks despite continued esketamine treatment. After discontinuation, the most common (>20%) new or worsening "withdrawal" symptoms were fatigue-lethargy/lack of energy and insomnia (Wajs et al. 2020). Intranasal esketamine also holds a risk for abuse and misuse as well as an increased risk of suicidal ideation and behavior in adolescents and children (Bahr et al. 2019); however, it is important to note that no abuse has been reported to date, and no patients requested an increase in dosing frequency in long-term studies of adults (Wajs et al. 2020). Recent findings also suggest that esketamine nasal spray appears to have neither short-term nor long-term adverse effects on nasal tolerability or olfactory function (Doty et al. 2021). However, cardiovascular and cerebrovascular conditions sensitive to increases in blood pressure should be treated as a contraindication or caution, as esketamine can increase blood pressure and heart rate.

2 Pharmacodynamics of Ketamine as Antidepressant

2.1 Central Mechanisms of Action

Ketamine's main pharmacological target is use-dependent antagonism of the ionotropic NMDAR and eventual decrease of calcium ion influx (Fig. 1). Crucially, the block is closely linked to prior α-amino-3-hydroxy-5-methyl-4-isoxazolepropionic acid receptor (AMPAR)-linked depolarization of the post-synapse and the removal of Mg2+ ions that block the channel pore (Hansen et al. 2018). At subanesthetic, antidepressant doses, ketamine facilitates glutamate release in the medial prefrontal cortex (mPFC), while higher doses decrease glutamate release (Moghaddam et al. 1997). This process occurs via the selective inhibition of NMDARs on gamma-aminobutyric acid (GABA)-ergic interneurons, which in turn disinhibits glutamatergic neurotransmission of pyramidal neurons in the mPFC (Duman et al. 2019; Miller et al. 2016). This hypothesis is supported by the finding that knockdown of the GluN2B subunit of NMDARs on GABAergic interneurons produced antidepressant effects in rodent models (Gerhard et al. 2020). The same effect was also found for somatostatin- and parvalbumin-expressing subtypes, but not glutamatergic pyramidal (CAMK2a) neurons (Gerhard et al. 2020).

An alternative hypothesis states that ketamine's rapid antidepressant effects are due to direct blocking of postsynaptic NMDAR activity in glutamatergic pyramidal neurons of the hippocampus and mPFC (Autry et al. 2011; Miller et al. 2016; Monteggia et al. 2013). This block, in turn, suppresses eukaryotic elongation factor 2 kinase (eEF2K), reducing the amount of phosphorylated elongation factor 2 (eEF2) and thus increasing brain-derived neurotrophic factor (BDNF) (Autry et al. 2011). This hypothesis is supported by evidence demonstrating that selective eEF2K inhibition in mice increased BDNF levels and led to subsequent antidepressant-like effects. Inhibition of protein synthesis – and, thus, BDNF synthesis – via anisomycin prevented ketamine-mediated rapid behavioral response (Autry et al. 2011).

In addition to its ability to modulate glutamate, ketamine affects a range of other neurotransmitter systems, including the GABAergic, dopaminergic, and serotonergic systems. With regard to GABAergic neurotransmission, ketamine was found to reverse reductions in GABA-synthesizing enzymes, proteins, and co-expressed neuropeptides due to chronic stress (Ghosal et al. 2017, 2020; Ren et al. 2016). In rats, a single intraperitoneal dose of ketamine led to a dose-dependent increase in GABA, glutamate, and glutamine cycling (Chowdhury et al. 2017). In addition, in mice, the prophylactic administration of ketamine prior to stress exposure was

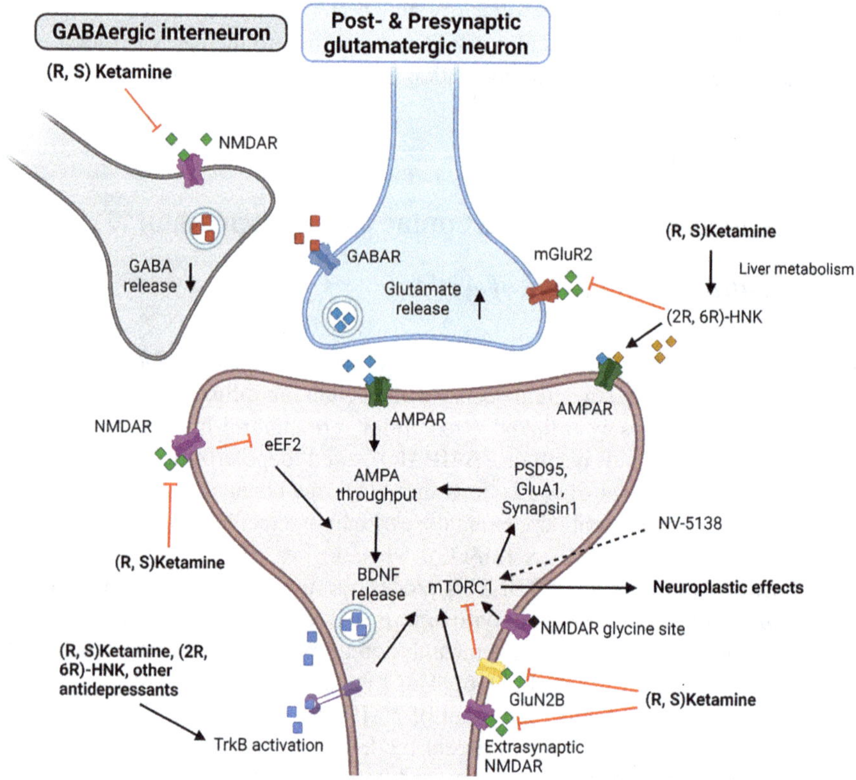

Fig. 1 Ketamine's general mechanism of action. Ketamine blocks the N-methyl-D-aspartate glutamate receptor (NMDAR) on gamma-aminobutyric acid (GABA)-ergic interneurons, thereby disinhibiting glutamate release by pyramidal neurons. Glutamate in turn binds to postsynaptic α-amino-3-hydroxy-5-methyl-4-isoxazolepropionic acid receptors (AMPARs) which mediates brain-derived neurotrophic factor (BDNF) release. BDNF binds primarily to tropomyosin-related kinase B (TrkB), leading to downstream activation of mammalian target of rapamycin complex 1 (mTORC1) and, consequently, neuroplastic effects. mTORC1 activation may, in turn, be associated with an increase in synaptic proteins such as postsynaptic density protein 95 (PSD95), synapsin 1, and the AMPAR subunit GluA, further increasing AMPAR throughput. Alternatively, ketamine may directly bind to postsynaptic NMDARs, reducing the amount of phosphorylated elongation factor 2 (eEF2) that, in turn, increases BDNF translation. Evidence also suggests that ketamine activates mTORC1 via direct TrkB activation and antagonism of extrasynaptic NMDARs. Phosphorylation of mTORC1 may also be increased by agonists binding to the glycine site of NMDARs or directly via NV-5138. In addition, the ketamine metabolite (2*R*,6*R*)-hydroxynorketamine (HNK) may increase downstream neuroplasticity by blocking mainly presynaptic metabotropic glutamate receptor 2 (mGluR2), thereby increasing glutamate release. Created with biorender.com

associated with an increase in precursors to inhibitory neurotransmitters and a decrease in precursors to excitatory neurotransmitters in response to a stressor (McGowan et al. 2018). This change in precursors did not occur when no stressor was presented, suggesting that ketamine might selectively enhance resilience to stressful events.

Ketamine's positive effects on motivation may be at least partially explained by its effect on the dopaminergic reward circuit (Abdallah et al. 2017; Mkrtchian et al. 2020). In rodents, ketamine increased activity in dopaminergic neurons in the ventral tegmental area (VTA) as well as extracellular dopamine in the nucleus accumbens and prefrontal cortex (PFC) (Witkin et al. 2016). The mediating role of AMPARs is underscored by the fact that the AMPAR antagonist NBQX (1,2,3,4-tetrahydro-6-nitro-2,3-dioxo- (9CI)-benzo[f]quinoxaline-7-sulfonamide) abolished both increased dopaminergic activity and antidepressant-like behavioral effects. Because activation of the D_1 receptor increases expression and excitability of NMDARs and AMPARs (Gao and Wolf 2008; Lavin and Grace 2001; Sun et al. 2005) as well as excitatory input to the PFC (Björkholm et al. 2017; Gonzalez-Islas and Hablitz 2003; Gurden et al. 2000), this effect might explain ketamine's impact on synaptogenesis and synaptic potentiation via increased BDNF and, thus, stimulation of the mechanistic target of rapamycin complex 1 (mTORC1). With regard to the behavioral effects of D1 receptor activity, repeated stress was found to reduce D1 receptor expression in the rodent mPFC, which in turn led to depression-related behaviors (Shinohara et al. 2018). Along these lines, D1 receptor-expressing pyramidal cells in the rodent mPFC produced a rapid antidepressant and anxiolytic response, while disruption of D1 receptor activity via the D1 receptor antagonist SCH39166 blocked ketamine's antidepressant-like effects (Hare et al. 2019).

Recent studies further suggest that serotonergic neurotransmission may be involved in ketamine's antidepressant effects. In rodents, subcutaneous ketamine injection increased prefrontal serotonin levels (Nishitani et al. 2014). This effect seemed to be mediated by AMPARs, given that injection of an AMPAR antagonist into the dorsal raphe nucleus (DRN) attenuated this effect while administration of an AMPAR agonist increased prefrontal serotonin. In addition, the prior infusion of a tryptophan hydroxylase inhibitor depleted serotonin levels, and direct application of the 5-HT1A antagonist WAY100635 into the mPFC abolished ketamine's antidepressant effects (Fukumoto et al. 2016, 2018; Gigliucci et al. 2013). Conversely, administration of the 5-HT1A agonist 8-OH-DPAT into the rodent mPFC mimicked the rapid antidepressant effects of ketamine and increased BDNF and subsequent mTORC1 signaling, ultimately increasing synaptic protein levels (Fukumoto et al. 2020). However, binding of the serotonin transporter (SERT) – as seen with SSRIs – is unlikely, given that positron emission tomography (PET) studies using the radioligand [^{11}C]N,N-dimethyl-2-(2-amino-4- cyanophenylthio)-benzylamine ([^{11}C]DASB) to measure SERT binding showed no measurable occupancy of the SERT after administering subanesthetic-dose ketamine (Spies et al. 2018).

Ketamine may also decrease the activity of extrasynaptic GluN2B containing NMDARs, which directly increases mTOR phosphorylation. Miller and colleagues demonstrated that deletion of GluN2B from principal cortical neurons abolished ketamine's antidepressant effects in mice (Miller et al. 2014). In addition, preclinical evidence suggests that ketamine's enantiomers are metabolically active. For instance, (2*R*,6*R*)-hydroxynorketamine (HNK) appears to have antidepressant-like and neuroplastic effects in animal models but lack ketamine's dissociative effects. Fewer dissociative side effects would be clinically beneficial and facilitate more widespread ketamine use. Interestingly, (2*R*,6*R*)-HNK's effects also seem to rely on AMPAR transmission, given that (2*R*,6*R*)-HNK administration led to both increased frequency and amplitude of AMPAR-mediated excitatory potentials and that its antidepressant effects were abolished by administering an AMPAR antagonist (Zanos et al. 2016). Increased glutamate release via presynaptic metabotropic glutamate receptor 2 (mGluR2) blockade by (2*R*,6*R*)-HNK further increased AMPAR transmission (Zanos et al. 2019). Additional evidence also suggests that (2*R*,6*R*)-HNK and ketamine – as well as traditional antidepressants – bind directly to tropomyosin receptor kinase B (TrkB), facilitating activation via BDNF and, thus, synaptic plasticity (Casarotto et al. 2021).

Beyond HNK, several trials have investigated novel glutamatergic compounds that might exhibit similar antidepressant effects to ketamine while avoiding its dissociative effects. Ketamine also possesses a glycine binding site in addition to its main NMDAR binding site, which presents a possible antidepressant target. In rats, GLYX-13, a partial glycine site agonist, was found to have antidepressant effects 24 h and seven days after administration (Burgdorf et al. 2013, 2015). However, recent trials with AV-101, a prodrug of 7-chlorokynurenic acid (a glycine site antagonist) failed to demonstrate antidepressant effects for patients with TRD (Park et al. 2020) or as adjunctive treatment in MDD (VistaGen Therapeutics 2019).

2.2 *Ketamine's Central and Peripheral Actions Beyond Glutamate*

Increasing evidence suggests that ketamine's biological effects transcend the glutamatergic synapse and affect other relevant neuropathological pathways, including neuroinflammatory pathways. One example is the kynurenine (KYN) pathway (Kadriu et al. 2019) (Fig. 2). In particular, overactivation of the neurotoxic branch of this pathway has been associated with mood disorders (Birner et al. 2017). Proinflammatory cytokines (e.g., interleukin-6 (IL-6) and quinolinic acid (QA), a key metabolite of the neurotoxic branch, were also found to be elevated in the cerebrospinal fluid of suicidal patients (Bay-Richter et al. 2015). Conversely, kynurenic acid (KYNA), a metabolite of the neuroprotective branch, was negatively associated with depressive symptoms (Bay-Richter et al. 2015).

More specifically, depression and psychological stress are hypothesized to lead to an increased immune response (e.g., IL-6, tumor necrosis factor alpha (TNF-α)), activating the enzyme indoleamine 2,3-dioxygenase (IDO) and thereby increasing the conversion of tryptophan (TRP) into KYN (Heisler and O'Connor 2015). In the neuroprotective branch of the KYN pathway, KYN is processed by astrocytes into KYNA, which acts as an NMDAR and α7nAChR antagonist and thus has positive effects on synaptic plasticity and neuronal protection, possibly via anti-inflammatory processes (Moroni et al. 2012) and the prevention of glutamate spill-over (Haroon

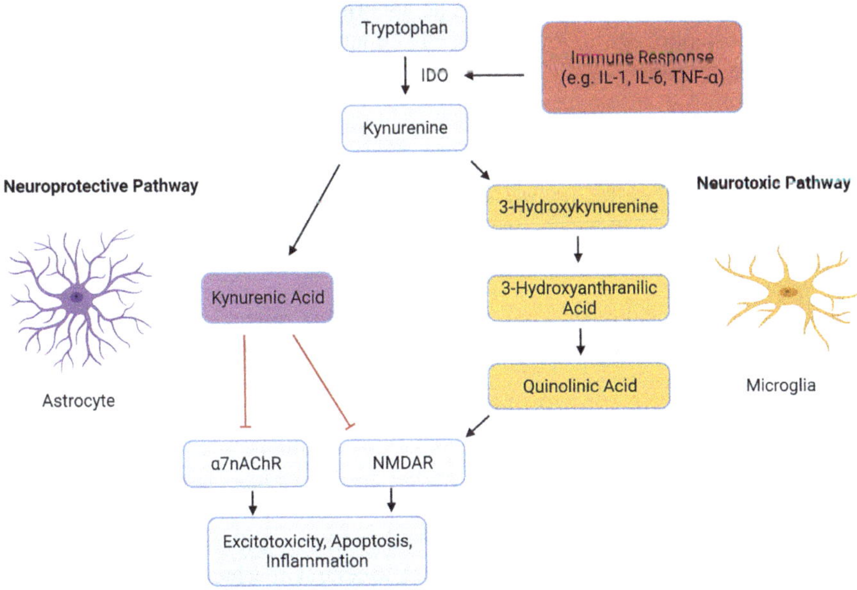

Fig. 2 The kynurenine pathway. The figure depicts the two branches of the kynurenine pathway. In the neuroprotective branch, mainly associated with astrocytes, kynurenine is transformed into kynurenic acid. Kynurenic acid acts as an N-methyl-D-aspartate glutamate receptor (NMDAR) and alpha 7 nicotinic acetylcholine receptor (α7nAChR) antagonist, thereby decreasing inflammation and excitotoxicity. In the neurotoxic pathway, mainly associated with microglia, kynurenine is metabolized into quinolinic acid, which acts as an NMDAR agonist, increasing excitotoxicity and neuronal apoptosis. Crucially, activation of rate-limiting enzymes such as indoleamine 2,3-dioxygenase (IDO), which transforms tryptophan into kynurenine via inflammation or psychological stress shifts the pathway toward its neurotoxic branch

et al. 2016; Vécsei et al. 2013). In the neurotoxic branch of the KYN pathway, KYN is processed by activated microglia into metabolites such as QA, which acts as a potent NMDAR agonist promoting excitotoxicity and neuronal apoptosis (Birner et al. 2017; Heyes et al. 1992).

Regarding ketamine, some evidence suggests that a single ketamine infusion may reduce circulating levels of proinflammatory cytokines in the blood of patients with TRD within hours of administration (Kiraly et al. 2017). Furthermore, in mice, a low dose of intraperitoneally-administered ketamine abolished depressive behaviors usually induced by lipopolysaccharide (LPS) via QA production (Walker et al. 2013). Similarly, in humans, a double-blind, placebo-controlled study of TRD patients found that KYN levels and the KYN/TRP ratio were decreased in ketamine responders 4 h post-ketamine infusion (Moaddel et al. 2018). Further corroborating this finding, Kadriu and colleagues found that KYNA levels were increased and IDO levels were decreased after a single ketamine infusion in patients with treatment-resistant bipolar disorder (Kadriu et al. 2019).

In addition, Kadriu et al. (2018) showed that ketamine also reduces markers of bone inflammation usually found in patients with long-term MDD. The study demonstrated that ketamine normalized an abnormal osteoprotegerin/RANKL ratio and plasma osteoprotegerin levels, possibly counteracting the loss of bone mineral density found in patients with MDD.

2.3 Common Downstream Mechanisms Engaged by Rapid-Acting Antidepressants

Spurred by the paradigm-shifting nature of ketamine research, investigators have examined other candidate drugs with potentially rapid antidepressant effects, including serotonergic psychedelics (SPs). These substances – most prominently psilocybin, lysergic acid diethylamide (LSD), and N,N-dimethyltryptamine (DMT) – have been shown to be potentially effective in treating conditions ranging from substance dependence (Johnson et al. 2014) to terminal cancer anxiety (Grob et al. 2011) as well as depression (Palhano-Fontes et al. 2019).

The exact underlying mechanisms of SPs are under investigation, but agonism of the 5-HT2a receptor has been proposed as a central mechanism underlying the efficacy of these compounds. For instance, administering the 5-HT2a receptor blocker ketanserin diminished the subjective psychedelic effects of LSD (Preller et al. 2017) and psilocybin (Vollenweider et al. 1998) in humans, as well as drug discrimination in animal models (Appel and Callahan 1989; Cunningham and Appel 1987). A blockade of the effects of DMT administered as ayahuasca proved only partially effective (Valle et al. 2016). This may be due to the additional monoamine oxidase (MAO) inhibition as well as beta-carboline alkaloids, which have been shown to interact with a variety of molecular targets (Deecher et al. 1992; Husbands et al. 2001). 5-HT2a receptors are expressed in a variety of brain regions, most notably the neocortex, amygdala, striatum, mammillary nucleus, and claustrum (Pasqualetti et al. 1996; Pazos et al. 1985; Weber and Andrade 2010). 5-HT2a agonism has been proposed to lead to the desynchronization of brain areas and, more specifically, the default mode network, leading to the psychedelic effects of these compounds (Carhart-Harris et al. 2014). However, recent evidence has also questioned the contribution of 5-HT2a receptors to the antidepressant effect of SPs, suggesting instead that 5-HT2a receptor-independent mechanisms such as induction of neuronal plasticity may play a role by directly binding to TrkB (Casarotto et al. 2021; Hesselgrave et al. 2021).

Another notable advance in clinical research regarding psychoactive substances is a recent Phase 3 study demonstrating the efficacy of 3,4-Methylenedioxymethamphetamine (MDMA)-assisted therapy in post-traumatic stress disorder (Mitchell et al. 2021). However, the mechanism of action and consequent effects of MDMA are distinct; this agent is thought to act primarily by increasing serotonin levels via binding of presynaptic serotonin transporters (Rudnick and Wall 1992). Given the monoamine hypothesis, this mechanism suggests a putative usefulness of MDMA for treating MDD, although little evidence for this indication presently exists (see Patel and Titheradge 2015 for a review).

SPs typically produce similarly rapid antidepressant effects despite possessing mechanistically and pharmacodynamically distinct properties (Muttoni et al. 2019). In particular, research on the molecular mechanisms that mediate rapid antidepressant effects expand our understanding of antidepressant mechanisms, with the ultimate goal of identifying future drug targets with similar or better rapid antidepressant effects than ketamine and a better side-effect profile. In this context, investigating the downstream mechanisms of other rapid-acting drugs could provide valuable insights. While common downstream mechanisms of action between ketamine and drugs such as SPs remain largely speculative, several mechanisms, discussed below, have been proposed that may account for the rapid antidepressant effects of both substance classes.

2.3.1 Rapid Restoration of Synaptic Deficits Due to Stress

Considerable evidence suggests that chronic stress and depression are associated with a decrease in dendritic spines, spine synapse connections, and activity in regions implicated in depression, such as the hippocampus and the PFC (Duman et al. 2016; Kang et al. 2012; McEwen et al. 2015; McEwen and Morrison 2013). One mechanism proposed to account for rapid antidepressant effects is the rapid and selective restoration of stress-induced neuronal deficits. In a mouse model, ketamine infusions reversed stress-related loss of dendritic spines (Moda-Sava et al. 2019). This effect was associated with a change in neuronal systems in the PFC two days post-treatment as well as more delayed antidepressant effects. Another study detected reversal of apical dendritic spine deficits in the hippocampal CA1 region of Flinders sensitive rats 60 min after ketamine treatment (Treccani et al. 2019). Höflich and colleagues similarly found changes in hippocampal subfield volumes after ketamine infusion in healthy controls, with peaking effects in CA1 (Höflich et al. 2021). These effects may be mediated by the aforementioned BDNF-dependent activation of mTORC1 and the consequent neuroplastic effects.

2.3.2 Rapid Glutamate Release

Both ketamine and SPs lead to rapid glutamate release post-administration (Razoux et al. 2007; Vollenweider and Kometer 2010). With regard to SPs, rodent studies found that both LSD and 2,5-Dimethoxy-4-iodoamphetamine (DOI) increased glutamate levels in the prefrontal and somatosensory cortices, an effect that was abolished by blocking the 5-HT2A receptor (Muschamp et al. 2004; Scruggs et al. 2003). Martin and Nichols (2016) further showed that DOI administration increased early-activation genes and the indicator for neuronal activity cFos, especially in cerebral regions with a high density of glutamate-releasing pyramidal neurons.

It should be noted that ketamine's ability to increase glutamate levels has been directly confirmed using carbon-13 magnetic resonance spectroscopy (Abdallah et al. 2018), but less evidence exists for SPs. However, one study found that, compared to placebo, psilocybin increased glutamate levels in the PFC (Mason et al. 2020).

2.3.3 Rapid Stimulation of AMPA Throughput

In the case of ketamine, the glutamate surge is known to stimulate AMPAR throughput, which mediates BDNF release and the downstream activation of mTORC1 (Duman 2018; Olson 2018). Administration of an AMPAR antagonist again abolished ketamine's rapid antidepressant effects and mTORC1 activation (Maeng et al. 2008; Moghaddam et al. 1997). While this link is less clear with SPs, in rodents DOI induced behavioral effects like head shakes that were subsequently blocked by administration of AMPAR antagonists (C. Zhang and Marek 2008).

2.3.4 Release of Neurotrophins

In preclinical studies, ketamine, LSD, and DOI all increased BDNF levels in the hippocampus and other cortical areas (Autry et al. 2011; Garcia et al. 2008; Silva Pereira et al. 2017; Ly et al. 2018). BDNF, in turn, is associated with neuronal growth and plasticity. Shifts in BDNF levels are known to increase both spinogenesis and neuritogenesis (Cohen-Cory et al. 2010). Conversely, in Val66Met knock-in mice, where BDNF messenger RNA transport is impaired, ketamine had neither synaptogenic nor antidepressant effects (Liu et al. 2012). Ly and colleagues further demonstrated that the TrkB antagonist ANA-12 abolished the neuroplastic effects of SPs (Ly et al. 2018). BDNF has a high affinity for binding TrkB, which in turn mediates neuronal plasticity via mTORC1 (Jaworski et al. 2005; Kumar et al. 2005). Thus, rapamycin administration – which inhibits mTORC1 – also abolished the neuroplastic effects of SPs (Ly et al. 2018). Finally, the selective 5-HT2A antagonist ketanserin also inhibited spinogenesis and neuritogenesis after SP administration, highlighting the importance of this receptor in initiating downstream effects. When directly compared to ketamine, some SPs proved to be more efficacious (e.g., MDMA) or potent (e.g., LSD) in promoting neuritogenesis, which may account for the prolonged antidepressant effect of these substances (Ly et al. 2018).

It should be noted, however, that the role of BDNF in rapidly decreasing depressive symptoms in humans is still under debate. Especially with regard to ketamine, the evidence has been mixed (Duncan et al. 2013; Haile et al. 2014; Machado-Vieira et al. 2009; Laje et al. 2012). Similarly mixed findings have also been found for SPs (de Almeida et al. 2019; Holze et al. 2020).

2.3.5 Stimulation of Intracellular Neuroplasticity Cascades

Both ketamine and SPs have been shown to increase neuroplasticity via dendritic growth and new synapse formation as well as strengthen preexisting synaptic connections (Kadriu et al. 2021). A single dose of ketamine increased levels of pre- and postsynaptic proteins such as PSD95, synapsin 1, and the AMPAR GluA subunit in the rodent PFC, which in turn was associated with the increased number and function of synapses (Li et al. 2010). The increase in synaptic proteins was detected as soon as 2 h after ketamine administration and lasted for at least 72 h, which may help explain ketamine's rapid antidepressant effects. Neuroplasticity genes, including for *Bdnf*, *Homer1a*, and activity related cytoskeletal protein (*Arc*) have also been shown to be activated via glutamatergic signaling (Bagot et al. 2017; De Bartolomeis et al. 2013).

In rodent models, both DOI and LSD administration similarly increased a wide array of neuroplasticity genes, including *Bdnf*, *Arc*, *egr-1*, *Nor1*, *Ania3*, *sgk*, *C/EBP-β*, and *Iκβ-α* (González-Maeso et al. 2007; Martin et al. 2014; Martin and Nichols 2016; Nichols et al. 2003; Nichols and Sanders-Bush 2002). However, in the case of SPs, gene expression is mediated via 5-HT2A receptor activation, which in turn affects neuroplasticity via G-protein-coupled receptor pathways.

Considerable similarities between SPs and ketamine also exist with regard to neuritogenesis and spinogenesis. Like ketamine, LSD, DOI, psilocybin, DMT, and noribogaine (a metabolite of ibogaine) all increased dendritic arbor complexity in cultured cortical neurons in vitro, both in terms of increasing the total length of arbors and the number of dendritic branches (Ly et al. 2018). LSD, DOI, and DMT also increased the number of dendritic spines (Ly et al. 2018). In vivo, treating Drosophila larvae with LSD and DOI increased dendritic branching (Ly et al. 2018). In rats, the intraperitoneal administration of DMT rapidly increased dendritic spines on cortical pyramidal neurons in the PFC 24 h after administration, comparable to an equivalent dose of ketamine (Ly et al. 2018). Although these effects remain poorly understood, they seem to be mediated by mTORC1, TrkB, and 5-HT2A signaling pathways. Because the effects were demonstrated in vivo in both vertebrate and invertebrate models, they suggest an evolutionarily conserved mechanism. Interestingly, in rodents, alterations in synaptic plasticity and ketamine's antidepressant effects have both been demonstrated to disappear when either mTORC1 or AMPARs are blocked (Li et al. 2010). Conversely, mTORC1 activation via NV-5138 produces rapid antidepressant effects in rodent models.

3 Conclusion

The evidence reviewed above describes the novel biochemical mechanisms of action that underlie ketamine's antidepressant effects as well as those of SPs. It should be noted, however, that recent attempts to identify or develop glutamatergic drugs that mimic ketamine's antidepressant qualities but lack its dissociative psychotomimetic effects have largely proven futile (Kadriu et al. 2020). Furthermore, while both the safety profile of ketamine and the lack of abuse potential associated with a single dose have been relatively well established (Acevedo-Diaz et al. 2020), only limited safety data exist regarding the long-term effects of repeated ketamine administration.

Similarly, few data exist regarding the development of tolerance to chronic antidepressant ketamine use. Given that the recommended dosing regimen of esketamine for depression is one to three times a week for the first two months and once every week or every two weeks thereafter, such long-term data are crucially needed (Canuso et al. 2018).

A key possible mechanism of action for the antidepressant effects of both ketamine and SPs is the ability of these compounds to increase neuroplasticity. Underscoring the plausibility of this hypothesis is that postmortem studies have demonstrated that individuals with depression have lower BDNF mRNA and protein levels (Castrén 2005; Thoenen 1995), serum BDNF protein levels (Sen et al. 2008; Shimizu et al. 2003) as well as decreased hippocampal volume (Sheline et al. 1996; Videbech and Ravnkilde 2004). However, neural plasticity may not be an exclusively beneficial state, regardless of the circumstances in which it occurs (Branchi 2011). For example, Belsky and colleagues demonstrated that polymorphisms of monoamine oxidase-A, dopamine receptor D4, or a 5-hydroxytryptamine-linked polymorphic region were linked to either beneficial or adverse outcomes depending on the environment (Belsky et al. 2009). Indeed, at least in the case of SPs, the importance of context or even psychotherapeutic care appears to be key for the beneficial effect of the neuroplastic state produced by these compounds to manifest (Carhart-Harris et al. 2018). Similarly, Chiarotti and colleagues identified a dose-dependent interaction between the SSRI escitalopram and the environmental context of patients (Chiarotti et al. 2017).

It is interesting to note that the US Food and Drug Administration (FDA) recently approved the rapid-acting antidepressant brexanolone (SAGE-547), a positive allosteric $GABA_A$ receptor modulator, for use in postpartum depression (U.S. Food and Drug Administration 2019). Research in mice further suggests that impairment of $GABA_A$Rs can lead to depressive-like behaviors as well as downregulation of both AMPA and NMDA receptors and consequent glutamatergic transmission, which can be normalized by ketamine infusion (Ren et al. 2016). In rats, the administration of brexanolone prevented a decrease in BDNF levels and consequent impairment of hippocampal neurogenesis due to stress (Evans et al. 2012).

The rapid-acting antidepressant effects of ketamine, SPs, and possibly also brexanolone/zuranolone suggest that downstream commonalities on glutamatergic systems may underlie the mechanism of action of all three agents, even if their initial targets differ. The emergence of rapid-acting antidepressants as treatments for depression offers investigators new and potentially fruitful avenues for exploration, both in terms of identifying novel biochemical mechanisms underlying the mechanism of action of successful rapid-acting antidepressants and developing novel therapeutics. Together, such evidence can be used to investigate and develop alternate and possibly more rapid-acting antidepressant therapies with the goal of helping numerous patients who suffer from treatment-resistant forms of mood disorders.

Acknowledgements Ioline Henter (NIMH) provided invaluable editorial assistance.

Disclosures Funding for this work was provided in part by the Intramural Research Program at the National Institute of Mental Health, National Institutes of Health (IRP-NIMH-NIH; ZIAMH002857). Dr. Zarate is listed as a co-inventor on a patent for the use of ketamine in major depression and suicidal ideation; as a co-inventor on a patent for the use of (2*R*,6*R*)-hydroxynorketamine, (*S*)-dehydronorketamine, and other stereoisomeric dehydroxylated and hydroxylated metabolites of (*R,S*)-ketamine metabolites in the treatment of depression and neuropathic pain; and as a co-inventor on a patent application for the use of (2*R*,6*R*)-hydroxynorketamine and (2*S*,6*S*)-hydroxynorketamine in the treatment of depression, anxiety, anhedonia, suicidal ideation, and post-traumatic stress disorders. He has assigned his patent rights to the U.-S. government but will share a percentage of any royalties that may be received by the government. C. Kraus has received honoraria from Janssen, LivaNova, AOP Orphan, and Roche Austria. Dr. Kadriu is full-time employee at Janssen Research & Development, LLC and owns stock and stock options in Johnson & Johnson.

All other authors have no conflict of interest to disclose, financial or otherwise.

References

Aarts AA, Anderson JE, Anderson CJ, Attridge PR, Attwood A, Axt J, Babel M, Bahník Š, Baranski E, Barnett-Cowan M, Bartmess E, Beer J, Bell R, Bentley H, Beyan L, Binion G, Borsboom D, Bosch A, Bosco FA et al (2015) Estimating the reproducibility of psychological science. Science 349(6251). https://doi.org/10.1126/science.aac4716

Abdallah CG, Jackowski A, Salas R, Gupta S, Sato JR, Mao X, Coplan JD, Shungu DC, Mathew SJ (2017) The nucleus accumbens and ketamine treatment in major depressive disorder. Neuropsychopharmacology 42(8):1739–1746. https://doi.org/10.1038/npp.2017.49

Abdallah CG, De Feyter HM, Averill LA, Jiang L, Averill CL, Chowdhury GMI, Purohit P, de Graaf RA, Esterlis I, Juchem C, Pittman BP, Krystal JH, Rothman DL, Sanacora G, Mason GF (2018) The effects of ketamine on prefrontal glutamate neurotransmission in healthy and depressed subjects. Neuropsychopharmacology 43(10):2154–2160. https://doi.org/10.1038/s41386-018-0136-3

Acevedo-Diaz EE, Cavanaugh GW, Greenstein D, Kraus C, Kadriu B, Zarate CA, Park LT (2020) Comprehensive assessment of side effects associated with a single dose of ketamine in treatment-resistant depression. J Affect Disord 263:568–575. https://doi.org/10.1016/j.jad.2019.11.028

Appel JB, Callahan PM (1989) Involvement of 5-HT receptor subtypes in the discriminative stimulus properties of mescaline. Eur J Pharmacol 159(1):41–46. https://doi.org/10.1016/0014-2999(89)90041-1

Autry AE, Adachi M, Nosyreva E, Na ES, Los MF, Cheng PF, Kavalali ET, Monteggia LM (2011) NMDA receptor blockade at rest triggers rapid behavioural antidepressant responses. Nature 475(7354):91–96. https://doi.org/10.1038/nature10130

Bagot RC, Cates HM, Purushothaman I, Vialou V, Heller EA, Yieh L, LaBonté B, Peña CJ, Shen L, Wittenberg GM, Nestler EJ (2017) Ketamine and imipramine reverse transcriptional signatures of susceptibility and induce resilience-specific gene expression profiles. Biol Psychiatry 81(4):285–295. https://doi.org/10.1016/j.biopsych.2016.06.012

Bahr R, Lopez A, Rey JA (2019) Intranasal esketamine (SpravatoTM) for use in treatment-resistant depression in conjunction with an oral antidepressant. Pharm Ther 44(6):340–375

Bay-Richter C, Linderholm KR, Lim CK, Samuelsson M, Träskman-Bendz L, Guillemin GJ, Erhardt S, Brundin L (2015) A role for inflammatory metabolites as modulators of the glutamate N-methyl-d-aspartate receptor in depression and suicidality. Brain Behav Immun 43:110–117. https://doi.org/10.1016/j.bbi.2014.07.012

Belsky J, Jonassaint C, Pluess M, Stanton M, Brummett B, Williams R (2009) Vulnerability genes or plasticity genes? Mol Psychiatry 14(8):746–754. https://doi.org/10.1038/mp.2009.44

Berman RM, Cappiello A, Anand A, Oren DA, Heninger GR, Charney DS, Krystal JH (2000) Antidepressant effects of ketamine in depressed patients. Biol Psychiatry 47(4):351–354. https://doi.org/10.1016/S0006-3223(99)00230-9

Birner A, Platzer M, Bengesser SA, Dalkner N, Fellendorf FT, Queissner R, Pilz R, Rauch P, Maget A, Hamm C, Herzog-Eberhard S, Mangge H, Fuchs D, Moll N, Zelzer S, Schütze G, Schwarz M, Reininghaus B, Kapfhammer HP, Reininghaus EZ (2017) Increased breakdown of kynurenine towards its neurotoxic branch in bipolar disorder. PLoS One 12(2):1–14. https://doi.org/10.1371/journal.pone.0172699

Björkholm C, Marcus MM, Konradsson-Geuken Å, Jardemark K, Svensson TH (2017) The novel antipsychotic drug brexpiprazole, alone and in combination with escitalopram, facilitates prefrontal glutamatergic transmission via a dopamine D1 receptor-dependent mechanism. Eur Neuropsychopharmacol 27(4):411–417. https://doi.org/10.1016/j.euroneuro.2017.01.014

Branchi I (2011) The double edged sword of neural plasticity: increasing serotonin levels leads to both greater vulnerability to depression and improved capacity to recover (for ketamine analogues). Psychoneuroendocrinology 36(3):339–351. https://doi.org/10.1016/j.psyneuen.2010.08.011

Burgdorf J, Zhang XL, Nicholson KL, Balster RL, David Leander J, Stanton PK, Gross AL, Kroes RA, Moskal JR (2013) GLYX-13, a NMDA receptor glycine-site functional partial agonist, induces antidepressant-like effects without ketamine-like side effects. Neuropsychopharmacology 38(5):729–742. https://doi.org/10.1038/npp.2012.246

Burgdorf J, Zhang XL, Weiss C, Gross A, Boikess SR, Kroes RA, Khan MA, Burch RM, Rex CS, Disterhoft JF, Stanton PK, Moskal JR (2015) The long-lasting antidepressant effects of rapastinel (GLYX-13) are associated with a metaplasticity process in the medial prefrontal cortex and hippocampus. Neuroscience 308:202–211. https://doi.org/10.1016/j.neuroscience.2015.09.004

Canuso CM, Singh JB, Fedgchin M, Alphs L, Lane R, Lim P, Pinter C, Hough D, Sanacora G, Manji H, Drevets WC (2018) Efficacy and safety of intranasal esketamine for the rapid reduction of symptoms of depression and suicidality in patients at imminent risk for suicide: results of a double-blind, randomized, placebo-controlled study. Am J Psychiatr 175(7): 620–630. https://doi.org/10.1176/appi.ajp.2018.17060720

Carhart-Harris RL, Leech R, Hellyer PJ, Shanahan M, Feilding A, Tagliazucchi E, Chialvo DR, Nutt D (2014) The entropic brain: a theory of conscious states informed by neuroimaging research with psychedelic drugs. Front Hum Neurosci 8(1 FEB):1–22. https://doi.org/10.3389/fnhum.2014.00020

Carhart-Harris RL, Roseman L, Haijen E, Erritzoe D, Watts R, Branchi I, Kaelen M (2018) Psychedelics and the essential importance of context. J Psychopharmacol 32(7):725–731. https://doi.org/10.1177/0269881118754710

Casarotto PC, Girych M, Fred SM, Kovaleva V, Moliner R, Enkavi G, Biojone C, Cannarozzo C, Sahu MP, Kaurinkoski K, Brunello CA, Steinzeig A, Winkel F, Patil S, Vestring S, Serchov T, Diniz CRAF, Laukkanen L, Cardon I et al (2021) Antidepressant drugs act by directly binding to TRKB neurotrophin receptors. Cell 184(5):1299–1313.e19. https://doi.org/10.1016/j.cell.2021.01.034

Castrén E (2005) Is mood chemistry? Nat Rev Neurosci 6(3):241–246

Chan KWS, Lee TMC, Siu AMH, Wong DPL, Kam C, Tsang SKM, Chan CCH (2013) Addictive Behaviors effects of chronic ketamine use on frontal and medial temporal cognition. Addict Behav 38(5):2128–2132. https://doi.org/10.1016/j.addbeh.2013.01.014

Chiarotti F, Viglione A, Giuliani A, Branchi I (2017) Citalopram ampli fi es the in fl uence of living conditions on mood in depressed patients enrolled in the STAR * D study. Transl Psychiatry 7(3):e1066. https://doi.org/10.1038/tp.2017.35

Chowdhury GMI, Zhang J, Thomas M, Banasr M, Ma X, Pittman B, Bristow L, Schaeffer E, Duman RS, Rothman DL, Behar KL, Sanacora G (2017) Transiently increased glutamate cycling in rat PFC is associated with rapid onset of antidepressant-like effects. Mol Psychiatry 22(1):120–126. https://doi.org/10.1038/mp.2016.34

Cohen BD, Rosenbaum G, Dobie SI, Gottlieb JS (1959) Sensory isolation: hallucinogenic effects of a brief procedure. J Nerv Ment Dis 129(5):486–491. https://doi.org/10.1097/00005053-195911000-00009

Cohen-Cory S, Kidane AH, Shirkey NJ, Marshak S (2010) Brain-derived neurotrophic factor and the development of structural neuronal connectivity. Dev Neurobiol 70(5):271–288. https://doi.org/10.1002/dneu.20774

Cunningham KA, Appel JB (1987) Neuropharmacological reassessment of the discriminative stimulus properties of d-lysergic acid diethylamide (LSD). Psychopharmacology 91(1):67–73. https://doi.org/10.1007/BF00690929

Curran HV, Monaghan L, Curran HV, Monaghan L (2001) In and out of the K-hole: a comparison of the acute and residual effects of ketamine in frequent and infrequent ketamine users. Addiction 96(5):749–760. https://doi.org/10.1080/09652140020039116

de Almeida RN, Galvão AC, Da Silva FS, Silva EA, Palhano-Fontes F, Maia-de-Oliveira JP, de Araújo DB, Lobão-Soares B, Galvão-Coelho NL (2019) Modulation of serum brain-derived neurotrophic factor by a single dose of ayahuasca: observation from a randomized controlled trial. Front Psychol 10(JUN):1–13. https://doi.org/10.3389/fpsyg.2019.01234

De Bartolomeis A, Sarappa C, Buonaguro EF, Marmo F, Eramo A, Tomasetti C, Iasevoli F (2013) Different effects of the NMDA receptor antagonists ketamine, MK-801, and memantine on postsynaptic density transcripts and their topography: role of Homer signaling, and implications for novel antipsychotic and pro-cognitive targets in psychosis. Prog Neuro-Psychopharmacol Biol Psychiatry 46:1–12. https://doi.org/10.1016/j.pnpbp.2013.06.010

Deecher DC, Teitler M, Soderlund DM, Bornmann WG, Kuehne ME, Glick SD (1992) Mechanisms of action of ibogaine and harmaline congeners based on radioligand binding studies. Brain Res 571(2):242–247. https://doi.org/10.1016/0006-8993(92)90661-R

Domino EF, Chodoff P, Corssen G (1965) Pharmacologic effects of CI-581, a new dissociative anesthetic, in man. Clin Pharmacol Ther 6(3):279–291. https://doi.org/10.1002/cpt196563279

Doty RL, Popova V, Wylie C, Fedgchin M, Daly E, Janik A, Ochs R, Rosanne R, Pilar L, Kim L, Rama C, Carol M, Jaskaran J (2021) Effect of esketamine nasal spray on olfactory function and nasal tolerability in patients with treatment - resistant depression: results from four multicenter, randomized, double - blind, placebo - controlled, phase III studies. CNS Drugs 35(7):781–794. https://doi.org/10.1007/s40263-021-00826-9

Duman RS (2018) Ketamine and rapid-acting antidepressants: a new era in the battle against depression and suicide [version 1; referees: 3 approved]. F1000Research 7(May):1–10. https://doi.org/10.12688/f1000research.14344.1

Duman RS, Aghajanian GK, Sanacora G, Krystal JH (2016) Synaptic plasticity and depression: new insights from stress and rapid-acting antidepressants. Nat Med 22(3):238–249. https://doi.org/10.1038/nm.4050

Duman RS, Shinohara R, Fogaça MV, Hare B (2019) Neurobiology of rapid-acting antidepressants: convergent effects on GluA1-synaptic function. Mol Psychiatry 24(12):1816–1832. https://doi.org/10.1038/s41380-019-0400-x

Duncan WC, Sarasso S, Ferrarelli F, Selter J, Riedner BA, Hejazi NS, Yuan P, Brutsche N, Manji HK, Tononi G, Zarate CA (2013) Concomitant BDNF and sleep slow wave changes indicate ketamine-induced plasticity in major depressive disorder. Int J Neuropsychopharmacol 16(2): 301–311. https://doi.org/10.1017/S1461145712000545

Evans J, Sun Y, McGregor A, Connor B (2012) Allopregnanolone regulates neurogenesis and depressive/anxiety-like behaviour in a social isolation rodent model of chronic stress. Neuropharmacology 63(8). https://doi.org/10.1016/j.neuropharm.2012.08.012

Findeis H, Sauer C, Cleare A, Bauer M, Ritter P (2020) Urothelial toxicity of esketamine in the treatment of depression. Psychopharmacology 237(11):3295–3302. https://doi.org/10.1007/S00213-020-05611-Y

Fukumoto K, Iijima M, Chaki S (2016) The antidepressant effects of an mGlu2/3 receptor antagonist and ketamine require AMPA receptor stimulation in the mPFC and subsequent activation of the 5-HT neurons in the DRN. Neuropsychopharmacology 41(4):1046–1056. https://doi.org/10.1038/npp.2015.233

Fukumoto K, Toki H, Iijima M, Hashihayata T, Yamaguchi JI, Hashimoto K, Chaki S (2017) Antidepressant potential of (R)-ketamine in rodent models: comparison with (S)-ketamine. J Pharmacol Exp Ther 361(1):9–16. https://doi.org/10.1124/jpet.116.239228

Fukumoto K, Iijima M, Funakoshi T, Chaki S (2018) Role of 5-HT 1A receptor stimulation in the medial prefrontal cortex in the sustained antidepressant effects of ketamine. Int J Neuropsychopharmacol 21(4):371–381. https://doi.org/10.1093/ijnp/pyx116

Fukumoto K, Fogaça MV, Liu RJ, Duman CH, Li XY, Chaki S, Duman RS (2020) Medial PFC AMPA receptor and BDNF signaling are required for the rapid and sustained antidepressant-like effects of 5-HT1A receptor stimulation. Neuropsychopharmacology 45(10):1725–1734. https://doi.org/10.1038/s41386-020-0705-0

Gallagher B, Neiman A, Slattery M-C, McLoughlin DM (2021) Online news media reporting of ketamine as a treatment for depression from 2000 to 2017. Ir J Psychol Med. https://doi.org/10.1017/IPM.2021.47

Gao C, Wolf ME (2008) Dopamine receptors regulate NMDA receptor surface expression in prefrontal cortex neurons. J Neurochem 106(6):2489–2501. https://doi.org/10.1111/j.1471-4159.2008.05597.x

Garcia LSB, Comim CM, Valvassori SS, Réus GZ, Barbosa LM, Andreazza AC, Stertz L, Fries GR, Gavioli EC, Kapczinski F, Quevedo J (2008) Acute administration of ketamine induces antidepressant-like effects in the forced swimming test and increases BDNF levels in the rat hippocampus. Prog Neuro-Psychopharmacol Biol Psychiatry 32(1):140–144. https://doi.org/10.1016/j.pnpbp.2007.07.027

Gaynes BN, Warden D, Trivedi MH, Wisniewski SR, Fava M, Rush AJ (2009) What did STAR*D teach us? Results from a large-scale, practical, clinical trial for patients with depression. Psychiatr Serv 60(11):1439–1445. https://doi.org/10.1176/ps.2009.60.11.1439

Gerhard DM, Pothula S, Liu RJ, Wu M, Li XY, Girgenti MJ, Taylor SR, Duman CH, Delpire E, Picciotto M, Wohleb ES, Duman RS (2020) GABA interneurons are the cellular trigger for ketamine's rapid antidepressant actions. J Clin Investig 130(3):1336–1349. https://doi.org/10.1172/JCI130808

Ghosal S, Hare BD, Duman RS (2017) Prefrontal cortex GABAergic deficits and circuit dysfunction in the pathophysiology and treatment of chronic stress and depression. Curr Opin Behav Sci 14:1–8. https://doi.org/10.1016/j.cobeha.2016.09.012

Ghosal S, Duman CH, Liu RJ, Wu M, Terwilliger R, Girgenti MJ, Wohleb E, Fogaca MV, Teichman EM, Hare B, Duman RS (2020) Ketamine rapidly reverses stress-induced impairments in GABAergic transmission in the prefrontal cortex in male rodents. Neurobiol Dis 134:104669. https://doi.org/10.1016/j.nbd.2019.104669

Gigliucci V, O'Dowd G, Casey S, Egan D, Gibney S, Harkin A (2013) Ketamine elicits sustained antidepressant-like activity via a serotonin-dependent mechanism. Psychopharmacology 228(1):157–166. https://doi.org/10.1007/s00213-013-3024-x

Goethe JW, Woolley SB, Cardoni AA, Woznicki BA, Piez DA (2007) Selective serotonin reuptake inhibitor discontinuation: side effects and other factors that influence medication adherence. J Clin Psychopharmacol 27(5):451–458. https://doi.org/10.1097/jcp.0b013e31815152a5

Gonzalez-Islas C, Hablitz JJ (2003) Dopamine enhances EPSCs in layer II-III pyramidal neurons in rat prefrontal cortex. J Neurosci 23(3):867–875. https://doi.org/10.1523/jneurosci.23-03-00867.2003

González-Maeso J, Weisstaub NV, Zhou M, Chan P, Ivic L, Ang R, Lira A, Bradley-Moore M, Ge Y, Zhou Q, Sealfon SC, Gingrich JA (2007) Hallucinogens recruit specific cortical 5-HT2A receptor-mediated signaling pathways to affect behavior. Neuron 53(3):439–452. https://doi.org/10.1016/j.neuron.2007.01.008

Grob CS, Danforth AL, Chopra GS, Hagerty M, McKay CR, Halberstad AL, Greer GR (2011) Pilot study of psilocybin treatment for anxiety in patients with advanced-stage cancer. Arch Gen Psychiatry 68(1):71–78. https://doi.org/10.1001/archgenpsychiatry.2010.116

Gurden H, Takita M, Jay TM (2000) Essential role of D1 but not D2 receptors in the NMDA receptor-dependent long-term potentiation at hippocampal-prefrontal cortex synapses in vivo. J Neurosci 20(22):1–5. https://doi.org/10.1523/jneurosci.20-22-j0003.2000

Haile CN, Murrough JW, Iosifescu DV, Chang LC, Al Jurdi RK, Foulkes A, Iqbal S, Mahoney JJ, De La Garza R, Charney DS, Newton TF, Mathew SJ (2014) Plasma brain derived neurotrophic factor (BDNF) and response to ketamine in treatment-resistant depression. Int J Neuropsychopharmacol 17(2):331–336. https://doi.org/10.1017/S1461145713001119

Hansen KB, Yi F, Perszyk RE, Furukawa H, Wollmuth LP, Gibb AJ, Traynelis SF (2018) Structure, function, and allosteric modulation of NMDA receptors. J Gen Physiol 150(8):1081–1105. https://doi.org/10.1085/jgp.201812032

Hare BD, Shinohara R, Liu RJ, Pothula S, DiLeone RJ, Duman RS (2019) Optogenetic stimulation of medial prefrontal cortex Drd1 neurons produces rapid and long-lasting antidepressant effects. Nat Commun 10(1):1–12. https://doi.org/10.1038/s41467-018-08168-9

Haroon E, Fleischer CC, Felger JC, Chen X, Woolwine BJ, Patel T, Hu XP, Miller AH (2016) Conceptual convergence: increased inflammation is associated with increased basal ganglia glutamate in patients with major depression. Mol Psychiatry 21(10):1351–1357. https://doi.org/10.1038/mp.2015.206

Heisler JM, O'Connor JC (2015) Indoleamine 2,3-dioxygenase-dependent neurotoxic kynurenine metabolism mediates inflammation-induced deficit in recognition memory. Brain Behav Immun 50:115–124. https://doi.org/10.1016/j.bbi.2015.06.022

Hesselgrave N, Troppoli TA, Wulff AB, Cole AB, Thompson SM (2021) Harnessing psilocybin: antidepressant-like behavioral and synaptic actions of psilocybin are independent of 5-HT2R activation in mice. Proc Natl Acad Sci U S A 118(17):1–7. https://doi.org/10.1073/pnas.2022489118

Heyes MP, Saito K, Crowley JS, Davis LE, Demitrack MA, Der M, Dilling LA, Elia J, Kruesi MJP, Lackner A, Larsen SA, Lee K, Leonard HL, Markey SP, Martin A, Milstein S, Mouradian MM, Pranzatelli MR, Quearry BJ et al (1992) Quinolinic acid and kynurenine pathway metabolism in inflammatory and non-inflammatory neurological disease. Brain 115(5):1249–1273. https://doi.org/10.1093/brain/115.5.1249

Höflich A, Kraus C, Pfeiffer RM, Seiger R, Rujescu D, Zarate CA, Kasper S, Winkler D, Lanzenberger R (2021) Translating the immediate effects of S-ketamine using hippocampal subfield analysis in healthy subjects-results of a randomized controlled trial. Transl Psychiatry 11(1). https://doi.org/10.1038/s41398-021-01318-6

Holze F, Vizeli P, Müller F, Ley L, Duerig R, Varghese N, Eckert A, Borgwardt S, Liechti ME (2020) Distinct acute effects of LSD, MDMA, and D-amphetamine in healthy subjects. Neuropsychopharmacology 45(3):462–471. https://doi.org/10.1038/s41386-019-0569-3

Husbands SM, Glennon RA, Gorgerat S, Gough R, Tyacke R, Crosby J, Nutt DJ, Lewis JW, Hudson AL (2001) B-carboline binding to imidazoline receptors. Drug Alcohol Depend 64(2): 203–208. https://doi.org/10.1016/S0376-8716(01)00123-5

Jaworski J, Spangler S, Seeburg DP, Hoogenraad CC, Sheng M (2005) Control of dendritic arborization by the phosphoinositide-3′-kinase- Akt-mammalian target of rapamycin pathway. J Neurosci 25(49):11300–11312. https://doi.org/10.1523/JNEUROSCI.2270-05.2005

Jhang J, Hsu Y, Kuo H (2015) Possible pathophysiology of ketamine-related cystitis and associated treatment strategies. Int J Urol 22(9):816–825. https://doi.org/10.1111/iju.12841

Johnson MW, Garcia-Romeau A, Cosimano MP, Griffiths RR (2014) Pilot study of the 5-HT2A R agonist psilocybin in the treatment of tobacco addiction. J Psychopharmacol 28(11):983–992. https://doi.org/10.1177/0269881114548296.Pilot

Kadriu B, Gold PW, Luckenbaugh DA, Lener MS, Ballard ED, Niciu MJ, Henter ID, Park LT, De Sousa RT, Yuan P, Machado-Vieira R, Zarate CA (2018) Acute ketamine administration corrects abnormal inflammatory bone markers in major depressive disorder. Mol Psychiatry 23(7):1626–1631. https://doi.org/10.1038/mp.2017.109

Kadriu B, Farmer CA, Yuan P, Park LT, De Deng Z, Moaddel R, Henter ID, Shovestul B, Ballard ED, Kraus C, Gold PW, Machado-Vieira R, Zarate CA (2019) The kynurenine pathway and bipolar disorder: intersection of the monoaminergic and glutamatergic systems and immune response. Mol Psychiatry. https://doi.org/10.1038/s41380-019-0589-8

Kadriu B, De Deng Z, Kraus C, Henter ID, Lisanby SH, Zarate CA (2020) Not so fast: recent successes and failures in treating depression. J Clin Psychiatry 81(4). https://doi.org/10.4088/JCP.19ac13138

Kadriu B, Greenwald M, Henter ID, Gilbert JR, Kraus C, Park LT, Zarate CA (2021) Ketamine and serotonergic psychedelics: common mechanisms underlying the effects of rapid-acting antidepressants. Int J Neuropsychopharmacol 24(1):8–21. https://doi.org/10.1093/ijnp/pyaa087

Kang HJ, Voleti B, Hajszan T, Rajkowska G, Stockmeier CA, Licznerski P, Lepack A, Majik MS, Jeong LS, Banasr M, Son H, Duman RS (2012) Decreased expression of synapse-related genes and loss of synapses in major depressive disorder. Nat Med 18(9):1413–1417. https://doi.org/10.1038/nm.2886

Kiraly DD, Horn SR, Van Dam NT, Costi S, Schwartz J, Kim-Schulze S, Patel M, Hodes GE, Russo SJ, Merad M, Iosifescu DV, Charney DS, Murrough JW (2017) Altered peripheral immune profiles in treatment-resistant depression: response to ketamine and prediction of treatment outcome. Transl Psychiatry 7(3). https://doi.org/10.1038/tp.2017.31

Kumar V, Zhang MX, Swank MW, Kunz J, Wu GY (2005) Regulation of dendritic morphogenesis by Ras-PI3K-Akt-mTOR and Ras-MAPK signaling pathways. J Neurosci 25(49): 11288–11299. https://doi.org/10.1523/JNEUROSCI.2284-05.2005

Laje G, Lally N, Mathews D, Brutsche N, Chemerinski A, Akula N, Kelmendi B, Simen A, McMahon FJ, Sanacora G, Zarate C (2012) Brain-derived neurotrophic factor Val66Met polymorphism and antidepressant efficacy of ketamine in depressed patients. Biol Psychiatry 72(11):e27–e28. https://doi.org/10.1016/j.biopsych.2012.05.031

Lapidus KAB, Levitch CF, Perez AM, Brallier JW, Parides MK, Soleimani L, Feder A, Iosifescu DV, Charney DS, Murrough JW (2014) A randomized controlled trial of intranasal ketamine in major depressive disorder. Biol Psychiatry 76(12):970–976. https://doi.org/10.1016/j.biopsych.2014.03.026

Lavin A, Grace AA (2001) Stimulation of D1-type dopamine receptors enhances excitability in prefrontal cortical pyramidal neurons in a state-dependent manner. Neuroscience 104(2): 335–346. https://doi.org/10.1016/S0306-4522(01)00096-3

Li N, Boyoung L, Rong-Jian L, Mounira B, Jason MD, Masaaki I, Xiao-Yuan L, Aghajanian G, Duman RS (2010) mTOR-dependent synapse formation underlies the rapid antidepressant effects of NMDA antagonists. Science 329(August):959–964

Li Y, Shen R, Wen G, Ding R, Du A, Zhou J (2017) Effects of ketamine on levels of inflammatory cytokines IL-6, IL-1 β, and TNF- α in the hippocampus of mice following acute or chronic administration. Front Pharmacol 8:139. https://doi.org/10.3389/fphar.2017.00139

Liu RJ, Lee FS, Li XY, Bambico F, Duman RS, Aghajanian GK (2012) Brain-derived neurotrophic factor Val66Met allele impairs basal and ketamine-stimulated synaptogenesis in prefrontal cortex. Biol Psychiatry 71(11):996–1005. https://doi.org/10.1016/j.biopsych.2011.09.030

Ly C, Greb AC, Cameron LP, Wong JM, Barragan EV, Wilson PC, Burbach KF, Soltanzadeh Zarandi S, Sood A, Paddy MR, Duim WC, Dennis MY, McAllister AK, Ori-McKenney KM, Gray JA, Olson DE (2018) Psychedelics promote structural and functional neural plasticity. Cell Rep 23(11):3170–3182. https://doi.org/10.1016/j.celrep.2018.05.022

Machado-Vieira R, Yuan P, Brutsche N, DiazGranados N, Luckenbaugh D, Manji HK, Zarate CA (2009) Brain-derived neurotrophic factor and initial antidepressant response to an N-methyl-D-aspartate antagonist. J Clin Psychiatry 70(12):1662–1666. https://doi.org/10.4088/JCP.08m04659

Maeng S, Zarate CA, Du J, Schloesser RJ, McCammon J, Chen G, Manji HK (2008) Cellular mechanisms underlying the antidepressant effects of ketamine: role of α-amino-3-hydroxy-5-

methylisoxazole-4-propionic acid receptors. Biol Psychiatry 63(4):349–352. https://doi.org/10.1016/j.biopsych.2007.05.028

Malhotra S, Santosh PJ (1996) Loading dose imipramine - new approach to pharmacotherapy of melancholic depression. J Psychiatr Res 30(1):51–58. https://doi.org/10.1016/0022-3956(95)00042-9

Martin DA, Nichols CD (2016) Psychedelics recruit multiple cellular types and produce complex transcriptional responses within the brain. EBioMedicine 11:262–277. https://doi.org/10.1016/j.ebiom.2016.08.049

Martin DA, Marona-Lewicka D, Nichols DE, Nichols CD (2014) Chronic LSD alters gene expression profiles in the mPFC relevant to schizophrenia. Neuropharmacology 83:1–8. https://doi.org/10.1016/j.neuropharm.2014.03.013

Mason NL, Kuypers KPC, Müller F, Reckweg J, Tse DHY, Toennes SW, Hutten NRPW, Jansen JFA, Stiers P, Feilding A, Ramaekers JG (2020) Me, myself, bye: regional alterations in glutamate and the experience of ego dissolution with psilocybin. Neuropsychopharmacology 45(12):2003–2011. https://doi.org/10.1038/s41386-020-0718-8

McCarthy DA, Chen G, Kaump DH, Ensor C (1965) General anesthetic and other pharmacological properties of 2-(O-chlorophenyl)-2-methylamino cyclohexanone HCl (CI-581). J New Drugs 5(1):21–33. https://doi.org/10.1002/j.1552-4604.1965.tb00219.x

McEwen BS, Morrison JH (2013) The brain on stress: vulnerability and plasticity of the prefrontal cortex over the life course. Neuron 79(1):16–29. https://doi.org/10.1016/j.neuron.2013.06.028

McEwen BS, Bowles NP, Gray JD, Hill MN, Hunter RG, Karatsoreos IN, Nasca C (2015) Mechanisms of stress in the brain. Nat Neurosci 18(10):1353–1363. https://doi.org/10.1038/nn.4086

McGowan JC, Hill C, Mastrodonato A, Lagamma CT, Kitayev A, Brachman RA, Narain NR, Kiebish MA, Denny CA (2018) Prophylactic ketamine alters nucleotide and neurotransmitter metabolism in brain and plasma following stress. Neuropsychopharmacology 43(9):1813–1821. https://doi.org/10.1038/s41386-018-0043-7

McIntyre RS, Rosenblat JD, Nemeroff CB, Sanacora G, Murrough JW, Berk M, Brietzke E, Dodd S, Gorwood P, Ho R, Iosifescu DV, Jaramillo CL, Kasper S, Kratiuk K, Lee JG, Lee Y, Lui LMW, Mansur RB, Papakostas GI et al (2021) Synthesizing the evidence for ketamine and esketamine in treatment-resistant depression: an international expert opinion on the available evidence and implementation. Am J Psychiatr 178(5):383–399. https://doi.org/10.1176/APPI.AJP.2020.20081251

Miller OH, Yang L, Wang CC, Hargroder EA, Zhang Y, Delpire E, Hall BJ (2014) GluN2B-containing NMDA receptors regulate depression-like behavior and are critical for the rapid antidepressant actions of ketamine. elife 2014(3):1–3. https://doi.org/10.7554/eLife.03581

Miller OH, Moran JT, Hall BJ (2016) Two cellular hypotheses explaining the initiation of ketamine's antidepressant actions: direct inhibition and disinhibition. Neuropharmacology 100:17–26. https://doi.org/10.1016/j.neuropharm.2015.07.028

Mitchell JM, Bogenschutz M, Lilienstein A, Harrison C, Kleiman S, Parker-Guilbert K, Ot'alora GM, Garas W, Paleos C, Gorman I, Nicholas C, Mithoefer M, Carlin S, Poulter B, Mithoefer A, Quevedo S, Wells G, Klaire SS, van der Kolk B et al (2021) MDMA-assisted therapy for severe PTSD: a randomized, double-blind, placebo-controlled phase 3 study. Nat Med 27(6):1025–1033. https://doi.org/10.1038/s41591-021-01336-3

Mkrtchian A, Evans JW, Kraus C, Yuan P, Kadriu B, Nugent AC, Roiser JP, Zarate CA (2020) Ketamine modulates fronto-striatal circuitry in depressed and healthy individuals. Mol Psychiatry. https://doi.org/10.1038/s41380-020-00878-1

Moaddel R, Shardell M, Khadeer M, Lovett J, Kadriu B, Ravichandran S, Morris PJ, Yuan P, Thomas CJ, Gould TD, Ferrucci L, Zarate CA (2018) Plasma metabolomic profiling of a ketamine and placebo crossover trial of major depressive disorder and healthy control subjects. Psychopharmacology 235(10):3017–3030. https://doi.org/10.1007/s00213-018-4992-7

Moda-Sava RN, Murdock MH, Parekh PK, Fetcho RN, Huang BS, Huynh TN, Witztum J, Shaver DC, Rosenthal DL, Alway EJ, Lopez K, Meng Y, Nellissen L, Grosenick L, Milner TA,

Deisseroth K, Bito H, Kasai H, Liston C (2019) Sustained rescue of prefrontal circuit dysfunction by antidepressant-induced spine formation. Science 364(6436). https://doi.org/10.1126/science.aat8078

Moghaddam B, Adams B, Verma A, Daly D (1997) Activation of glutamatergic neurotransmission by ketamine: a novel step in the pathway from NMDA receptor blockade to dopaminergic and cognitive disruptions associated with the prefrontal cortex. J Neurosci 17(8):2921–2927. https://doi.org/10.1523/jneurosci.17-08-02921.1997

Monteggia L, Gideons E, Kavalali E (2013) The role of eEF2 kinase in rapid antidepressant action of ketamine. Biol Psychiatry 73(12):1199–1203. https://doi.org/10.1016/j.biopsych.2012.09.006.The

Moroni F, Cozzi A, Sili M, Mannaioni G (2012) Kynurenic acid: a metabolite with multiple actions and multiple targets in brain and periphery. J Neural Transm 119(2):133–139. https://doi.org/10.1007/s00702-011-0763-x

Muschamp JW, Regina MJ, Hull EM, Winter JC, Rabin RA (2004) Lysergic acid diethylamide and [−]-2,5-dimethoxy-4-methylamphetamine increase extracellular glutamate in rat prefrontal cortex. Brain Res 1023(1):134–140. https://doi.org/10.1016/j.brainres.2004.07.044

Muttoni S, Ardissino M, John C (2019) Classical psychedelics for the treatment of depression and anxiety: a systematic review. J Affect Disord 258(July):11–24. https://doi.org/10.1016/j.jad.2019.07.076

Narendran R, Frankle WG, Keefe R, Ph D, Gil R, Martinez D, Slifstein M, Ph D, Kegeles LS, Ph D, Talbot PS, Huang Y, Ph D, Hwang D, Ph D, Khenissi L, Cooper TB, Laruelle M, Abi-dargham A (2005) Altered prefrontal dopaminergic function in chronic recreational ketamine users. Am J Psychiatr 162(12):2352–2359

Nichols CD, Sanders-Bush E (2002) A single dose of lysergic acid diethylamide influences gene expression patterns within the mammalian brain. Neuropsychopharmacology 26(5):634–642. https://doi.org/10.1016/S0893-133X(01)00405-5

Nichols CD, Garcia EE, Sanders-Bush E (2003) Dynamic changes in prefrontal cortex gene expression following lysergic acid diethylamide administration. Mol Brain Res 111(1–2):182–188. https://doi.org/10.1016/S0169-328X(03)00029-9

Nishitani N, Nagayasu K, Asaoka N, Yamashiro M, Shirakawa H, Nakagawa T, Kaneko S (2014) Raphe AMPA receptors and nicotinic acetylcholine receptors mediate ketamine-induced serotonin release in the rat prefrontal cortex. Int J Neuropsychopharmacol 17(8):1321–1326. https://doi.org/10.1017/S1461145714000649

Olson DE (2018) Psychoplastogens: a promising class of plasticity-promoting neurotherapeutics. J Exp Neurosci 12. https://doi.org/10.1177/1179069518800508

Palhano-Fontes F, Barreto D, Onias H, Andrade KC, Novaes MM, Pessoa JA, Mota-Rolim SA, Osório FL, Sanches R, Dos Santos RG, Tófoli LF, De Oliveira Silveira G, Yonamine M, Riba J, Santos FR, Silva-Junior AA, Alchieri JC, Galvão-Coelho NL, Lobão-Soares B et al (2019) Rapid antidepressant effects of the psychedelic ayahuasca in treatment-resistant depression: a randomized placebo-controlled trial. Psychol Med 49(4):655–663. https://doi.org/10.1017/S0033291718001356

Park LT, Kadriu B, Gould TD, Zanos P, Greenstein D, Evans JW, Yuan P, Farmer CA, Oppenheimer M, George JM, Adeojo LW, Snodgrass HR, Smith MA, Henter ID, MacHado-Vieira R, Mannes AJ, Zarate CA (2020) A randomized trial of the N-methyl-d-aspartate receptor glycine site antagonist prodrug 4-chlorokynurenine in treatment-resistant depression. Int J Neuropsychopharmacol 23(7):417–425. https://doi.org/10.1093/IJNP/PYAA025

Pasqualetti M, Nardi I, Ladinsky H, Marazziti D, Cassano GB (1996) Comparative anatomical distribution of serotonin 1A, 1Dα and 2A receptor mRNAs in human brain postmortem. Mol Brain Res 39(1–2):223–233. https://doi.org/10.1016/0169-328X(96)00026-5

Patel R, Titheradge D (2015) MDMA for the treatment of mood disorder: all talk no substance? Ther Adv Psychopharmacol 5(3):179–188. https://doi.org/10.1177/2045125315583786

Pazos A, Cortés R, Palacios JM (1985) Quantitative autoradiographic mapping of serotonin receptors in the rat brain. II. Serotonin-2 receptors. Brain Res 346(2):231–249. https://doi.org/10.1016/0006-8993(85)90857-1

Preller KH, Herdener M, Pokorny T, Planzer A, Kraehenmann R, Stämpfli P, Liechti ME, Seifritz E, Vollenweider FX (2017) The fabric of meaning and subjective effects in LSD-induced states depend on serotonin 2A receptor activation. Curr Biol 27(3):451–457. https://doi.org/10.1016/j.cub.2016.12.030

Razoux F, Garcia R, Léna I (2007) Ketamine, at a dose that disrupts motor behavior and latent inhibition, enhances prefrontal cortex synaptic efficacy and glutamate release in the nucleus accumbens. Neuropsychopharmacology 32(3):719–727. https://doi.org/10.1038/sj.npp.1301057

Ren Z, Pribiag H, Jefferson SJ, Shorey M, Fuchs T, Stellwagen D, Luscher B (2016) Bidirectional homeostatic regulation of a depression-related brain state by gamma-aminobutyric acidergic deficits and ketamine treatment. Biol Psychiatry 80(6):457–468. https://doi.org/10.1016/j.biopsych.2016.02.009

Rosenbaum G (1959) Comparison of sernyl with other drugs. A.M.A. Arch Gen Psychiatry 1(6). https://doi.org/10.1001/archpsyc.1959.03590060113013

Rudnick G, Wall SC (1992) The molecular mechanism of "ecstasy" [3,4-methylenedioxy-methamphetamine (MDMA)]: serotonin transporters are targets for MDMA-induced serotonin release. Proc Natl Acad Sci U S A 89(5):1817–1821. https://doi.org/10.1073/pnas.89.5.1817

Sallee FR, Vrindavanam NS, Deas-Nesmith D, Carson SW, Sethuraman G (1997) Pulse intravenous clomipramine for depressed adolescents: double- blind, controlled trial. Am J Psychiatr 154(5):668–673. https://doi.org/10.1176/ajp.154.5.668

Scruggs JL, Schmidt D, Deutch AY (2003) The hallucinogen 1-[2,5-dimethoxy-4-iodophenyl]-2-aminopropane (DOI) increases cortical extracellular glutamate levels in rats. Neurosci Lett 346(3):137–140. https://doi.org/10.1016/S0304-3940(03)00547-0

Sen S, Duman R, Sanacora G (2008) Serum brain-derived neurotrophic factor, depression, and antidepressant medications: meta-analyses and implications. Biol Psychiatry 64(6):527–532. https://doi.org/10.1016/j.biopsych.2008.05.005

Sheline YI, Wang PW, Gado MH, Csernansky JG, Vannier MW (1996) Hippocampal atrophy in recurrent major depression. Proc Natl Acad Sci U S A 93(9):3908–3913. https://doi.org/10.1073/pnas.93.9.3908

Shimizu E, Hashimoto K, Okamura N, Koike K, Komatsu N, Kumakiri C, Nakazato M, Watanabe H, Shinoda N, Okada SI, Iyo M (2003) Alterations of serum levels of brain-derived neurotrophic factor (BDNF) in depressed patients with or without antidepressants. Biol Psychiatry 54(1):70–75. https://doi.org/10.1016/S0006-3223(03)00181-1

Shinohara R, Taniguchi M, Ehrlich AT, Yokogawa K, Deguchi Y, Cherasse Y, Lazarus M, Urade Y, Ogawa A, Kitaoka S, Sawa A, Narumiya S, Furuyashiki T (2018) Dopamine D1 receptor subtype mediates acute stress-induced dendritic growth in excitatory neurons of the medial prefrontal cortex and contributes to suppression of stress susceptibility in mice. Mol Psychiatry 23(8):1717–1730. https://doi.org/10.1038/mp.2017.177

Short B, Fong J, Galvez V, Shelker W, Loo CK (2018) Side-effects associated with ketamine use in depression: a systematic review. Lancet Psychiatry 5(1):65–78. https://doi.org/10.1016/S2215-0366(17)30272-9

Silva Pereira V, Elfving B, Joca SRL, Wegener G (2017) Ketamine and aminoguanidine differentially affect Bdnf and Mtor gene expression in the prefrontal cortex of adult male rats. Eur J Pharmacol 815(May):304–311. https://doi.org/10.1016/j.ejphar.2017.09.029

Skolnick P, Layer RT, Popik P, Nowak G, Paul IA, Trullas R (1996) Adaptation of N-methyl-D-aspartate (NMDA) receptors following antidepressant treatment: implications for the pharmacotherapy of depression. Pharmacopsychiatry 29(1):23–26. https://doi.org/10.1055/s-2007-979537

Spies M, James GM, Berroterán-Infante N, Ibeschitz H, Kranz GS, Unterholzner J, Godbersen M, Gryglewski G, Hienert M, Jungwirth J, Pichler V, Reiter B, Silberbauer L, Winkler D,

Mitterhauser M, Stimpfl T, Hacker M, Kasper S, Lanzenberger R (2018) Assessment of ketamine binding of the serotonin transporter in humans with positron emission tomography. Int J Neuropsychopharmacol 21(2):145–153. https://doi.org/10.1093/ijnp/pyx085

Sun X, Zhao Y, Wolf ME (2005) Dopamine receptor stimulation modulates AMPA receptor synaptic insertion in prefrontal cortex neurons. J Neurosci 25(32):7342–7351. https://doi.org/10.1523/JNEUROSCI.4603-04.2005

Thoenen H (1995) Neurotrophins and neuronal plasticity. Science 270(5236):593–598. https://doi.org/10.1126/science.270.5236.593

Treccani G, Ardalan M, Chen F, Musazzi L, Popoli M, Wegener G, Nyengaard JR, Müller HK (2019) S-ketamine reverses hippocampal dendritic spine deficits in flinders sensitive line rats within 1 h of administration. Mol Neurobiol 56(11):7368–7379. https://doi.org/10.1007/s12035-019-1613-3

U.S. Food and Drug Administration (2019) FDA approves first treatment for post-partum depression

Valle M, Maqueda AE, Rabella M, Rodríguez-Pujadas A, Antonijoan RM, Romero S, Alonso JF, Mañanas MT, Barker S, Friedlander P, Feilding A, Riba J (2016) Inhibition of alpha oscillations through serotonin-2A receptor activation underlies the visual effects of ayahuasca in humans. Eur Neuropsychopharmacol 26(7):1161–1175. https://doi.org/10.1016/j.euroneuro.2016.03.012

Vécsei L, Szalárdy L, Fülöp F, Toldi J (2013) Kynurenines in the CNS: recent advances and new questions. Nat Rev Drug Discov 12(1):64–82. https://doi.org/10.1038/nrd3793

Videbech P, Ravnkilde B (2004) Reviews and overviews hippocampal volume and depression: a meta-analysis of MRI studies. Am J Psychiatr 161:1957–1966

VistaGen Therapeutics (2019) VistaGen reports topline phase 2 results for AV-101 as an adjunctive treatment of major depressive disorder. https://www.prnewswire.com/news-releases/vistagen-reports-topline-phase-2-results-for-av-101-as-an-adjunctive-treatment-of-major-depressive-disorder-300958317.html

Vollenweider FX, Kometer M (2010) The neurobiology of psychedelic drugs: implications for the treatment of mood disorders. Nat Rev Neurosci 11(9):642–651. https://doi.org/10.1038/nrn2884

Vollenweider FX, Vollenweider-Scherpenhuyzen MFI, Bäbler A, Vogel H, Hell D (1998) Psilocybin induces schizophrenia-like psychosis in humans via a serotonin-2 agonist action. Neuroreport 9(17):3897–3902. https://doi.org/10.1097/00001756-199812010-00024

Wajs E, Aluisio L, Holder R, Daly EJ, Lane R, Lim P, George JE, Morrison RL, Sanacora G, Young AH (2020) Esketamine nasal spray plus oral antidepressant in patients with treatment-resistant depression: assessment of long-term safety in a phase 3, open-label study (SUSTAIN-2). J Clin Psychiatry 81(3). https://doi.org/10.4088/JCP.19M12891

Walker AK, Budac DP, Bisulco S, Lee AW, Smith RA, Beenders B, Kelley KW, Dantzer R (2013) NMDA receptor blockade by ketamine abrogates lipopolysaccharide-induced depressive-like behavior in C57BL/6J mice. Neuropsychopharmacology 38(9):1609–1616. https://doi.org/10.1038/npp.2013.71

Weber ET, Andrade R (2010) Htr2a gene and 5-HT2A receptor expression in the cerebral cortex studied using genetically modified mice. Front Neurosci 4(AUG):1–12. https://doi.org/10.3389/fnins.2010.00036

Whiteford HA, Degenhardt L, Rehm J, Baxter AJ, Ferrari AJ, Erskine HE, Charlson FJ, Norman RE, Flaxman AD, Johns N, Burstein R, Murray CJL, Vos T (2013) Global burden of disease attributable to mental and substance use disorders: findings from the global burden of disease study 2010. Lancet 382(9904):1575–1586. https://doi.org/10.1016/S0140-6736(13)61611-6

Witkin JM, Monn JA, Schoepp DD, Li X, Overshiner C, Mitchell SN, Carter G, Johnson B, Rasmussen K, Rorick-Kehn LM (2016) The rapidly acting antidepressant ketamine and the mGlu2/3 receptor antagonist LY341495 rapidly engage dopaminergic mood circuits. In. J Pharmacol Exp Ther 358(1). https://doi.org/10.1124/jpet.116.233627

World Health Organization, Geneva (2019) World Health Organization model list of essential medicines, 21st List, 2019. https://www.who.int/publications/i/item/WHOMVPEMPIAU2019.06

Zanos P, Moaddel R, Morris PJ, Georgiou P, Fischell J, Elmer GI, Alkondon M, Yuan P, Pribut HJ, Singh NS, Dossou KSS, Fang Y, Huang XP, Mayo CL, Wainer IW, Albuquerque EX, Thompson SM, Thomas CJ, Zarate CA, Gould TD (2016) NMDAR inhibition-independent antidepressant actions of ketamine metabolites. Nature 533(7604):481–486. https://doi.org/10.1038/nature17998

Zanos P, Highland JN, Stewart BW, Georgiou P, Jenne CE, Lovett J, Morris PJ, Thomas CJ, Moaddel R, Zarate CA, Gould TD (2019) (2R,6R)-hydroxynorketamine exerts mGlu2 receptordependent antidepressant actions. Proc Natl Acad Sci U S A 116(13):6441–6450. https://doi.org/10.1073/pnas.1819540116

Zarate CA, Singh JB, Carlson PJ, Brutsche NE, Ameli R, Luckenbaugh DA, Charney DS, Manji HK (2006) A randomized trial of an N-methyl-D-aspartate antagonist in treatment-resistant major depression. Arch Gen Psychiatry 63(8). https://doi.org/10.1001/archpsyc.63.8.856

Zhang C, Marek GJ (2008) AMPA receptor involvement in 5-hydroxytryptamine2A receptor-mediated pre-frontal cortical excitatory synaptic currents and DOI-induced head shakes. Prog Neuro-Psychopharmacol Biol Psychiatry 32(1):62–71. https://doi.org/10.1016/j.pnpbp.2007.07.009

Zhang JC, Li SX, Hashimoto K (2014) R (−)-ketamine shows greater potency and longer lasting antidepressant effects than S (+)-ketamine. Pharmacol Biochem Behav 116:137–141. https://doi.org/10.1016/j.pbb.2013.11.033

The Potential of Psychedelics for End of Life and Palliative Care

David B. Yaden, Sandeep M. Nayak, Natalie Gukasyan, Brian T. Anderson, and Roland R. Griffiths

Contents

Abstract End of life and palliative care has improved in recent decades but the psychopharmacological options available to clinicians and patients in these contexts remain limited. In particular, psychological factors such as depression, existential distress, and well-being remain challenging to address with current medications. Here, we review recent research on the use of psychedelics in clinical settings with a particular focus on patients with life-threatening diagnoses. We propose that psychedelics may provide clinicians with an additional psychopharmacological treatment in the context of end of life and palliative care.

Keywords Psychedelics · Psilocybin · End of Life · Palliative care · Psychiatry

---

D. B. Yaden (✉), S. M. Nayak, and N. Gukasyan
Department of Psychiatry and Behavioral Sciences, Center for Psychedelic and Consciousness Research, Johns Hopkins University School of Medicine, Baltimore, MD, USA
e-mail: dyaden1@jhmi.edu; smn@jhmi.edu; gukasyan@jhmi.edu

B. T. Anderson
Department of Psychiatry & Behavioral Sciences, Zuckerberg San Francisco General Hospital, UCSF, San Francisco, CA, USA
e-mail: brian.anderson@ucsf.edu

R. R. Griffiths
The Oliver Lee McCabe, III Professor in the Neuropsychopharmacology of Consciousness, Department of Psychiatry and Behavioral Sciences, Center for Psychedelic and Consciousness Research, Department of Neuroscience, Johns Hopkins University School of Medicine, Baltimore, MD, USA
e-mail: rgriff@jhmi.edu

Curr Topics Behav Neurosci (2022) 56: 169–184
https://doi.org/10.1007/7854_2021_278
Published Online: 28 December 2021

Medical advances have made it possible to better manage many of the discomforts involved in dying, but there are still few medications available to address the accompanying psychological distress. In *Being Mortal* (2014), physician Atul Gawande argues that contemporary society may be somewhat historically anomalous insofar as there are few well-known contemporary norms or guidelines to dealing with the psychological side of dying. Gawande points to the historical examples of books called *Ars Moriendi* (Art of Dying; Shinners 1997) that were popular in medieval Europe as well as the *Tibetan Book of the Dead* and the *Egyptian Book of the Dead* that each provide cultural forms of instruction on how to accept the psychological aspects of dying. Gawande (2014, 2016) observes that the modern hospice movement provides psychological support at end of life through an interdisciplinary clinical team who are tasked with holistically addressing the biopsychosocial aspects of dying.

Such services both increase the quality of life and extend life in several terminal illnesses (Connor et al. 2007), contrary to impressions that hospice care reduces longevity (i.e., "giving up"). Despite their demonstrated value (for a review, see Connor et al. 2007), these services are chronically underutilized (Gawande 2016). The proliferation of hospice services resulted in the broader palliative care movement, a specialty focused on reducing suffering and improving well-being for patients with serious, chronic, or life-threatening illnesses or injuries in general (WHO 2011). End of life and palliative care (EOLPC) is a quickly growing medical specialty, which addresses pain and symptom management among other biopsychosocial concerns (Aziz et al. 2013).

EOLPC, while valuable, is limited by the psychopharmacological treatments available. Here, we review research on the efficacy of psychedelic treatments in the context EOLPC. A number of clinical trials with psilocybin have found decreased depression and anxiety as well as increased well-being in psychologically distressed patients who had a life-threatening diagnosis. In this chapter, we suggest that psychedelics could provide a novel psychopharmacological treatment capable of reducing psychological distress and supporting the psychological well-being of actively dying patients and, more generally, in those receiving palliative care.

1 Contemporary End of Life and Palliative Care

End of life care represents a serious economic issue in contemporary healthcare and poor care can be a source of needless suffering for patients (Gawande 2014). In response to these issues, the now worldwide hospice movement was created by English nurse Cicely Saunders to address the psychological suffering of actively dying patients (Connor 1998; Saunders 1978). Despite their value, most patients are either not being referred or failing to avail themselves of these services until mere days before death (Finucane 1999). Some of the many institutional, cultural, and psychological reasons for failing to utilize these services may be due to an inability of the clinician to accept the seriousness of the diagnosis (or to see one's role as intervening to prolong life in all cases). Likewise, patients may fail to understand the implications of their diagnosis or to appreciate their ability to manage aspects of their own death. Patients may also believe that requesting such services would let their family down (for a thoughtful discussion, see Feudtner 2009).

The subject of how one wants to die is frequently avoided, but, when asked, people express definite preferences. In general, people want to be relatively free from pain, be surrounded by loved ones, and to feel a degree of meaning and well-being throughout the dying process (e.g., Singer et al. 1999). When patients, family members, physicians, and other care providers were surveyed about what is valued most while dying, all four of these groups indicated at rates above 90% that freedom from both pain and anxiety were important attributes of end of life care (Steinhauser et al. 2000). A substantial subset of these patients also indicated the importance of addressing religious, spiritual, and existential concerns and well-being (Steinhauser et al. 2000). Additionally, most people say that they want to engage in meaningful discussions with loved ones and feel a sense of meaning, but many people experience psychological suffering that prevents such interpersonal connection (e.g., Gruneir et al. 2007).

There are a number of psychological services available to address distress in palliative care patients, such as psychotherapy, social services, access to chaplains, and integrative medicine modalities (Gawande 2014, 2016). In addition to psychotherapies like cognitive-behavioral therapy (CBT) and acceptance and commitment therapy (ACT), there are several manualized, evidence-based psychotherapies available that are tailored to the needs of EOLPC patients, such as existential interventions focused on meaning and purpose (Bauereiß et al. 2018; Park et al. 2019) and dignity therapy (Li et al. 2020). While effective to varying extents, these psychosocial treatments could be complemented by psychopharmacological treatments to enhance outcomes.

In addition to psychosocial therapies, psychopharmacological treatment currently provides an important but understudied part of end of life care. Standard of care calls for individualized assessment of treatment needs in terms of pain management, depression, anxiety, appetite, nausea, and drowsiness (Bruera et al. 1991). Opioids are routinely used for pain management (Quigley 2008). Cannabinoids such as dronabinol may be used to stimulate appetite, reduce nausea, and mitigate anxiety (Mücke et al. 2018). Commonly used drugs for these indications include serotonergic antidepressants, sedative hypnotics, stimulants, and neuroleptics (Candy et al. 2012; Ostuzzi et al. 2018) (for a review, see Grassi and Riba 2014). Studies of antidepressants in palliative care populations demonstrate small to moderate effect sizes (e.g., Rayner et al. 2011).

In general, existing pharmacotherapies can produce some symptomatic relief for patients, but treatment often involves unwanted side effects such as decreased levels of alertness, memory problems, and impaired coordination (Grassi and Riba 2014). While the pharmacotherapies available in end-of-life contexts are largely effective at managing pain, they are less effective at managing depression (Grassi and Riba 2014). Furthermore, existential distress and patient well-being are currently only indirectly impacted by existing medications to a small degree. Additional psychopharmacological treatments would be valuable in the EOLPC context.

2 Classic Psychedelics

Classic psychedelic drugs (previously called "hallucinogens") may prove capable of effectively addressing psychological needs in end of life care. The psychedelics are a group of compounds whose action are mediated at the 5-HT_{2A} receptor and that produce substantial changes in perception, affect, and cognition, often accompanied by a profound sense of personal meaning (Nichols 2016; Vollenweider and Preller 2020). The best known of these compounds are psilocybin, LSD, DMT (and the DMT-containing plant brew ayahuasca), and mescaline. Psychedelics have been used in ritual and religious contexts across a number of cultures over hundreds if not thousands of years (Schultes and Hofmann 1992). Psychedelics were studied in the 1950s and 1960s before burdensome governmental regulations halted research until around the year 2000 (Johnson et al. 2019).

In this review, we will focus on psilocybin which has been studied more than other classic psychedelics in clinical trials. Psilocybin (4-phosphoryloxy-N,N-dimethyltryptamine) is generally similar to serotonin (5-hydroxytryptamine) with regard to chemical structure and binding activity (Nichols 2016). Psilocybin is generally safe, well-tolerated and has limited addiction or abuse potential (Johnson et al. 2019; Nutt et al. 2010; Vollenweider and Preller 2020; Nichols 2016). Although physically safe, psilocybin experiences can be extremely psychologically challenging. Some people rate their psilocybin experiences among the most challenging of their life; however, these same individuals may nevertheless claim that the experience was meaningful and beneficial (Carbonaro et al. 2016). In clinical settings with therapeutic support, persisting adverse effects have been very limited (Johnson et al. 2019).

The subjective states associated with psilocybin have been characterized a variety of different ways, with increasing convergence across psychometric self-report instruments and qualitative research. The Five-Dimensional States of Consciousness (5-DASC; Dittrich et al. 2010; Studerus et al. 2011) assesses several dimensions of changes to subjectivity that occur from psilocybin, including oceanic boundlessness, anxious ego dissolution, and complex imagery. Among the most therapeutically relevant mental states that psilocybin produces can be more parsimoniously described as a "mystical-type" experience, an altered state of consciousness classically described by William James (1902) and elaborated by other scholars such as Stace (1960; for a review, see Yaden et al. 2017a). The mystical experience is most frequently measured in psychedelic research using the Mystical Experience Questionnaire (MEQ30: Barrett et al. 2015), which includes sub-scales to measure a sense of unity, reverence, and authoritative truth, positive emotions, transcendence of time/space, and ineffability.

Psilocybin may also produce experiences of therapeutic relevance characterized as psychological insight (Carbonaro et al. 2020; Davis et al. 2020a, b) and which can be assessed with the Psychological Insight Questionnaire (Davis et al. 2020c). Notably, the majority of participants at Johns Hopkins who have experienced high-dose psilocybin in clinical research report that this experience is among the most meaningful of their entire lives (Griffiths et al. 2006, 2011, 2016, 2018). Overall, there is substantial evidence that the subjective effects of psychedelics are an important factor in their therapeutic effects (see Yaden and Griffiths 2020).

Among the first studies in the contemporary era of clinical psychedelic research was one that administered psilocybin to healthy psychedelic-naïve participants and examined changes to well-being (Griffiths et al. 2006). This study compared psilocybin to an active control (methylphenidate) condition in a randomized controlled trial (RCT). Results from this study showed large improvements in various measures of well-being such as mood, life satisfaction, relationships, and meaning, which persisted for more than a year, and were mediated by the degree of psilocybin-associated mystical experience (Griffiths et al. 2006, 2008).

In addition to the aspects of well-being mentioned above, in a number of clinical trials with psilocybin, participants have reported enhanced spiritual well-being as a persisting positive effect from their experience (Griffiths et al. 2006, 2008, 2011, 2016, 2018). The improvements to spiritual forms of well-being is of particular relevance for end of life contexts, as patients report preferring that this psychological domain is addressed while dying (Steinhauser et al. 2000). Spirituality has been defined in a number of ways, and while religious and otherwise supernatural concepts are commonly part of such definitions, supernatural beliefs need not necessarily be part of spirituality (Yaden et al. 2018, 2021a, b). For example, the Death Transcendence scale (Hood and Morris 1983) measures the extent to which one believes that one's self will survive beyond bodily death through several different possible means: the memories of family and friends, the work that one has contributed to society, by becoming part of nature, religious/spiritual conceptions of the afterlife, and/or through a sense of unity with all things. This measure is one way of conceptualizing well-being and a healthy cognitive mindset regarding one's own death in a way that could be considered broadly "spiritual" but without necessarily including supernatural concepts. Griffiths et al. (2011, 2018) showed that a measure of death transcendence was increased after psilocybin.

Psilocybin has shown promise for treating several disorders spanning several diagnostic categories. An open-label trial ($N = 26$) demonstrated initial safety and feasibility of addressing treatment-resistant depression with up to 25 mg of oral psilocybin with psychological support (Carhart-Harris et al. 2018). A subsequent RCT ($N = 24$) showed marked decreases in depression among moderately to severely depressed participants compared to a waitlist control using a similar intervention (Davis et al. 2020b). A more recent head-to-head RCT ($N = 59$) provided data suggesting that psychological support plus two doses of psilocybin 25 mg was not superior than psychological support plus daily escitalopram (a widely used serotonin reuptake inhibitor) on the primary endpoint assessment (Carhart-Harris et al. 2021). While the secondary outcomes favored psilocybin, these were not corrected for multiple comparisons so must be cautiously interpreted as exploratory findings (Carhart-Harris et al. 2021). There appears to be some trans-diagnostic efficacy with psilocybin, as preliminary data also suggest the potential for demonstrating efficacy in the treatment of substance use disorders (Johnson et al. 2014; Bogenschutz et al. 2015) and possibly obsessive-compulsive disorder (OCD) (Moreno et al. 2006).

In summary, psychedelics are a class of generally well-tolerated and largely non-addictive psychoactive substances that have demonstrated therapeutic or otherwise positive effects under a number of experimental conditions. There is evidence of potential efficacy across a range of psychiatric disorders and psychedelics are currently being tested for a wider range of applications, including EOLPC contexts.

3 Psychedelics in Palliative Care and End of Life Contexts

Beyond increasing well-being in healthy volunteers and reducing mood and substance use disorders in clinical populations, psychedelics have been specifically examined in the context of coping with a life-threatening cancer diagnosis (Grob et al. 2011; Griffiths et al. 2016; Ross et al. 2016). Thus, there is evidence bearing directly on our primary topic, which we review in more detail below.

Palliative care and end of life contexts were among the first considered clinical use cases of psychedelics in the previous wave of research (Kast 1962, 1964; Cohen 1965; Pahnke 1969; Fisher 1970; Grof et al. 1973; Richards et al. 1977). Kast (1966, 1967) conducted the first two studies of LSD in patients who were terminally ill. In the first, 80 patients with terminal cancer and a life expectancy of weeks to months were administered 100 mcg IM LSD under open-label conditions (Kast 1966). Patients reported an improvement in mood that persisted about 10 days before declining again. A follow-up study involved treatment of 128 patients with similar inclusion criteria and design (Kast 1967). Several were quite ill inpatients, with six dying in the one-week observation period before drug administration. This follow-up study showed a transient elevation of mood, and improved attitudes toward death that were evident at 3 days, but not 10 days (Kast 1967).

Grof et al. (1973) reported on 31 cancer patients with at least 3 months life expectancy who received open-label LSD PO 200–500 mcg under supportive conditions with preparatory and integration sessions. This report showed statistically significant baseline to post-treatment improvements in depression, fear of death, and isolation following the experience, but did not assess how durable these were.

In the contemporary era of psychedelic research (see Table 1), Gasser et al. (2014) performed the only modern trial of LSD in patients with diagnoses of life-threatening illnesses. Patients were required to have an advanced-stage potentially fatal illness with a probability of survival >6 months and meet criteria for a DSM-IV anxiety disorder or score ≥40 on either the state or trait scale of the State-Trait Anxiety Inventory. Of 12 patients, five met criteria for GAD, 6 for MDD, 1 for dysthymia, 1 for PTSD, and 2 for panic disorder (these were not mutually exclusive). Patients

Table 1 Psychedelic studies relevant to end of life and palliative care

Study	Participants	Design	Drug condition (s)	Primary outcomes
Grob et al. (2011)	$N = 12$ Axis 1 diagnoses: acute stress disorder, GAD, anxiety disorder due to cancer, adjustment disorder Axis 3 diagnoses: advanced-stage cancer (100%)	Double-blind, randomized placebo-controlled crossover	Psilocybin (0.2 mg/kg) Niacin 250 mg	Feasibility and safety: mild increases in heart rate and blood pressure during psilocybin session. No significant differences before crossover in BDI, STAI, or POMS at 2 weeks
Gasser et al. (2014)	$N = 12$ Axis 1 diagnoses: GAD (54.5%); MDD (63.6%); panic disorder (27.3%); dysthymia (18.2%); PTSD (8.3%) Axis 3 diagnoses: life-threatening diseases including metastatic carcinomas ($N = 50\%$), other malignancy ($N = 17\%$), celiac disease (8%), Parkinson's disease (8%), Bechterew's disease (8%)	Double-blind, randomized active placebo-controlled crossover	LSD 200 μg LSD 20 μg	At 2 months post first session, 200 μg LSD group had: reduction in trait anxiety ($p = 0.03$, $d = 1.1$), reduction in state anxiety ($p = 0.021$, $d = 1.2$)
Ross et al. (2016)	$N = 29$ Axis 1 diagnoses: adjustment disorder (90%); GAD (10%) Axis 3 diagnoses: stage III or IV cancer (62%); other malignancy (38%)	Double-blind, randomized active placebo-controlled crossover	Psilocybin 21 mg/70 kg Niacin 250 mg	Prior to crossover, reductions in the following measures compared to placebo group (Cohen's d effect sizes at 1-day and 7-weeks post session): HADS ($d = 1.4$ and 1.4); BDI ($d = 1.1$ and 0.8); STAI state ($d = 1.2$ and 1.2); STAI trait ($d = 1.0$ and 1.3)
Griffiths et al. (2016)	$N = 51$ Axis 1 diagnoses: adjustment disorder (22/51); dysthymia (5/51); GAD (5/51); MDD (14/51); dual	Randomized, double-blind, crossover	Psilocybin low dose (1 or 3 mg/70 kg) psilocybin high dose (22 or 30 mg/70 kg)	At 1 week post session 1, high-dose first group had significant decreases in all measures including GRID-HAMD

(continued)

Table 1 (continued)

Study	Participants	Design	Drug condition (s)	Primary outcomes
	diagnosis GAD + depressive disorder (5/51) Axis 3 diagnoses: stage III or IV cancer			(d = 1.0), BDI (d = 1.4), HAM-A (d = 1.2) at 6 months post baseline, the entire group (collapsed across conditions) had significant decreases in all measures including GRID-HAMD (d = 3.0), BDI (d = 1.6), HAM-A (d = 3.4)
Anderson et al. (2020)	$N = 18$ (all men ≥50 years old) DSM-5 diagnoses: GAD (7/18); MDD (5/18); panic disorder (3/18); borderline personality disorder (3/18) Current general medical condition: HIV (18/18), metastatic malignancy (1/18)	Open-label, single-arm. Preparation and integration done as group therapy	Psilocybin 0.3 mg/kg or 0.36 mg/kg po	17/18 completed intervention with 1 participant discontinuing treatment due to a study-related adverse event. Zero study-related serious adverse events were detected. Pre-post (baseline to 3 months) resulted in a clinically significant change in demoralization ($\eta_p^2 = 0.47$, 90% CI 0.21–0.60)

were randomized to two sessions of 200 mcg or 20 mcg LSD. Those who received active placebo had the option to later receive open-label 200 mcg LSD. The study showed significant decreases in anxiety within the high-dose group ($n = 8$) from pretreatment to the 2-month follow-up with a large effect size of 1.1. In contrast, the low dose group ($n = 4$) demonstrated an increase in anxiety over that same time period. Between the two groups, state anxiety was statistically significantly lower in the high-dose group and trait anxiety was non-significantly lower at 2 months.

In contemporary research with psilocybin, at UCLA, Grob et al. (2011) conducted a study with patients who had been diagnosed with advanced-stage terminal cancers (prognoses of 6 months to 1 year) who also had DSM-IV diagnoses of acute stress disorder, GAD, anxiety disorder due to cancer or adjustment disorder with anxiety. The study was a placebo-controlled RCT within-subject crossover with 12 participants. Niacin was used as placebo and 0.2 mg/kg psilocybin (14 mg for a

70 kg person – a modest dose) was used as the active dose. While this pilot study established the safety of treating anxiety in advanced cancer patients with oral psilocybin, there were no statistically significant group differences in anxiety or depression at follow-up timepoints.

At Johns Hopkins, Griffiths et al. (2016) conducted a larger study ($N = 51$) also in a population of patients who had received a life-threatening cancer diagnosis. This trial included patients with an active cancer (e.g., stage III or IV) with a poor prognosis or disease progression or recurrence ($n = 33$) or the possibility of recurrence ($n = 18$). In addition, they had to have a DSM-IV diagnosis of GAD, acute stress disorder, PTSD, mild or moderate MDD, dysthymic disorder, or adjustment disorder (with a variety of qualifiers). This study compared a very low placebo-like dose of psilocybin (1–3 mg/70 kg) to a large dose of psilocybin (22–30 mg/70 kg of body weight). Participants in the high-dose psilocybin group, compared to the placebo-like control group, reported higher levels of well-being as well as lower levels of anxiety and depression at 5 weeks. For the majority of the sample (80%), these changes persisted for 6 months. This study also included observer ratings of the participant who were blinded to condition, and these observers (e.g., friends, neighbors) reported improvements in participants who had received psilocybin. As has been reported in several psilocybin trials, self-reported subjective qualities of the drug administration session predicted positive persisting effects (Yaden and Griffiths 2020).

At NYU, Ross et al. (2016) conducted a study with participants ($N = 29$) who had a life-threatening cancer diagnosis and a DSM-IV diagnosis of acute stress disorder, GAD, or adjustment disorder with anxiety and/or depression. This study initially began recruiting terminally ill patients with stage IV cancer, but later broadened the inclusion criteria to include participants in remission. Ninety percent of patients met criteria for Adjustment Disorder, and the remaining 10% did for GAD (Ross et al. 2016). In this randomized placebo-controlled crossover study, participants who received a single session of psilocybin 0.3 mg/kg (e.g., 21 mg for a 70 kg person) showed reduced anxiety, depression, and cancer-related demoralization, compared to a niacin placebo group. These findings persisted at 6-month follow-up. This study also found improvements in demoralization and hopelessness, constructs highly relevant to end of life contexts. At about 6 months after the study, the majority of the sample (70%) indicated that their psilocybin session was among the top five most meaningful experiences of their life (Agin-Liebes et al. 2020). Many of these findings persisted at 4.5-year follow-up (Agin-Liebes et al. 2020).

At UCSF, Anderson et al. (2020) conducted a pilot study ($N = 18$, in 3 cohorts of 6) of psilocybin-assisted group therapy for older long-term AIDS survivor (LTAS) men with moderate-to-severe demoralization. Such individuals live with a chronic life-threatening illness (i.e., HIV), and many have been acutely ill at various times in their disease course. Of the enrolled participants, baseline evaluation found that 7 met SCID-5 criteria for GAD, 5 for MDD, 3 for borderline personality disorder, and 6 had a history of a life-threatening malignancy. Participants underwent 4 pre-drug and 4–6 post-drug group therapy sessions; psilocybin was administered individually (without other group members present) at 0.3 mg/kg po to 7 participants, and then 0.36 mg/kg to the remaining 11 participants. Feasibility was demonstrated and the intervention was found to be relatively safe with no psilocybin-related serious adverse events detected in the trial, although 2 unexpected adverse reactions occurred, 1 participant discontinued treatment due to an adverse reaction, and 14 participants experienced adverse reactions that were at least moderate in severity. Exploratory pre-post analysis found an improvement in demoralization from baseline to 3-month follow-up with a mean difference of −5.8 (SD 6.0) and an effect size of $\eta_p^2 = 0.47$, 90% CI 0.21–0.60.

The safety, feasibility, and clinical potential demonstrated in these three recent studies with psilocybin and one recent study with LSD continue to be evaluated in ongoing research. It will be important to better understand therapeutic mechanisms, contraindications, and optimal dosing and psychological context conditions. Because psilocybin and LSD administration may produce an intense and challenging psychological experiences with low probability but significant risks (Johnson et al. 2008; Carbonaro et al. 2016), it is important to proceed with caution. It nevertheless appears likely, assuming that additional studies result in similar findings, that psilocybin may be an effective medication for palliative care and end of life contexts.

4 Clinical Considerations for Psychedelics in End of Life and Palliative Care

Psychedelic substances have the potential to be a powerful tool in the context of EOLPC. However, the nature of these substances raises a number of clinical considerations including both opportunities and challenges.

One area of concern is the risks associated with psychedelics in the context of common physical symptoms and medical conditions in palliative care populations. Much of the research on psychedelic-assisted treatment to date, even in patients with advanced cancer, has been in relatively medically stable individuals who are able to engage in outpatient care. In a hospice setting, it is possible that psychedelics may exacerbate nausea or diarrhea, breathlessness, or insomnia. Of particular importance is whether psychedelics may worsen or precipitate delirium in vulnerable patients. Barrett et al. (2018) found that global cognitive impairment was not observed in healthy volunteers at doses of psilocybin up to 30 mg/70 kg. However, deficits in individual cognitive domains were present and dose-dependent. Such impairments may be more pronounced in palliative care populations who are more at risk of developing delirium.

A second area of concern is the safety of psychedelics when co-administered with other medications commonly used at the end of life. Serotonergic antidepressants are typically contraindicated for co-administration with psychedelics due to the potential for blunting of subjective effects, as well as a theoretical risk of serotonin syndrome. Thus, most currently approved protocols require an antidepressant washout period of 4–5 half-lives prior to administration of a psychedelic. Relatively little is known about effects of co-administration with other psychotropic drugs but it is likely that other clinically significant interactions exist. In healthy volunteers, for example, haloperidol co-administration with psilocybin was associated with derealization experiences associated with arousal and anxiety (Vollenweider et al. 1998), and administration of psychedelics to individuals using lithium has been associated with seizures (Nayak et al. 2021). Other drugs commonly used in this population that may be problematic when co-administered with psychedelics include corticosteroids and stimulants given their risk of precipitating mania, as well as serotonergic agonists such as ondansetron, since they may theoretically contribute to serotonin syndrome. While these possible risks have not been systematically evaluated, they are nonetheless worth considering.

A number of other areas of concern remain regarding the generalizability of psychedelic treatments. While safety guidelines have been provided (Johnson et al. 2008), ongoing research has generally not included individuals with a family history of psychotic or bipolar disorders. The end of life context and the stressors involved may provide a particularly stressful context which may increase the likelihood of adverse responses, although this has not yet been studied. Additionally, there may be an increased tendency to pair psychedelic treatments with non-evidence-based and fringe treatments in this context, which should be cautioned against in favor of more evidence-based treatments (Yaden et al. 2020).

Findings suggest that psychedelics may have analgesic properties, which may have important implications in palliative care (Castellanos et al. 2020; Kast 1964). Current mainstay analgesia treatments such as opioid medications have the risk of sedation and other side effects. Ramaekers et al. (2021) found that LSD acutely reduced subjective discomfort and pain ratings in healthy volunteers, and that this effect was achieved with relatively low doses (10–20 μg), which might have the added benefit of lower risk of cognitive impairment when compared to high doses.

Psychedelics have been delivered in the context of various psychotherapeutic modalities and have the potential to be integrated with existing evidence-based psychotherapies specific to palliative care (Nayak and Johnson 2021). Therapeutic life review (Keall et al. 2015) and meaning centered therapies (Rosenfeld et al. 2017), for example, closely resemble the life review process commonly done during preparatory visits in many psychedelic clinical trials.

The end of life context usually involves religious and existential contemplation. While psychedelics are sometimes claimed to facilitate such contemplations, it is possible that psychedelics could interfere with this process or appear to some to interfere with what might be considered a "natural" process. Indeed, the end of life context offers a number of bioethical considerations and the possibility of psychedelic treatments may further complicate this process. Relatedly, psychedelics present a number of informed consent concerns. Smith and Sisti (2020) argue that informed consent for psychedelic treatment should include the possibility of changes to one's belief system and sense of identity. There are reasons for concern regarding the possibility of religious/spiritual belief change as a result of using psychedelics from some self-report surveys (Griffiths et al. 2019; Yaden et al. 2017b), although these do not represent population base rates and other samples have not found such associations (Yaden and Anderson 2021).

While risks must be weighed, it is also important to consider the positive potential for psychedelic experiences and the costs of preventing patients from having such experiences. Earp (2018) argues that psychedelic experiences constitute an important form of enhancement that goes beyond the reduction of suffering. Specifically, Earp proposes that psychedelics could promote improvements in one's relationships. There is good evidence for this, as a number of psychedelic studies report improvements in social relationships (e.g., Griffiths et al. 2006, 2008, 2011, 2016, 2018; Pahnke et al. 1970). The relational enhancements provided by psychedelic treatments could open an important window for interpersonal connection with family and friends during a time that will be the last opportunity for patients to have such meaningful moments with loved ones. Additionally, Earp (2018) points to other kinds of enhancements, such as experiences relevant to one's belief system or worldview. Empirical evidence indicates that experiences resulting from psychedelic substances are among the most meaningful of one's entire life (e.g., Griffiths et al. 2006, 2008, 2011, 2016, 2018). Denying such experiences to individuals when, perhaps, they need them most is a significant ethical issue to consider.

There are other clinical considerations regarding how psychedelics could be safely and ethically administered in end of life settings, such as whether there are evidence-based protocols to safely administer such treatments in end of life contexts. While it is likely the case that aspects of evidence-based psychotherapies apply generally to psychedelic treatments (Nayak and Johnson 2021), it remains unclear how such therapies should be modified when applied to administering psychedelics in palliative care and end of life contexts.

An important unresolved issue with the prospect of administering psychedelics at end of life has to do with their specific indication. For example, three of the recent studies cited above involving psilocybin (Grob et al. 2011; Griffiths et al. 2016; Ross et al. 2016) use medical and psychiatric inclusion criteria that do not fully overlap. More work is needed in order to specify the indications that are clinically appropriate and specific for end of life care, and will be acceptable for approval by regulatory authorities.

5 Conclusion

In this review, we find evidence suggesting possible efficacy of classic psychedelics in treating a variety of psychiatric conditions including end of life distress. Psychedelic treatments can provide experiences of meaning and well-being amidst the process of dying that are highly valued by patients and their families. For some, this treatment could potentially provide little benefit and add additional stress to an already difficult time, so further research is needed in order to minimize such risks. For others, such experiences may be among the most important of their entire lives and could represent a positive intervention with immense psychological value amidst one of life's most difficult moments – its end.

Acknowledgements *Funding*: Support for Drs. D. Yaden, S. Nayak, N. Gukasyan, and R. Griffiths through the Johns Hopkins Center for Psychedelic and Consciousness Research was provided by Tim Ferriss, Matt Mullenweg, Blake Mycoskie, Craig Nerenberg, and the Steven and Alexandra Cohen Foundation as well as a grant from the Y.C. Ho/Helen and Michael Chiang Foundation.

References

Agin-Liebes GI, Malone T, Yalch MM, Mennenga SE, Ponté KL, Guss J, Bossis AP, Grigsby J, Fischer S, Ross S (2020) Long-term follow-up of psilocybin-assisted psychotherapy for psychiatric and existential distress in patients with life-threatening cancer. J Psychopharmacol 34 (2):155–166

Anderson BT, Danforth A, Daroff R, Stauffer C, Ekman E, Agin-Liebes G et al (2020) Psilocybin-assisted group therapy for demoralized older long-term AIDS survivor men: an open-label safety and feasibility pilot study. EClinicalMedicine 27:100538

Aziz NM, Grady PA, Curtis JR (2013) Training and career development in palliative care and end-of-life research: opportunities for development in the US. J Pain Symptom Manage 46 (6):938–946

Barrett FS, Carbonaro TM, Hurwitz E, Johnson MW, Griffiths RR (2018) Double-blind comparison of the two hallucinogens psilocybin and dextromethorphan: effects on cognition. Psychopharmacology (Berl) 235(10):2915–2927

Barrett FS, Johnson MW, Griffiths RR (2015) Validation of the revised mystical experience questionnaire in experimental sessions with psilocybin. J Psychopharmacol 29(11):1182–1190

Bauereiß N, Obermaier S, Özünal SE, Baumeister H (2018) Effects of existential interventions on spiritual, psychological, and physical well-being in adult patients with cancer: systematic review and meta-analysis of randomized controlled trials. Psychooncology 27(11):2531–2545

Bogenschutz MP, Forcehimes AA, Pommy JA, Wilcox CE, Barbosa PCR, Strassman RJ (2015) Psilocybin-assisted treatment for alcohol dependence: a proof-of-concept study. J Psychopharmacol 29(3):289–299

Bruera E, Kuehn N, Miller MJ, Selmser P, Macmillan K (1991) The Edmonton symptom assessment system (ESAS): a simple method for the assessment of palliative care patients. J Palliat Care 7(2):6–9

Candy B, Jackson KC, Jones L, Tookman A, King M (2012) Drug therapy for symptoms associated with anxiety in adult palliative care patients. Cochrane Database Syst Rev (10)

Carbonaro TM, Bradstreet MP, Barrett FS, MacLean KA, Jesse R, Johnson MW, Griffiths RR (2016) Survey study of challenging experiences after ingesting psilocybin mushrooms: acute and enduring positive and negative consequences. J Psychopharmacol 30(12):1268–1278

Carbonaro TM, Johnson MW, Griffiths RR (2020) Subjective features of the psilocybin experience that may account for its self-administration by humans: A double-blind comparison of psilocybin and dextromethorphan. Psychopharmacology (Berl) 237:2293–2304

Carhart-Harris R, Giribaldi B, Watts R, Baker-Jones M, Murphy-Beiner A, Murphy R, Martell J, Blemings A, Erritzoe D, Nutt DJ (2021) Trial of psilocybin versus escitalopram for depression. New Eng J Med 384(15):1402–1411

Carhart-Harris RL, Bolstridge M, Day CMJ, Rucker J, Watts R, Erritzoe DE, Kaelen M, Giribaldi B, Bloomfield M, Pilling S, Rickard JA, Forbes B, Feilding A, Taylor D, Curran HV, Nutt DJ (2018) Psilocybin with psychological support for treatment-resistant depression: six-month follow-up. Psychopharmacology (Berl) 235(2):399–408

Castellanos JP, Woolley C, Bruno KA, Zeidan F, Halberstadt A, Furnish T (2020) Chronic pain and psychedelics: a review and proposed mechanism of action. Reg Anesth Pain Med 45 (7):486–494

Cohen S (1965) LSD and the anguish of dying. Harper's, pp 69–78

Connor SR (1998) Hospice: practice, pitfalls, and promise. Taylor & Francis, New York

Connor SR, Pyenson B, Fitch K, Spence C, Iwasaki K (2007) Comparing hospice and nonhospice patient survival among patients who die within a three-year window. J Pain Symptom Manage 33(3):238–246

Davis AK, Barrett FS, Griffiths RR (2020a) Psychological flexibility meditates the relations between acute psychedelic effects and subjective decreases in depression and anxiety. J Contextual Behav Sci 15:39–45

Davis AK, Barrett FS, May DG, Cosimano MP, Sepeda ND, Johnson MW, Finan PH, Griffiths RR (2020b) Effects of psilocybin-assisted therapy for major depressive disorder: a randomized clinical trial. JAMA Psychiatry 78(5):481–489

Davis AK, Barrett FS, So S, Gukasyan N, Swift TC, Griffiths RR (2020c) Development of the psychological insight questionnaire among a sample of people who have consumed psilocybin or LSD. J Psychopharmacol

Dittrich A, Lamparter D, Maurer M (2010) 5D-ASC: questionnaire for the assessment of altered states of consciousness. In: A short introduction. PSIN PLUS, Zurich

Earp BD (2018) Psychedelic moral enhancement. R Inst Philos Suppl 83:415–439

Feudtner C (2009) The breadth of hopes. N Engl J Med 361:2306–2307

Finucane TE (1999) How gravely ill becomes dying. A key to end-of-life care. JAMA 282:1670–1672

Fisher G (1970) Psychotherapy for the dying: principles and illustrative cases with special reference to the use of LSD. Omega 1:3–15

Gasser P, Holstein D, Michel Y, Doblin R, Yazar-Klosinski B, Passie T, Brenneisen R (2014) Safety and efficacy of lysergic acid diethylamide-assisted psychotherapy for anxiety associated with life-threatening diseases. J Nerv Ment Dis 202(7):513–520

Gawande A (2014) Being mortal: medicine and what matters in the end. Metropolitan Books

Gawande A (2016) Quantity and quality of life: duties of care in life-limiting illness. JAMA 315 (3):267–269

Grassi, L., & Riba, M. (Eds.). (2014). Psychopharmacology in oncology and palliative care: a practical manual. Springer

Griffiths RR, Richards WA, McCann U, Jesse R (2006) Psilocybin can occasion mystical-type experiences having substantial and sustained personal meaning and spiritual significance. Psychopharmacology (Berl) 187(3):268–283

Griffiths RR, Richards WA, Johnson MW, McCann UD, Jesse R (2008) Mystical-type experiences occasioned by psilocybin mediate the attribution of personal meaning and spiritual significance 14 months later. J Psychopharmacol 22(6):621–632

Griffiths RR, Johnson MW, Richards WA, Richards BD, McCann U, Jesse R (2011) Psilocybin occasioned mystical-type experiences: immediate and persisting dose-related effects. Psychopharmacology (Berl) 218(4):649–665

Griffiths RR, Johnson MW, Carducci MA, Umbricht A, Richards WA, Richards BD et al (2016) Psilocybin produces substantial and sustained decreases in depression and anxiety in patients with life-threatening cancer: a randomized double-blind trial. J Psychopharmacol 30 (12):1181–1197

Griffiths RR, Johnson MW, Richards WA, Richards BD, Jesse R, MacLean KA et al (2018) Psilocybin-occasioned mystical-type experience in combination with meditation and other spiritual practices produces enduring positive changes in psychological functioning and in trait measures of prosocial attitudes and behaviors. J Psychopharmacol 32(1):49–69

Griffiths RR, Hurwitz ES, Davis AK, Johnson MW, Jesse R (2019) Survey of subjective "God encounter experiences": comparisons among naturally occurring experiences and those occasioned by the classic psychedelics psilocybin, LSD, ayahuasca, or DMT. PLoS One 14(4):e0214377

Grob CS, Danforth AL, Chopra GS, Hagerty M, McKay CR, Halberstadt AL, Greer GR (2011) Pilot study of psilocybin treatment for anxiety in patients with advanced-stage cancer. Arch Gen Psychiatry 68(1):71–78

Grof S, Goodman LE, Richards WA, Kurland AA (1973) LSD-assisted psychotherapy in patients with terminal cancer. Int Pharmacopsychiatry 8:129–144

Gruneir A, Mor V, Weitzen S, Truchil R, Teno J, Roy J (2007) Where people die: a multilevel approach to understanding influences on site of death in America. Med Care Res Rev 64 (4):351–378. https://doi.org/10.1177/1077558707301810

Hood RW, Morris RJ (1983) Toward a theory of death transcendence. J Sci Stud Relig 22:353–365

James W (1902) The varieties of religious experience: a study in human nature. Harvard University Press

Johnson MW, Richards WA, Griffiths RR (2008) Human hallucinogen research: guidelines for safety. J Psychopharmacol 22(6):603–620

Johnson MW, Garcia-Romeu A, Cosimano MP, Griffiths RR (2014) Pilot study of the 5-HT2AR agonist psilocybin in the treatment of tobacco addiction. J of Psychopharm 28(11):983–992

Johnson MW, Hendricks PS, Barrett FS, Griffiths RR (2019) Classic psychedelics: an integrative review of epidemiology, therapeutics, mystical experience, and brain network function. Pharmacol Ther 197:83–102

Kast EC (1962) The measurement of pain, a new approach to an old problem. J New Drugs 2:344–351

Kast EC (1964) Lysergic acid diethylamide as an analgesic agent. Anesth Analg 43:285–291

Kast E (1966) LSD and the dying patient. Chic Med Sch Q 26(2):80–87

Kast E (1967) Attenuation of anticipation: a therapeutic use of lysergic acid diethylamide. Psychiatry Q 41(4):646–657

Keall RM, Clayton JM, Butow PN (2015) Therapeutic life review in palliative care: a systematic review of quantitative evaluations. J Pain Symptom Manage 49(4):747–761

Li Y, Li X, Hou L, Cao L, Liu G, Yang K (2020) Effectiveness of dignity therapy for patients with advanced cancer: a systematic review and meta-analysis of 10 randomized controlled trials. Depress Anxiety 37(3):234–246

Moreno FA, Wiegand CB, Taitano EK, Delgado PL (2006) Safety, tolerability, and efficacy of psilocybin in 9 patients with obsessive-compulsive disorder. J Clin Psychol 67(11):1735–1740

Mücke M, Weier M, Carter C, Copeland J, Degenhardt L, Cuhls H et al (2018) Systematic review and meta-analysis of cannabinoids in palliative medicine. J Cachexia Sarcopenia Muscle 9 (2):220–234

Nayak S, Johnson MW (2021) Psychedelics and psychotherapy. Pharmacopsychiatry 54 (4):167–175

Nayak S, Gukasyan N, Barrett FS, Erowid E, Erowid F, Griffiths RR (2021) Classic psychedelic coadministration with lithium, but not lamotrigine, is associated with seizures: an analysis of online psychedelic experience reports. Pharmacopsychiatry 54(5):240–245

Nichols DE (2016) Psychedelics. Pharm Rev 68(2):264–355

Nutt DJ, King LA, Phillips LD (2010) Drug harms in the UK: a multicriteria decision analysis. Lancet 376(9752):1558–1565

Ostuzzi G, Matcham F, Dauchy S, Barbui C, Hotopf M (2018, 2018) Antidepressants for the treatment of depression in people with cancer. Cochrane Database Syst Rev (4):CD011006

Pahnke WN (1969) The psychedelic mystical experience in the human encounter with death. Harv Theol Rev 62:1–21

Pahnke WN, Kurland A, Unger S, Savage C, Grof S (1970) The experimental use of psychedelic (LSD) psychotherapy. JAMA 212(11):1856–1856

Park CL et al (2019) Effects of psychosocial interventions on meaning and purpose in adults with cancer: A systematic review and meta-analysis. Cancer 125:2383–2393

Quigley C (2008) Opioids in people with cancer-related pain. BMJ Clin Evid 2008:2408

Ramaekers JG, Hutten N, Mason NL, Dolder P, Theunissen EL, Holze F et al (2021) A low dose of lysergic acid diethylamide decreases pain perception in healthy volunteers. J Psychopharmacol 35(4):398–405

Rayner L, Price A, Evans A, Valsraj K, Hotopf M, Higginson IJ (2011) Antidepressants for the treatment of depression in palliative care: systematic review and meta-analysis. Palliat Med 25 (1):36–51

Richards WA, Rhead JC, DiLeo FB, Yensen R, Kurland AA (1977) The peak experience variable in DPT-assisted psychotherapy with cancer patients. J Psychedelic Drugs 9:1–10

Rosenfeld B, Saracino R, Tobias K, Masterson M, Pessin H, Applebaum A et al (2017) Adapting meaning-centered psychotherapy for the palliative care setting: results of a pilot study. Palliat Med 31(2):140–146

Ross S, Bossis A, Guss J, Agin-Liebes G, Malone T, Cohen B et al (2016) Rapid and sustained symptom reduction following psilocybin treatment for anxiety and depression in patients with life-threatening cancer: a randomized controlled trial. J Psychopharmacol 30(12):1165–1180

Saunders CM (1978) The management of terminal malignant disease. Edward Arnold, London

Schultes RE, Hofmann A (1992) Plants of the gods: their sacred, healing and hallucinogenic powers. Healing Arts Press, Rochester

Shinners JR (1997) Medieval popular religion, 1000–1500: a reader. Broadview Press

Singer PA, Martin DK, Kelner M (1999) Quality end-of-life care: patients' perspectives. JAMA 281 (2):163–168

Smith WR, Sisti D (2020) Ethics and ego dissolution: the case of psilocybin. J Med Ethics

Steinhauser KE, Christakis NA, Clipp EC, McNeilly M, McIntyre L, Tulsky JA (2000) Factors considered important at the end of life by patients, family, physicians, and other care providers. JAMA *284*(19):2476–2482

Studerus E, Kometer M, Hasler F, Vollenweider FX (2011) Acute, subacute and long-term subjective effects of psilocybin in healthy humans: a pooled analysis of experimental studies. J Psychopharmacol 25(11):1434–1452

Vollenweider FX, Preller KH (2020) Psychedelic drugs: neurobiology and potential for treatment of psychiatric disorders. Nat Rev Neuro 21(11):611–624

Vollenweider FX, Vollenweider-Scherpenhuyzen MF, Bäbler A, Vogel H, Hell D (1998) Psilocybin induces schizophrenia-like psychosis in humans via a serotonin-2 agonist action. Neuroreport 9(17):3897–3902

World Health Organization (2011) WHO Definition of Palliative Care. (2011, accessed online February 2011)

Yaden DB, Anderson DE (2021) The psychology of philosophy: associating philosophical views with psychological traits in professional philosophers. Philos Psychol 34(5):721–755

Yaden DB, Griffiths RR (2020) The subjective effects of psychedelics are necessary for their enduring therapeutic effects. ACS Pharmacol Transl Sci 4(2):568–572

Yaden DB, Haidt J, Hood RW Jr, Vago DR, Newberg AB (2017a) The varieties of self-transcendent experience. Rev Gen Psychol 21(2):143–160

Yaden DB, Le Nguyen KD, Kern ML, Belser AB, Eichstaedt JC, Iwry J et al (2017b) Of roots and fruits: a comparison of psychedelic and nonpsychedelic mystical experiences. J Humanist Psychol 57(4):338–353

Yaden DB, Eichstaedt JC, Kern ML, Smith LK, Buffone A, Stillwell DJ, Kosinski M, Ungar LH, Seligman MEP, Schwartz HA (2018) The language of religious affiliation: social, emotional, and cognitive differences. Soc Psychol Personal Sci 9(4):444–452

Yaden DB, Yaden ME, Griffiths RR (2020) Psychedelics in psychiatry – keeping the renaissance from going off the rails. JAMA Psychiatry 78(5):469–470

Yaden DB, Berghella AP, Regier PS, Garcia-Romeu A, Johnson MW, Hendricks PS (2021a) Classic psychedelics in the treatment of substance use disorder: potential synergies with twelve-step programs. Int J Drug Policy 98:103380

Yaden DB, Giorgi S, Kern ML, Adler A, Ungar LH, Seligman ME, Eichstaedt JC (2021b) Beyond beliefs: multidimensional aspects of religion and spirituality in language. Psychol Relig Spiritual. Online first https://psycnet.apa.org/doiLanding?doi=10.1037%2Frel0000408

Part III
Disruptive Psychopharmacology for Other Disorders

Psychedelic-Assisted Therapy for Substance Use Disorders and Potential Mechanisms of Action

Nathalie M. Rieser, Marcus Herdener, and Katrin H. Preller

Contents

Abstract Substance use disorders (SUD) represent a significant public health issue with a high need for novel and efficacious treatment options. In light of this high unmet need, recent results reporting beneficial outcomes of psychedelic-assisted therapy in SUD are particularly relevant. However, several questions remain with

Nathalie M. Rieser, Marcus Herdener, and Katrin H. Preller contributed equally to this work.

N. M. Rieser (✉) and K. H. Preller
Pharmaco-Neuroimaging and Cognitive-Emotional Processing, Department of Psychiatry, Psychotherapy and Psychosomatics, University of Zurich, Psychiatric University Hospital Zurich, Zurich, Switzerland
e-mail: nathalie.rieser@bli.uzh.ch

M. Herdener
Center for Addictive Disorders, Department of Psychiatry, Psychotherapy and Psychosomatics, University of Zurich, Psychiatric University Hospital Zurich, Zurich, Switzerland

Curr Topics Behav Neurosci (2022) 56: 187–212
https://doi.org/10.1007/7854_2021_284
Published Online: 15 December 2021

regard to this treatment approach. The clinical mechanisms of action of psychedelic substances in the treatment of SUD are not well understood. Closing this knowledge gap is critical to inform and optimize the psychotherapeutic embedding of the acute substance administration. In this chapter, we discuss potential mechanisms that have implications on psychotherapeutic approaches including induced neuroplasticity, alterations in brain network connectivity, reward and emotion processing, social connectedness, insight, and mystical experiences. Furthermore, we outline considerations and approaches that leverage these mechanisms in order to optimize the therapeutic embedding by maximizing synergy between substance effects and psychotherapy. Understanding the mechanisms of action, developing psychotherapeutic approaches accordingly, and evaluating their synergistic efficacy in scientific studies will be critical to advance the framework of psychedelic-assisted therapy for addiction, create evidence-based approaches, and achieve the best treatment outcome for patients with SUD.

Keywords Addiction · Hallucinogen · LSD · Mechanism of action · Psilocybin · Psychedelic · Psychedelic-assisted therapy · Substance use disorder · SUD

1 Introduction

Serotonergic psychedelic substances, also called hallucinogens, have been studied extensively in the 1950s and 1960s for the treatment of various psychiatric disorders (Vollenweider and Preller 2020). One major area of interest was the use of LSD in the treatment of substance use disorders (SUD) (Dyck 2006). While most of these early clinical studies do not comply with current scientific standards, a meta-analysis of six randomized trials testing LSD for alcohol dependence showed consistent treatment effects on drinking outcomes (Krebs and Johansen 2012). These early results remain of great interest given that alcohol addiction still accounts for a substantial proportion of global health burden and alcohol addicted patients lose more than 20 years in average life expectancy compared to the normal population (GBD 2018). Pharmacological treatment approaches for alcohol use disorder (AUD) have limited effectiveness with up to 50% relapse rates (Moos and Moos 2006). Currently, research on the therapeutic effects of psychedelics is gaining momentum again and various studies investigating the efficacy of psychedelics in the treatment of alcohol (NCT04620759, NCT04410913, NCT04141501), nicotine (NCT01943994), and cocaine addiction (NCT02037126) are ongoing. At this time, two modern clinical trials have been published, reporting beneficial long-term effects of psilocybin for nicotine and alcohol use disorder for up to 6 months (Bogenschutz et al. 2015; Johnson et al. 2017). Although these studies are small and lack appropriate control conditions, the results are promising. However, potential mechanisms of actions of psilocybin in patients with SUD have not been investigated.

Importantly, psychedelics are usually not administered without psychotherapeutic embedding. In the two published studies testing the effects of psychedelics in the treatment of SUD, one or two administrations of psilocybin were combined with various preparation and integration sessions based on cognitive-behavioral or motivational enhancement therapeutic approaches (Bogenschutz et al. 2015; Johnson et al. 2017). The use of psychedelics for treating psychiatric illnesses is therefore considered to be best described as pharmacologically-assisted psychotherapy (PAT) that leverages psychedelic-induced neurobiological and psychological effects (Vollenweider and Preller 2020). However, two major questions remain regarding this approach:

1. What is the clinically relevant mechanism of action of psychedelics in SUD? In other words, what are the psychedelic-induced neurobiological and psychological effects that can be leveraged in psychotherapy?
2. What is the best way to conduct psychedelic-assisted psychotherapy in SUD?

These two questions are not unrelated. As discussed in detail below, understanding the clinical mechanisms of action will help to inform, optimize, and advance currently implemented psychotherapeutic approaches in the context of psychedelic-assisted therapy. Testing these approaches empirically is essential to be able to provide evidence-based psychedelic-assisted therapy and maximize the clinical potential of psychedelic substances in the treatment of SUD.

In this chapter, we discuss potential clinically relevant mechanisms of action of psychedelics and highlight current knowledge gaps. We use the term "psychedelics" to refer to 5-HT_{2A} receptor agonists, i.e., "classic psychedelics" such as psilocybin, LSD, ayahuasca/DMT, and mescaline. If other substances are discussed, this is noted in the text. Please note that the effects discussed here do not represent an exhaustive list of potential mechanisms of action, but rather focus on insights derived from studies with psychedelics that may be relevant for the treatment of SUD. Furthermore, we discuss the implications these mechanistic hypotheses have for the content and conduction of accompanying psychotherapy. While some of these mechanisms and considerations may be applicable for the treatment of other disorders as well, this chapter focuses specifically on SUD.

2 Potential Mechanisms of Action of Psychedelic Substances in the Treatment of Substance Use Disorders

2.1 *Induced Neuroplasticity*

Maladaptive learning and memory processes are major characteristics of SUD (Hyman 2005). Alcohol and other drugs of abuse have been suggested to hijack synaptic plasticity systems and create persistent links between reinforcing aspects of the drug-induced experience and associated stimuli (Kauer and Malenka 2007).

These associations between cues and drug experiences can contribute to relapse even after prolonged periods of abstinence (Lüscher 2016). Furthermore, chronic exposure to alcohol has been shown to reduce neuroplasticity as evidenced by dysregulations in long-term potentiation/long-term depression and Brain-derived neurotrophic factor (BDNF) signaling (Lovinger and Abrahao 2018). These long-term plasticity impairments may underlie learning deficits and reductions in cognitive flexibility observed in alcohol addicted patients (Beylergil et al. 2017). These neuroadaptations pose a major challenge for psychotherapeutic treatments in AUD. Treatment success may be severely limited by reductions in neuroplasticity as these treatment approaches require the patient to learn new strategies and skills that help them to abstain from drinking alcohol.

Classical psychedelics may open a window of opportunity for "therapeutic learning" by inducing neuroplasticity that potentially underlies the long-lasting effects of a single administration and could potentially be leveraged in the therapeutic process. In vivo and in vitro studies in rodents revealed that psychedelics increase neuritogenesis, spinogenesis, and synaptogenesis as well as gene expression of BDNF and Immediate Early Genes associated with plasticity (Catlow et al. 2013; Ly et al. 2018; Nichols 2016; Shao et al. 2021; Zhang et al. 2013). These neuroplastic changes appear to be mediated through activation of the 5-HT_{2A} receptor, tyrosine receptor kinase B (TrkB), and mammalian target of rapamycin (mTOR) signaling pathways (Ly et al. 2018). Another recent study in pigs reports significantly higher Synaptic vesicle glycoprotein 2A density – a marker of presynaptic density – in the hippocampus and the prefrontal cortex 7 days after a single dose of psilocybin (Raval et al. 2021). Importantly, animal studies also provide evidence for psychedelic-induced functional neuroplastic changes as low doses of psilocybin lead to significantly faster extinction learning in a cued fear conditioning paradigm than saline in mice (Catlow et al. 2013). In a recent study, psilocybin evoked dendritic spinal density, widths of spinal heads, and spine protrusion lengths in the medial prefrontal cortex in mice (Shao et al. 2021). Furthermore, half of the newly formed spines remained stable until day 7 (Shao et al. 2021).

It still needs to be investigated if these neuroplastic effects observed in animals translate to humans. So far, only indirect evidence has been presented such as increased levels of medial prefrontal cortical glutamate during the acute experience (Mason et al. 2020) and post-acute reductions of glutamate + glutamine in the posterior cingulate cortex (Sampedro et al. 2017). If future studies find evidence that psychedelics increase neuroplasticity and facilitate (extinction) learning, this may have important implications for the therapeutic embedding of a psychedelic experience and may open up various opportunities to beneficially leverage the synergy between drug-induced effects and psychotherapy.

2.2 Alterations in Brain Networks Connectivity

Previous studies report widespread changes in the functional architecture of the brain in patients suffering from SUD as measured with connectivity metrics during resting state. Specifically, reductions in functional connectivity of the precuneus, postcentral gyrus, insula, and visual cortex were reported in AUD (Vergara et al. 2017). Changes within the dorsal anterior cingulate cortex-striatum circuit have repeatedly been reported in nicotine dependence (Scarlata et al. 2021). Salience and default mode network dysregulation has been shown to predict treatment outcome in chronic cocaine users (Geng et al. 2017). Additionally, converging evidence points to changes between and within cortical midline structures as a common mechanism underlying various SUD such as alcohol, nicotine, cannabis, and heroin addiction (Ersche et al. 2020; Zhang and Volkow 2019).

Acutely, psychedelics have been shown to alter thalamo-cortical connectivity patterns (Mueller et al. 2017; Preller et al. 2018). Specifically, increased functional connectivity between the thalamus and sensory brain areas and increased effective connectivity from the thalamus to the posterior cingulate cortex have been reported (Preller et al. 2018, 2019). These changes, together with results obtained with neurophysiological measurements such as prepulse inhibition (Quednow et al. 2012; Riba et al. 2002; Schmid et al. 2015; Vollenweider et al. 2007), point to changes in thalamic gating of internal and external sensory and cognitive information. Additionally, increased synchronization of sensory brain areas (including the visual cortex, postcentral gyrus, and precuneus) together with a de-synchronization of associative brain regions under the acute influence of LSD and psilocybin suggests a disruption of cortical information processing and changes in information integration in the psychedelic state (Barrett et al. 2020; Preller et al. 2018; Preller et al. 2020). In line with this, Mason et al. (2021) reported increased ratings of creativity after psilocybin administration, which were predicted by acutely decreased resting-state functional MRI connectivity of the default mode network. These changes in information integration could enable patients to escape rigid thinking patterns and potentially help develop novel insights into problems during the therapeutic work after a psychedelic dose.

Unfortunately, little is known yet about long-lasting effects of psychedelics on brain connectivity. Two studies report psilocybin-induced changes in resting-state functional connectivity measured 1 week and 1 month after administration (Barrett et al. 2020; McCulloch et al. 2021). Furthermore, post-acute changes in default mode network connectivity were detectable 1 day after ayahuasca administration (Pasquini et al. 2020; Sampedro et al. 2017). Although only investigated in healthy participants so far, this result could implicate lasting functional re-organization effects that may normalize pathological connectivity patterns specifically with regard to connectivity with and between cortical midline structures in clinical populations.

2.3 Alterations in Emotion Processing

Patients with SUD show difficulties in regulating negative emotions and believe it to be less socially acceptable to express negative feelings, compared to healthy controls (Dingle et al. 2018). Suppression or avoidance of negative emotions is amplifying negative mood (Bastian et al. 2012) which may in turn increase maladaptive drinking pattern to reduce negative emotional states. Several key motivation factors for substance misuse, such as escape or avoidance of negative affect (Baker et al. 2004), chronic irritability, emotional pain, dysphoria, or alexithymia (Koob and Le Moal 2005), point to the inability to regulate negative emotions as an important mechanistic candidate in the pathogenesis of SUD. An additional maladaptive strategy to deal with negative emotions is the engagement in rumination, which is the repetitive focusing on one's emotion and its underlying cause (Aldao et al. 2010). Rumination has been positively associated with psychopathology in anxiety disorders, depression, eating disorders, and SUD (Aldao et al. 2010). Additionally, patients with AUD show difficulties in decoding other people's emotions (Le Berre 2019).

Recent studies report that psilocybin and LSD acutely attenuate the recognition of negative facial expressions in healthy participants (Bershad et al. 2019; Dolder et al. 2016; Kometer et al. 2012). Furthermore, psilocybin and LSD reduced the neural response to negative stimuli in the amygdala (Kraehenmann et al. 2015; Mueller et al. 2017), an effect that was sustained for 1 week after administration (Barrett et al. 2020). However, one study reported increased amygdala reactivity in depressed patients the morning after psilocybin administration, suggesting increased emotional responsiveness in this patient group before therapeutic integration work had started (Roseman et al. 2018). Additionally, psilocybin has been shown to enhance autobiographical memory recall, which was associated with subjective well-being in healthy participants (Carhart-Harris et al. 2012).

Together, these results suggest a modulatory effect of psychedelics on emotion processing, in particular the perception of negative stimuli. The clinical importance of these effects is supported by Watts et al. (2017) who report beneficial changes in emotion processing characterized by reduced avoidance and concurrently increased acceptance of emotions in depressed patients after psilocybin-supported treatment. Increased acceptance may also decrease ruminative thoughts, as patients accept the current mood state instead of repetitively thinking about its underlying causes and consequences. Given the deficits in regulating negative emotions reported in SUD patients, psychedelics could have a positive effect on the maladaptive processes by reducing negative affect, enhancing the ability to regulate negative affect, and strengthening adaptive problem solving, which may in turn decrease substance use.

2.4 Alterations in Reward and Stress Processing

Reinforcing effects are key characteristics of psychoactive substances and motivate their use. Neurobiologically, these reinforcing effects depend on dopamine release in the ventral tegmental area (VTA) and subsequently in the Nucleus Accumbens (NAcc) (Wise 2008). Dopamine release from VTA to NAcc plays a central role in the reward circuit and motivational processes (Salamone and Correa 2012; Volkow et al. 2019). Reinforcing effects are also caused by environmental factors, as substance craving and use is often triggered by sensitization, stress, priming dose, or drug-related cues (Belin et al. 2013; Jasinska et al. 2014). These triggers can drive drug users to compulsive drug taking (Volkow et al. 2011b). Repeated perturbation of reward systems leads to long-lasting decreases in dopamine D2 receptors and decreased dopamine cell activity (Volkow et al. 2003). Specifically, chronic drug use induces neuroadaptations in the dopamine striato-thalamo-cortical and limbic pathways (Volkow et al. 2019). Various brain imaging studies have shown that these neuroadaptations cause a long-lasting decrease in sensitivity to natural reinforcers, such as monetary reward and social interaction (e.g., Goldstein et al. 2007; Preller et al. 2014; Tobler et al. 2016). Furthermore, obsessive-compulsive thoughts related to cocaine use and lifetime cocaine consumption were associated with impaired reward sensitivity for natural reinforcers in cocaine users (Kirschner et al. 2018). Therefore, modifying maladaptive reward sensitivity and reinstating natural reward responsiveness is a promising therapeutic strategy in SUD.

An additional neurotransmitter influencing the cycle of addiction is serotonin. Serotonin is involved in the regulation of stress, anxiety, cognitive functions, social behavior, and reinforcement of properties of drugs (Belmer et al. 2016). Serotonin neurotransmission is reduced in patients with AUD after alcohol-withdrawal leading to increased stress-induced anxiety, which in turn reinforces craving and relapse (Belmer et al. 2016). Furthermore, frontolimbic 5-HT_{2A} receptor binding correlates with anxiety and difficulties in the regulation of stress (Frokjaer et al. 2008). Therefore, stimulation of serotonergic neurotransmission by the administration of psychedelics may decrease stress-induced anxiety, improve mood, and reduce attentional bias, and in turn decrease craving (Bogenschutz and Pommy 2012). This is in line with participants being treated with psilocybin for smoking cessation reporting reduced withdrawal symptoms and craving (Noorani et al. 2018).

2.5 *Increased Social Connectedness*

Potentially related to impaired responsiveness to natural rewards, patients with SUD show dysfunctional social cognition and interaction, subsequently leading to decreased social contact and support. This may result in increased social isolation (Quednow 2020; Tobler et al. 2016). For example, cocaine users showed decreased social interaction and brain activity in regions related to reward processing. This decreased brain activity was associated with a decreased social network size, suggesting that reduced social reward processing is leading to impairments of social behavior in real life (Preller et al. 2014). Similarly, patients with AUD showed reduced theory of mind skills (Le Berre 2019). Such difficulties may increase the probability of relapse and hinder treatment efficacy. Therefore, the training of social and emotional skills should be implemented in therapy of SUD.

Studies with healthy controls have shown that classic psychedelics acutely decrease feelings of social exclusion and increase emotional empathy, prosocial behavior, and the desire to be with other people (Dolder et al. 2016; Pokorny et al.

2017; Preller et al. 2016). Positive social effects and interpersonal closeness were increased up to 1 year after the psychedelic experience (Griffiths et al. 2008, 2018; Schmid and Liechti 2018). Furthermore, participants receiving psilocybin in the treatment of smoking cessation reported feelings of love and a sense of unity and interconnection with their environment, which they identified to be important factors for staying abstinent from smoking (Noorani et al. 2018). Additionally, participants engaged more in social activities after the therapy (Noorani et al. 2018). After the administration of a single dose of (±)3,4-methylendioxymethamphetamine (MDMA), the reopening of a critical period for social reward learning was shown in mice (Nardou et al. 2019). However, MDMA is not a classic psychedelic, but rather an "entactogen" and these effects may therefore not translate to classic psychedelics. Classic psychedelics may counteract social withdrawal associated with SUD by decreasing social anxiety and increasing emotional empathy, prosociality, and interpersonal closeness. Psychedelics may not only improve social ties within daily life, but also the patient-therapist relationship. This may suggest that a group setting for discussing the experience or including a person close to the patient into the preparation or follow-up therapy sessions may be beneficial.

2.6 *Subjective Experiences and Personal Meaning of the Experiences*

Self-awareness, interoception, and insight are related constructs that describe the ability to identify one's own behaviors, emotions, and mental states (David et al. 2012). Patients suffering from SUD often show difficulties in self-awareness, interoception, and insight (Goldstein et al. 2009). Specifically, patients with SUD fail to perceive the personal importance of stimuli or situations that have a meaning to the self. This leads to drug-biased attention, deviant processing of non-drug-related cues, and abnormalities in social cognition (Moeller and Goldstein 2014). A dysfunctional neural circuit involving the insula, anterior cingulate cortices, probably orbitofrontal cortex (Goldstein et al. 2009), and ventromedial prefrontal cortex (Moeller and Goldstein 2014) may underlie these difficulties. Consequently, patients with SUD show difficulties in recognizing and understanding their feelings, and gaining insight into potential dysfunctional behavior. Furthermore, impaired self-awareness is often associated with increased illness severity, poorer prognosis, and decreased insight into the severity of the disorder (David et al. 2012). Therefore, increasing self-awareness and improving insight into problematic behavior could represent a promising therapeutic approach.

Some studies propose that mystical experiences are essential to attribute personal meaning to the psychedelic-induced experience and a positive treatment outcome (e.g., Davis et al. 2020; Garcia-Romeu et al. 2019; Griffiths et al. 2016, 2018; Ross et al. 2016). Mystical experience refers to the first-person experience during a psychedelic session such as feeling of unity, blissful state, insightfulness, and feeling of awe. In a study in nicotine addiction, mystical experiences were associated with smoking cessation at 6-month and 12-month follow-up (Garcia-Romeu et al. 2014; Johnson et al. 2017). Similarly, in patients with AUD, changes in drinking pattern were associated with mystical experiences, but also the general intensity of subjective effects as measured with the Altered State of Consciousness Scale (Bogenschutz et al. 2015). Why and how mystical experiences influence treatment outcomes is not well understood, but Bogenschutz and Pommy (2012) suggest that the high personal meaning of the experience is associated with increased self-efficacy to stay abstinent and decrease temptation to use the substance. In line with this, feeling of awe has been suggested as a driving mechanism in long-term effects of classic psychedelic-assisted therapy by influencing unitive experience and ego dissolution, which in turn affects ineffability, sacredness, positive mood, and insight (Hendricks 2018).

Additionally, insightfulness – partially overlapping with the concept of mystical experiences – is often an essential characteristic of the psychedelic experience (Bogenschutz et al. 2018; Noorani et al. 2018). Insight can refer to a patient's understanding of their disorder, problematic behavior, emotions, and feelings. Such insight has been suggested to lead to enhanced self-efficacy and motivation to change (Bogenschutz and Pommy 2012). A previous online survey showed that increased insight in a non-clinical setting correlates with decreased alcohol use (Garcia-Romeu et al. 2019) and may therefore represent an important mediator of treatment success in SUD.

Associated with increased insight, psychedelics may also (1) increase realization and understanding of negative consequences of problematic behavior patterns and substance use, which may lead to a heightened desire and motivation to change one's behavior, (2) alter the prospect of life, (3) influence one's conviction that change is possible, (4) reduce ambivalence, and (5) facilitate personality change (Bogenschutz and Pommy 2012). However, divergent results were reported for personality change after psychedelic admission. For example, psilocybin increased openness, but did not influence neuroticism, extraversion, agreeableness, or conscientiousness in healthy participants post-acutely (MacLean et al. 2011; Madsen et al. 2020). Similarly, patients with treatment resistant depression showed increased openness, but also increased extraversion and decreased neuroticism following psilocybin sessions (Erritzoe et al. 2018). On the other hand, LSD in healthy participants did not affect openness, but increased conscientiousness in one study (Schmid and Liechti 2018) but not another (Carhart-Harris et al. 2016).

Importantly, SUD patients also reported post-acute alterations of self-perception and more control over their own choices and behavior after a psychedelic experience within a clinical setting (Bogenschutz et al. 2018; Noorani et al. 2018). Together, these results suggest that the acute experience of altered self-perception facilitates treatment success in SUD and has a long-lasting impact on perception of self-efficacy.

Further research is needed to investigate how subjective experiences relate to positive treatment outcomes and whether specific subjective experiences are more beneficial than others to promote treatment success (Bogenschutz et al. 2015). Studies in patients with SUD have not yet investigated the potential relationship between other subjective experiences and therapeutic outcome measures, such as visual/auditive alterations, anxiety, or alterations in cognition and control.

In contrast to results pointing to the importance of subjective effects for beneficial clinical effects, two studies in mice reported neuroplasticity was induced by psilocybin with ketanserin pretreatment (Shao et al. 2021) and tabernanthalog (Cameron et al. 2021), a non-hallucinogenic 5-HT_{2A} agonist, which also reduced alcohol-seeking. These results may challenge the idea that subjective effects are necessary for treatment success in AUD. However, these effects still need to be replicated in humans.

3 Therapeutic Implications

It is widely accepted that psychedelics should be administered in combination with psychotherapeutic treatment that includes preparation and integration sessions, and transfer to daily life. The psychotherapeutic embedding is considered vital to ensure safety and efficacy (e.g., Watts and Luoma 2020). However, it remains unknown which psychotherapeutic strategies are best suited to be combined with psychedelic interventions in the context of SUD. Here, we briefly describe some of the main psychotherapeutic interventions for SUD often applied in clinical settings, and discuss the potential benefits of their combination with psychedelics based on their mechanisms of action as outlined above.

Existing evidence-based psychotherapeutic interventions for SUD include mainly contingency management (CM), cognitive-behavioral therapy (CBT), and motivational interviewing (MI), or variants and combinations thereof. CM is a behavioral intervention in which non-drug taking behavior and other positive behavioral changes are "reinforced" or rewarded, therefore supporting a shift from substance-related to non-substance related behaviors based on principles of reward learning (Petry et al. 2001, 2013). It therefore seeks to reduce the imbalance in reward processing that is at the core of SUD. Although there is good evidence for the application of CM in general, one of the major challenges is to achieve sustained effects that have a positive impact on substance use that outlasts the reinforcement intervention (Prendergast et al. 2006). Here, we discuss how, e.g., neuroplasticity as induced by psychedelics might contribute to enhanced and more sustained effects of CM interventions (see Sects. 3.1, 3.2, and 3.3). Strengthening engagement in behaviors not related to substance use is also at the heart of other psychotherapies like MI. MI focuses on resolving ambivalence related to behavioral decisions between (continued) substance use and other activities, and enhancing motivational engagement in non-substance-related "rewarding" activities (Rollnick and Miller 1995). Accordingly, it has been suggested that MI remodels maladaptive reward learning associated with addiction-related behaviors and leads to changes within the corresponding brain circuitries (Feldstein Ewing et al. 2011). However, effect sizes for MI are rather low (Stein et al. 2009). We expect that fostering brain connectivity changes by psychedelics may enhance the therapeutic potential of MI interventions (see Sect. 2.2). Moreover, as in MI it sometimes proves to be challenging for patients to develop new attractive goals in life that are unrelated to substance use, and, in addition, to generate creative novel ways to achieve them, psychedelics might support MI approaches by enabling "out-of-the-box-thinking" not restricted by rigid thinking patterns. In addition, increased insight into problematic behavior and changes in self-perception and self-efficacy that both have been related to the mystical experiences induced by psychedelics are considered key factors contributing to the clinical effects of MI. Therefore, we expect that MI based interventions and psychedelic therapy are ideally suited to complement each other.

For example, we are currently conducting a study investigating clinical and mechanistic effects of psilocybin in AUD (NCT04141501). Our accompanying psychotherapeutic interventions are based on the BRENDA-approach that is rooted in the principles of MI, which involves "(1) a biopsychosocial evaluation; (2) a report of findings from the evaluation given to the patient; (3) empathy; (4) addressing patient needs; (5) providing direct advice; and (6) assessing patient reaction to advice and adjusting the treatment plan as needed" (p.1, Starosta et al. 2006). The BRENDA psychosocial therapeutic approach has specifically been designed to be combined with pharmacological interventions in AUD (Starosta et al. 2006). It has successfully been implemented in various clinical trials testing the efficacy of pharmacotherapies in SUD (Monterosso et al. 2001; Pettinati et al. 2000). In our study, we extended the BRENDA-approach to include therapeutic elements specific for psychedelic-assisted therapy. After assessing the patients' medical history and their reasons for drinking alcohol in a first visit, patients are prepared for the substance session in a second visit. Patients' previous experiences with psychedelics, their expectations and concerns, handling of challenging situations and potential effects of psilocybin are discussed. Furthermore, patients define intentions for the substance session and goals for their future. On the day of psilocybin administration (visit 3), we instruct patients to focus on themselves, to immerse themselves in whatever may come up even if it may be challenging or unpleasant and to avoid judging the situation or the content. In a fourth visit, the patients revisit the substance session together with their therapist. This includes the discussion of challenging situations and emotions during the day. The interpretation and integration of the experience takes place in the fifth and sixth visits. Here, we discuss how potential insights can lead to beneficial changes in daily life.

We suggest that leveraging synergies between psychedelic-specific mechanisms of action and psychotherapeutic elements in future trials will contribute to optimizing the success of the treatment. Targeted approaches informed by the underlying mechanisms of action will need to be developed to achieve the goal of delivering the best psychedelic-assisted therapy possible. Based on the potential mechanisms of action in SUD outlined above, we are discussing potential therapeutic elements below (for an overview see Fig. 1) and suggest testing the efficacy of these targeted approaches in future studies.

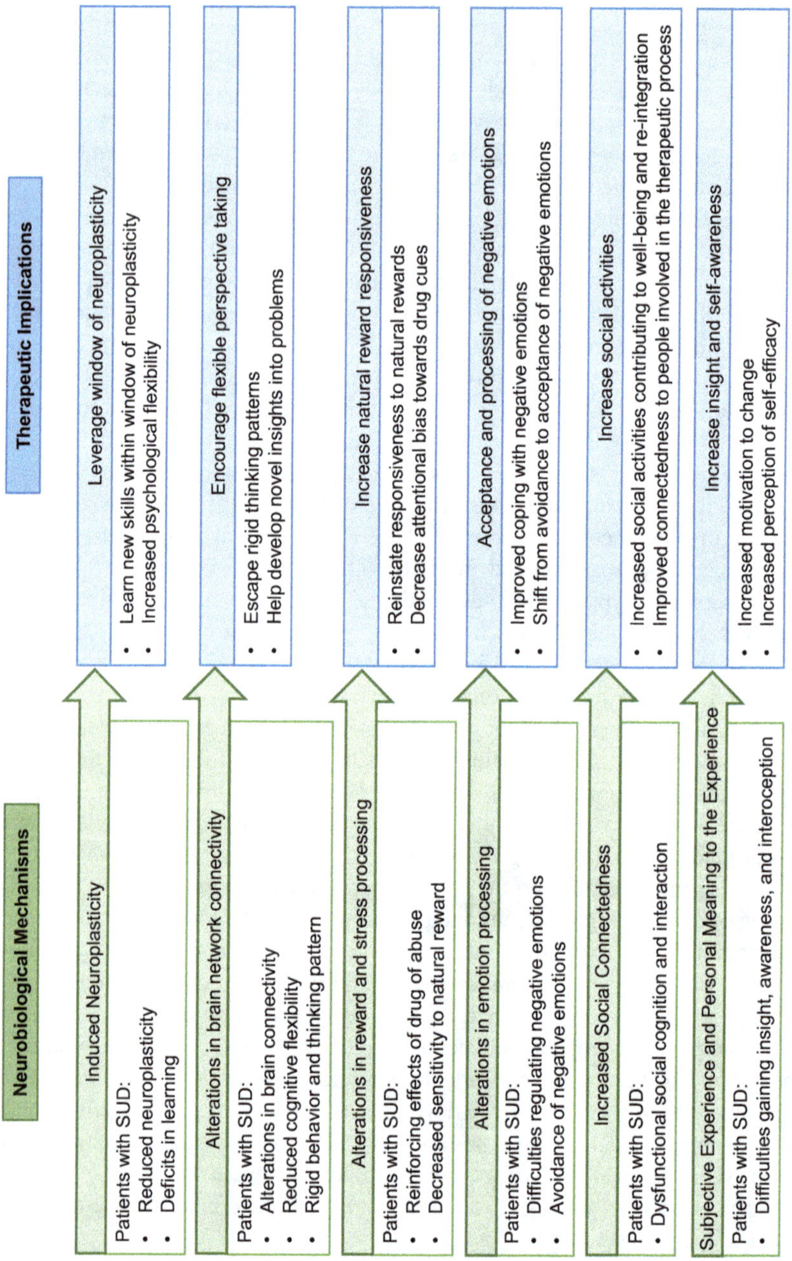

Fig. 1 Overview of neurobiological mechanisms and their therapeutic implications

3.1 Implications of Induced Neuroplasticity

As outlined in Sect. 2.1 psychedelics may open a neuroplastic window of opportunity during which learning (including extinction learning) is facilitated. Unfortunately, we currently do not know if these results translate to humans. Furthermore, we do not have a clear understanding of when and how long this neuroplastic effect can be leveraged. Nevertheless, as outlined above, patients suffering from SUD benefit from learning new behaviors that can replace drug use habits. This may help patients to stay abstinent and weaken the association between drug cues and craving. It is conceivable that psychedelic-induced neuroplasticity may greatly facilitate these processes.

For example, CM is mainly based on operant conditioning, a form of associative learning, where a voluntary behavior is modified or strengthened by reinforcement. Patients usually receive incentives to reinforce drug abstinence. While effect sizes for this intervention are large as long as the target behavior is reinforced, these effects are often not well sustained beyond the therapeutic intervention. Therefore, psychedelic-induced neuroplasticity may enhance and extend the learning effects associated with CM and other psychotherapeutic learning interventions.

Moreover, in synergy with psychedelic-induced plasticity, therapeutic approaches based on extinction learning (i.e., exposure to drug stimuli without providing the drug of abuse itself) may also prove to be more effective in SUD. Leveraging these approaches in combination with immersive technology such as Virtual Reality (e.g., Segawa et al. 2019) may further support the translation to real life as it allows to create realistic scenes (e.g., visiting a bar, seeing other people drink alcohol) that have been shown to trigger craving in AUD patients. Additionally, neurofeedback approaches are helpful tools to specifically train self-regulation of brain activity, and, in combination with psychedelics, could support patients in either learning to reduce their reactions to drug cues, or to enhance their sensitivity to non-drug related reinforcers in a very precise manner (Kirschner et al. 2018).

Increased psychological flexibility resulting from induced neuroplasticity may furthermore be beneficial in supporting creative problem solving specifically with regard to developing novel strategies to avoid relapse or deal with negative emotions. The therapist may encourage the patient to engage in "out-of-the-box" thinking to come up with alternative behaviors that support the patients in staying abstinent. It has to be noted though that LSD did not enhance cognitive flexibility acutely in one study (Pokorny et al. 2020). It is therefore important to consider the timing of these interventions.

While these approaches are likely to be effective after the acute effects have subsided, the finding of increased acute plasticity raises the question of how the acute experience shapes treatment outcomes. If induced neuroplasticity is already present while the patient experiences psychedelic effects, the quality and content of these experiences is likely to have a long-lasting impact. It may therefore be important to avoid deeply frightening experiences during the acute effects as this could potentially lead to enduring anxiety. Positive experiences, on the other hand, may support treatment success. Therefore, a "framing" towards positive and non-drug-related experiences during preparation sessions might be beneficial for subsequent behavioral change. Moreover, the environment could be adapted to the specific needs of the patients, for example by using a group setting if reductions of social problems are a treatment goal.

3.2 *Implications of Alterations in Brain Network Connectivity*

Acute changes in brain connectivity reveal a pattern of increased connectivity between sensory and decreased connectivity between associative brain regions under the influence of LSD and psilocybin (Preller et al. 2018, 2020). These effects point to alterations in information integration and cognitive control that may allow patients to gain a new perspective on themselves and their lives by escaping rigid thinking patterns. Therapists could leverage this effect by specifically discussing novel perspectives with the patient and encouraging the patient to challenge old ideas and habits. Therefore, psychedelics appear ideally suited to support "change talk", i.e. self-expressed speech that argues for change, which is considered one of the key elements of MI.

It is currently unclear if these changes in brain connectivity – and associated with this, information processing – persist beyond the acute drug effects. It may therefore be critical to (re-)engage in these discussions as soon as possible after the peak effects have subsided. This challenges the idea of therapists not interfering with the acute experience and emphasizes directing the content of the experience towards specific, personalized topics that have ideally been discussed in the preparation sessions. Depending on the dose administered, this may not be possible while the patient is experiencing a strong altered state of consciousness, but may be feasible towards the end of the acute effects. Alternatively, lower doses may suffice to induce changes in connectivity between associative brain regions and, together with a targeted intervention that encourages a change of perspective, may promote beneficial clinical outcomes. Future studies investigating changes in brain connectivity induced by psychedelics at various doses and time points may help to uncover the optimal dose and time point for more targeted psychotherapeutic interventions under the influence of or in combination with psychedelics.

Additionally, the strength of psilocybin-induced changes in brain connectivity was predicted by baseline connectivity in healthy participants (Preller et al. 2020). It is therefore possible that the functional architecture of an individual's brain as measured with resting-state functional magnetic resonance imaging may represent a predictive biomarker for acute psychedelic effects and treatment success. Future studies need to explore if this measure can serve as a stratifying factor predicting who is likely/unlikely to benefit from psychedelic-assisted therapy.

3.3 *Implications of Alterations in Emotion, Stress, and Reward Processing*

During the acute and post-acute phase of psychedelic drug action, the neural response to negative stimuli is decreased. Therefore, psychedelics may provide the patient with the opportunity to face negative memories and emotions instead of suppressing or avoiding them. Dealing with the negative emotions possibly coming up during acute psychedelic experience is often more comfortable than pushing them away (Watts and Luoma 2020). Negative emotions may be less hurtful, thus accessible and possible to process (Kraehenmann et al. 2015; Mueller et al. 2017). With the increased processing of difficult situations, rumination may decrease concurrently. As the reduced neural response to negative stimuli is prolonged up to 1 week after psychedelic administration (Barrett et al. 2020), it may be beneficial to process negative life events within this timeframe. This way, psychedelic interventions may complement and enhance the effects of psychotherapeutic interventions like CBT or mindfulness-based relapse prevention that aim at developing skills to cope with heightened stress levels that are often associated with craving and drug use (Bowen et al. 2014). Mindfulness techniques used in relapse prevention are often used to increase tolerance of negative emotional and cognitive states, thereby decreasing the need to alleviate discomfort by engaging in substance use.

Furthermore, the stimulation of serotonin receptors by psychedelics may normalize serotonin neurotransmission and in turn decrease craving by increasing mood and reducing stress and attentional bias (Bogenschutz and Pommy 2012). With heightened mood and reduced attentional bias, drug cues may become less important and drug rewards less worth striving for. Therefore, integration sessions could focus on the normalization of natural rewards processing. To sustain decreased attentional bias, drug-cue reactivity combined with neurofeedback or virtual reality practice may support long-term changes and reinstate natural reward sensitivity.

However, as one study in depressed individuals showed that the neural response to negative emotions was increased the day after psilocybin administration before any integration therapy had taken place (Roseman et al. 2018), it is necessary that the process of negative emotion processing is carefully embedded in therapy and closely accompanied by a trained therapist. Especially shortly after the psychedelic experience, patients may be more vulnerable and may need more support from the therapists. Future research should investigate when best to discuss negative life events, whether patients need more support right after psychedelic administration, and what type of integration is necessary to achieve long-term improvements in emotion processing.

3.4 Implications of Social Connectedness

Increased emotional empathy induced by psilocybin may benefit the treatment of SUD by increasing social activity and decreasing social isolation. While SUD are often associated with severe social harms (Nutt et al. 2010; Volkow et al. 2011a) in combination with more general anhedonia (Garfield et al. 2014), patients with SUD receiving psychedelics may feel more connected with their environment and the people close to them especially during acute effects. This could not only improve the patient-therapist relationship but also make social interactions in general more meaningful and valuable. As long-term increases in prosocial behavior and interaction with the environment have been reported after exposure to a psychedelic, therapists should support patients in engaging in social activities that contribute to the patients' well-being. More specifically, psychedelics seem well suited to be combined with and enhance therapeutic benefits of approaches like behavioral activation treatment that aim at increasing social engagement of patients (Daughters et al. 2018). We also propose to evaluate carefully if group therapy sessions have an additional beneficial effect on therapeutic outcomes as interaction with other people is necessary to deepen social activities and social connectedness (Anderson et al. 2020; Kettner et al. 2021). Another approach could be to include family members or people close to the patients in the therapeutic process, which is already being done in MDMA-assisted treatment of PTSD-patients (e.g., Monson et al. 2020). However, therapists need to be aware that the experience of increased social connectedness induced by psychedelic substances requires them to be particularly mindful to keep their professional distance.

3.5 Implications of Subjective Experiences and Personal Meaning

The personal meaning of the experience seems to be essential for psychedelics to induce self-awareness, create insight, and increase interoception. Thus, it ought to be beneficial for patients to attribute personal meaning to the experience, gain insight and a better understanding of their disorder, increase self-awareness, and motivation. Furthermore, insight may give patients a new perspective on their dysfunctional behavior (i.e., avoidance, suppression, or rumination) and may lead to new ideas about the rationale for their addiction-related behaviors and potential possibility of change (i.e., acceptance). It has to be noted that excessive self-awareness has been associated with poor mental health as it can induce anxiety and rumination (Nezlek 2002). Psychedelic-induced self-awareness and facilitated insights might therefore need to be guided towards the development of new goals in life. This way, they could provide a rich source of ideas that can be further developed and applied in well-established semi-structured therapeutic approaches like MI. An increased range of possibilities and more flexible thinking together with altered self-perception may lead to the perception of increased self-efficacy to change their (substance use) behavior. Although self-efficacy is a good predictor of treatment outcome in SUD, there is no clear consensus on how to therapeutically enhance the perception of self-efficacy in SUD patients. Psychedelics may provide an important contribution in this context. Therefore, it may be essential to focus on topics like personal meaning, insight, awareness, and self-efficacy during preparation, the acute psychedelic experience, and the following integration sessions. These processes could potentially be augmented by practicing mindfulness techniques (see also Sect. 3.3) before or after the psychedelic experience to re-induce some aspects of the experience. Additionally, Hendricks (2018) suggests increasing the feeling of awe during the acute experience via exposure to nature, art, and music. Furthermore, gained insight needs to be transferred and implemented into daily life in order to facilitate long-term changes.

One main mechanism of action of psychedelics is gaining insight. Therefore, it seems essential to outline dysfunctional behavior patterns and define goals for the future during the preparation phase and also encourage changing of behavior and thinking patterns during integration sessions. Furthermore, new habits need to be strengthened to support long-term change. Current clinical studies in humans suggest that the subjective experience is necessary to gain insight and support positive clinical outcomes (e.g., Bogenschutz et al. 2015; Garcia-Romeu et al. 2014). However, recent research in mice contradicts these findings (Cameron et al. 2021). Further research is needed to answer this question.

3.6 Implications for Dose-Finding and Dosing Regimen

It is currently unknown which dose or dosing regimen is leading to the most positive treatment outcomes. In in-vivo experiments in mice, significant faster extinction learning in a cued fear conditioning paradigm was observed with low doses of psilocybin compared to high doses (Catlow et al. 2013). This study furthermore showed a dose-dependent decrease in hippocampal neurogenesis; with a trend towards increased neurogenesis by low doses of psilocybin (Catlow et al. 2013). However, there are currently no studies investigating the effect of psychedelics on learning in humans. Translating these animal findings into human studies is necessary to understand the impact of different doses on potential clinical mechanisms of action.

While these results point to the beneficial effect of low doses, Kurland et al. (1967) and Chwelos et al. (1959) proposed the administration of relatively high doses of LSD in patients with AUD to elicit a "peak-psychedelic" or mystical experience of ego loss. Additionally, Bogenschutz et al. (2015) reported lower acute effects in patients with AUD (measured with mystical experience questionnaire (MEQ) and hallucinogen rating scale (HRS)) than in healthy volunteers and, therefore, suggest that these patients require higher doses to have a strong acute experience. Given the assumption that mystical experiences are essential for positive treatment outcomes (Bogenschutz et al. 2015; Garcia-Romeu et al. 2014), higher doses of the psychedelic may be needed for clinical efficacy in this patient population. Yet, anecdotal reports also suggest a beneficial effect of very low, sub-threshold doses of psychedelics, so-called microdoses. Future research therefore needs to investigate whether microdoses of psychedelics (1) indeed have clinical efficacy and (2) if these effects are induced by a different mechanism of action as compared to those described in Sect. 2. Another question that has not been resolved yet is the most promising dosing regimen, i.e. is a single dose enough or are repeated doses necessary for a positive long-term treatment outcome? Our currently ongoing study will provide some answers to this question as it involves the administration of only a single dose and allows for results to be compared with previous studies using repeated dosing in patients with AUD and smoking cessation (Bogenschutz et al. 2015; Garcia-Romeu et al. 2014). Furthermore, it needs to be investigated if long-lasting results can be fostered by repeating the treatment cycle after some months.

4 Conclusion

In this chapter, we have outlined various potential mechanisms of action of psychedelics that could prove to have significant synergies with psychotherapeutic approaches. Today, most clinical trials that involve the administration of a psychedelic substance for the treatment of SUD include multiple sessions of preparation and integration based on anecdotal reports of therapists working with these substances – usually some form of talk therapy – combined with motivational enhancement approaches. However, given the momentum this research has gained including efforts to test psychedelic-assisted therapy in larger phase III trials in the future, it will be important to establish evidence-based psychotherapeutic interventions and optimize therapeutic approaches in order to achieve maximally beneficial treatment effects for patients. Understanding the clinical mechanisms of action of psychedelics will be critical to guide this optimization process as leveraging synergies between pharmacological mechanisms and non-pharmacological interventions represent a promising approach to achieve optimal treatment effects. In this chapter, we have outlined knowledge gaps, but also make concrete suggestions about possible psychotherapeutic interventions that seem well suited for a psychedelic-assisted therapy of SUD based on our current understanding of the neurobiology of these substances. We suggest exploring the benefits of fostering creative thinking styles, including family members or friends in the therapeutic process, and enable changes of perspectives and behaviors within an MI framework. We further propose to enhance learning strategies as conceptualized in CM or extinction learning approaches for SUD by psychedelic-induced neuroplasticity. Psychedelics have the potential to support behavioral therapies such as behavioral activation aiming at increasing social engagement. Additionally, we propose to leverage novel technologies, for example VR-based approaches and neurofeedback, to promote and sustain these effects. The efficacy of these combined approaches as well as optimal dose and timing needs to be investigated in future studies. Furthermore, the potential of personalizing these approaches as well as predictive markers for treatment success need to be studied. If these knowledge gaps can be filled, psychedelic-assisted therapy has a great potential to become an efficacious, evidence-based, and modern treatment for patients suffering from SUD.

References

Aldao A, Nolen-Hoeksema S, Schweizer S (2010) Emotion-regulation strategies across psychopathology: a meta-analytic review. Clin Psychol Rev 30:217–237. https://doi.org/10.1016/j.cpr.2009.11.004

Anderson BT, Danforth A, Daroff R, Stauffer C, Ekman E, Agin-Liebes G, Trope A, Boden MT, Dilley J, Mitchell J, Woolley J (2020) Psilocybin-assisted group therapy for demoralized older long-term AIDS survivor men: an open-label safety and feasibility pilot study. EClinicalMedicine 27:1–11. https://doi.org/10.1016/j.eclinm.2020.100538

Baker TB, Piper ME, McCarthy DE, Majeskie MR, Fiore MC (2004) Addiction motivation reformulated: an affective processing model of negative reinforcement. Psychol Rev 111:33–51. https://doi.org/10.1037/0033-295X.111.1.33

Barrett FS, Doss MK, Sepeda ND, Pekar JJ, Griffiths RR (2020) Emotions and brain function are altered up to one month after a single high dose of psilocybin. Sci Rep 10:1–14. https://doi.org/10.1038/s41598-020-59282-y

Bastian B, Kuppens P, Hornsey MJ, Park J, Koval P, Uchida Y (2012) Feeling bad about being sad: the role of social expectancies in amplifying negative mood. Emotion 12:69–80. https://doi.org/10.1037/a0024755

Belin D, Belin-Rauscent A, Murray JE, Everitt BJ (2013) Addiction: failure of control over maladaptive incentive habits. Curr Opin Neurobiol 23:564–572. https://doi.org/10.1016/j.conb.2013.01.025

Belmer A, Patkar OL, Pitman KM, Bartlett SE (2016) Serotonergic neuroplasticity in alcohol addiction. Brain Plast 1:177–206. https://doi.org/10.3233/BPL-150022

Bershad AK, Schepers ST, Bremmer MP, Lee R, de Wit H (2019) Acute subjective and behavioral effects of microdoses of lysergic acid diethylamide in healthy human volunteers. Biol Psychiatry 86:792–800. https://doi.org/10.1016/j.biopsych.2019.05.019

Beylergil SB, Beck A, Deserno L, Lorenz RC, Rapp MA, Schlagenhauf F, Heinz A, Obermayer K (2017) Dorsolateral prefrontal cortex contributes to the impaired behavioral adaptation in alcohol dependence. Neuroimage Clin 15:80–94. https://doi.org/10.1016/j.nicl.2017.04.010

Bogenschutz MP, Pommy JM (2012) Therapeutic mechanisms of classic hallucinogens in the treatment of addictions: from indirect evidence to testable hypotheses. Drug Test Anal 4:543–555. https://doi.org/10.1002/dta.1376

Bogenschutz MP, Forcehimes AA, Pommy JA, Wilcox CE, Barbosa PCR, Strassman RJ (2015) Psilocybin-assisted treatment for alcohol dependence: a proof-of-concept study. J Psychopharmacol (Oxf) 29:289–299. https://doi.org/10.1177/0269881114565144

Bogenschutz MP, Podrebarac SK, Duane JH, Amegadzie SS, Malone TC, Owens LT, Ross S, Mennenga SE (2018) Clinical interpretations of patient experience in a trial of psilocybin-assisted psychotherapy for alcohol use disorder. Front Pharmacol 9:1–10. https://doi.org/10.3389/fphar.2018.00100

Bowen S, Witkiewitz K, Clifasefi SL, Grow J, Chawla N, Hsu SH, Carroll HA, Harrop E, Collins SE, Lustyk MK, Larimer ME (2014) Relative efficacy of mindfulness-based relapse prevention, standard relapse prevention, and treatment as usual for substance use disorders: a randomized clinical trial. JAMA Psychiatry 71:547–556. https://doi.org/10.1001/jamapsychiatry.2013.4546

Cameron LP, Tombari RJ, Lu J, Pell AJ, Hurley ZQ, Ehinger Y, Vargas MV, McCarroll MN, Taylor JC, Myers-Turnbull D, Liu T, Yaghoobi B, Laskowski LJ, Anderson EI, Zhang G, Viswanathan J, Brown BM, Tjia M, Dunlap LE, Rabow ZT, Fiehn O, Wulff H, McCorvy JD,

Lein PJ, Kokel D, Ron D, Peters J, Zuo Y, Olson DE (2021) A non-hallucinogenic psychedelic analogue with therapeutic potential. Nature 589:474–479. https://doi.org/10.1038/s41586-020-3008-z

Carhart-Harris RL, Leech R, Williams TM, Erritzoe D, Abbasi N, Bargiotas T, Hobden P, Sharp DJ, Evans J, Feilding A, Wise RG, Nutt DJ (2012) Implications for psychedelic-assisted psychotherapy: functional magnetic resonance imaging study with psilocybin. Br J Psychiatry 200: 238–244. https://doi.org/10.1192/bjp.bp.111.103309

Carhart-Harris RL, Kaelen M, Bolstridge M, Williams TM, Williams LT, Underwood R, Feilding A, Nutt DJ (2016) The paradoxical psychological effects of lysergic acid diethylamide (LSD). Psychol Med 46:1379–1390. https://doi.org/10.1017/S0033291715002901

Catlow BJ, Song S, Paredes DA, Kirstein CL, Sanchez-Ramos J (2013) Effects of psilocybin on hippocampal neurogenesis and extinction of trace fear conditioning. Exp Brain Res 228:481–491. https://doi.org/10.1007/s00221-013-3579-0

Chwelos N, Blewett DB, Smith CM, Hoffer A (1959) Use of d -lysergic acid diethylamide in the treatment of alcoholism. Q J Stud Alcohol 20:577–590. https://doi.org/10.15288/qjsa.1959.20.577

Daughters SB, Magidson JF, Anand D, Seitz-Brown CJ, Chen Y, Baker S (2018) The effect of a behavioral activation treatment for substance use on post-treatment abstinence: a randomized controlled trial. Addiction 113:535–544. https://doi.org/10.1111/add.14049

David AS, Bedford N, Wiffen B, Gilleen J (2012) Failures of metacognition and lack of insight in neuropsychiatric disorders. Philos Trans R Soc Lond Ser B Biol Sci 367:1379–1390. https://doi.org/10.1098/rstb.2012.0002

Davis AK, Barrett FS, Griffiths RR (2020) Psychological flexibility mediates the relations between acute psychedelic effects and subjective decreases in depression and anxiety. J Contextual Behav Sci 15:39–45. https://doi.org/10.1016/j.jcbs.2019.11.004

Dingle GA, Da Neves DC, Alhadad SSJ, Hides L (2018) Individual and interpersonal emotion regulation among adults with substance use disorders and matched controls. Br J Clin Psychol 57:186–202. https://doi.org/10.1111/bjc.12168

Dolder PC, Schmid Y, Müller F, Borgwardt S, Liechti ME (2016) LSD acutely impairs fear recognition and enhances emotional empathy and sociality. Neuropsychopharmacology 41: 2638–2646. https://doi.org/10.1038/npp.2016.82

Dyck E (2006) 'Hitting highs at rock bottom': LSD treatment for alcoholism, 1950–1970. Soc Hist Med 19:313–329. https://doi.org/10.1093/shm/hkl039

Erritzoe D, Roseman L, Nour MM, MacLean K, Kaelen M, Nutt DJ, Carhart-Harris RL (2018) Effects of psilocybin therapy on personality structure. Acta Psychiatr Scand 138:368–378. https://doi.org/10.1111/acps.12904

Ersche KD, Meng C, Ziauddeen H, Stochl J, Williams GB, Bullmore ET, Robbins TW (2020) Brain networks underlying vulnerability and resilience to drug addiction. Proc Natl Acad Sci U S A 117:15253–15261. https://doi.org/10.1073/pnas.2002509117

Feldstein Ewing SW, Filbey FM, Hendershot CS, McEachern AD, Hutchison KE (2011) Proposed model of the neurobiological mechanisms underlying psychosocial alcohol interventions: the example of motivational interviewing. J Stud Alcohol Drugs 72:903–916. https://doi.org/10.15288/jsad.2011.72.903

Frokjaer VG, Mortensen EL, Nielsen FA, Haugbol S, Pinborg LH, Adams KH, Svarer C, Hasselbalch SG, Holm S, Paulson OB, Knudsen GM (2008) Frontolimbic serotonin 2A receptor binding in healthy subjects is associated with personality risk factors for affective disorder. Biol Psychiatry 63:569–576. https://doi.org/10.1016/j.biopsych.2007.07.009

Garcia-Romeu A, Griffiths RR, Johnson MW (2014) Psilocybin-occasioned mystical experiences in the treatment of tobacco addiction. Curr Drug Abuse Rev 7:157–164. https://doi.org/10.2174/1874473708666150107121331

Garcia-Romeu A, Davis AK, Erowid F, Erowid E, Griffiths RR, Johnson MW (2019) Cessation and reduction in alcohol consumption and misuse after psychedelic use. J Psychopharmacol (Oxf) 33:1088–1101. https://doi.org/10.1177/0269881119845793

Garfield JBB, Lubman DI, Yücel M (2014) Anhedonia in substance use disorders: a systematic review of its nature, course and clinical correlates. Aust N Z J Psychiatry 48:36–51. https://doi.org/10.1177/0004867413508455

GBD 2017 Risk Factor Collaborators (2018) Global, regional, and national comparative risk assessment of 84 behavioural, environmental and occupational, and metabolic risks or clusters of risks for 195 countries and territories, 1990–2017: a systematic analysis for the Global Burden of Disease Study 2017. Lancet 392:1923–1994. https://doi.org/10.1016/S0140-6736(18)32225-6

Geng X, Hu Y, Gu H, Salmeron BJ, Adinoff B, Stein EA, Yang Y (2017) Salience and default mode network dysregulation in chronic cocaine users predict treatment outcome. Brain 140:1513–1524. https://doi.org/10.1093/brain/awx036

Goldstein RZ, Alia-Klein N, Tomasi D, Zhang L, Cottone LA, Maloney T, Telang F, Caparelli EC, Chang L, Ernst T, Samaras D, Squires NK, Volkow ND (2007) Is decreased prefrontal cortical sensitivity to monetary reward associated with impaired motivation and self-control in cocaine addiction? Am J Psychiatry 164:43–51. https://doi.org/10.1176/ajp.2007.164.1.43

Goldstein RZ, Craig ADB, Bechara A, Garavan H, Childress AR, Paulus MP, Volkow ND (2009) The neurocircuitry of impaired insight in drug addiction. Trends Cogn Sci 13:372–380. https://doi.org/10.1016/j.tics.2009.06.004

Griffiths R, Richards W, Johnson M, McCann U, Jesse R (2008) Mystical-type experiences occasioned by psilocybin mediate the attribution of personal meaning and spiritual significance 14 months later. J Psychopharmacol (Oxf) 22:621–632. https://doi.org/10.1177/0269881108094300

Griffiths RR, Johnson MW, Carducci MA, Umbricht A, Richards WA, Richards BD, Cosimano MP, Klinedinst MA (2016) Psilocybin produces substantial and sustained decreases in depression and anxiety in patients with life-threatening cancer: a randomized double-blind trial. J Psychopharmacol (Oxf) 30:1181–1197. https://doi.org/10.1177/0269881116675513

Griffiths RR, Johnson MW, Richards WA, Richards BD, Jesse R, MacLean KA, Barrett FS, Cosimano MP, Klinedinst MA (2018) Psilocybin-occasioned mystical-type experience in combination with meditation and other spiritual practices produces enduring positive changes in psychological functioning and in trait measures of prosocial attitudes and behaviors. J Psychopharmacol (Oxf) 32:49–69. https://doi.org/10.1177/0269881117731279

Hendricks PS (2018) Awe: a putative mechanism underlying the effects of classic psychedelic-assisted psychotherapy. Int Rev Psychiatry 30:331–342. https://doi.org/10.1080/09540261.2018.1474185

Hyman SE (2005) Addiction: a disease of learning and memory. Am J Psychiatry 162:1414–1422. https://doi.org/10.1176/appi.ajp.162.8.1414

Jasinska AJ, Stein EA, Kaiser J, Naumer MJ, Yalachkov Y (2014) Factors modulating neural reactivity to drug cues in addiction: a survey of human neuroimaging studies. Neurosci Biobehav Rev 38:1–16. https://doi.org/10.1016/j.neubiorev.2013.10.013

Johnson MW, Garcia-Romeu A, Griffiths RR (2017) Long-term follow-up of psilocybin-facilitated smoking cessation. Am J Drug Alcohol Abuse 43:55–60. https://doi.org/10.3109/00952990.2016.1170135

Kauer JA, Malenka RC (2007) Synaptic plasticity and addiction. Nat Rev Neurosci 8:844–858. https://doi.org/10.1038/nrn2234

Kettner H, Rosas FE, Timmermann C, Kärtner L, Carhart-Harris RL, Roseman L (2021) Psychedelic communitas: intersubjective experience during psychedelic group sessions predicts enduring changes in psychological wellbeing and social connectedness. Front Pharmacol 12:1–20. https://doi.org/10.3389/fphar.2021.623985

Kirschner M, Sladky R, Haugg A, Stämpfli P, Jehli E, Hodel M, Engeli E, Hösli S, Baumgartner MR, Sulzer J, Huys QJM, Seifritz E, Quednow BB, Scharnowski F, Herdener M (2018) Self-regulation of the dopaminergic reward circuit in cocaine users with mental imagery and neurofeedback. EBioMedicine 37:489–498. https://doi.org/10.1016/j.ebiom.2018.10.052

Kometer M, Schmidt A, Bachmann R, Studerus E, Seifritz E, Vollenweider FX (2012) Psilocybin biases facial recognition, goal-directed behavior, and mood state toward positive relative to negative emotions through different serotonergic subreceptors. Biol Psychiatry 72:898–906. https://doi.org/10.1016/j.biopsych.2012.04.005

Koob GF, Le Moal M (2005) Plasticity of reward neurocircuitry and the 'dark side' of drug addiction. Nat Neurosci 8:1442–1444. https://doi.org/10.1038/nn1105-1442

Kraehenmann R, Preller KH, Scheidegger M, Pokorny T, Bosch OG, Seifritz E, Vollenweider FX (2015) Psilocybin-induced decrease in amygdala reactivity correlates with enhanced positive mood in healthy volunteers. Biol Psychiatry 78:572–581. https://doi.org/10.1016/j.biopsych.2014.04.010

Krebs TS, Johansen P-Ø (2012) Lysergic acid diethylamide (LSD) for alcoholism: meta-analysis of randomized controlled trials. J Psychopharmacol (Oxf) 26:994–1002. https://doi.org/10.1177/0269881112439253

Kurland AA, Unger S, Shaffer JW, Savage C (1967) Psychedelic therapy utilizing LSD in the treatment of the alcoholic patient: a preliminary report. Am J Psychiatry 123:1202–1209. https://doi.org/10.1176/ajp.123.10.1202

Le Berre A-P (2019) Emotional processing and social cognition in alcohol use disorder. Neuropsychology 33:808–821. https://doi.org/10.1037/neu0000572

Lovinger DM, Abrahao KP (2018) Synaptic plasticity mechanisms common to learning and alcohol use disorder. Learn Mem 25:425–434. https://doi.org/10.1101/lm.046722.117

Lüscher C (2016) The emergence of a circuit model for addiction. Annu Rev Neurosci 39:257–276. https://doi.org/10.1146/annurev-neuro-070815-013920

Ly C, Greb AC, Cameron LP, Wong JM, Barragan EV, Wilson PC, Burbach KF, Soltanzadeh Zarandi S, Sood A, Paddy MR, Duim WC, Dennis MY, McAllister AK, Ori-McKenney KM, Gray JA, Olson DE (2018) Psychedelics promote structural and functional neural plasticity. Cell Rep 23:3170–3182. https://doi.org/10.1016/j.celrep.2018.05.022

MacLean KA, Johnson MW, Griffiths RR (2011) Mystical experiences occasioned by the hallucinogen psilocybin lead to increases in the personality domain of openness. J Psychopharmacol (Oxf) 25:1453–1461. https://doi.org/10.1177/0269881111420188

Madsen MK, Fisher PM, Stenbæk DS, Kristiansen S, Burmester D, Lehel S, Páleníček T, Kuchař M, Svarer C, Ozenne B, Knudsen GM (2020) A single psilocybin dose is associated with long-term increased mindfulness, preceded by a proportional change in neocortical 5-HT2A receptor binding. Eur Neuropsychopharmacol 33:71–80. https://doi.org/10.1016/j.euroneuro.2020.02.001

Mason NL, Kuypers KPC, Müller F, Reckweg J, Tse DHY, Toennes SW, Hutten NRPW, Jansen JFA, Stiers P, Feilding A, Ramaekers JG (2020) Me, myself, bye: regional alterations in glutamate and the experience of ego dissolution with psilocybin. Neuropsychopharmacology 45:2003–2011. https://doi.org/10.1038/s41386-020-0718-8

Mason NL, Kuypers KPC, Reckweg JT, Müller F, Tse DHY, Da Rios B, Toennes SW, Stiers P, Feilding A, Ramaekers JG (2021) Spontaneous and deliberate creative cognition during and after psilocybin exposure. Transl Psychiatry 11:1–13. https://doi.org/10.1038/s41398-021-01335-5

McCulloch DE-W, Madsen MK, Stenbæk DS, Kristiansen S, Ozenne B, Jensen PS, Knudsen GM, Fisher PM (2021) Lasting effects of a single psilocybin dose on resting-state functional connectivity in healthy individuals. J Psychopharmacol (Oxf):1–11. https://doi.org/10.1177/02698811211026454

Moeller SJ, Goldstein RZ (2014) Impaired self-awareness in human addiction: deficient attribution of personal relevance. Trends Cogn Sci 18:635–641. https://doi.org/10.1016/j.tics.2014.09.003

Monson CM, Wagner AC, Mithoefer AT, Liebman RE, Feduccia AA, Jerome L, Yazar-Klosinski-B, Emerson A, Doblin R, Mithoefer MC (2020) MDMA-facilitated cognitive-behavioural conjoint therapy for posttraumatic stress disorder: an uncontrolled trial. Eur J Psychotraumatol 11:1–7. https://doi.org/10.1080/20008198.2020.1840123

Monterosso JR, Flannery BA, Pettinati HM, Oslin DW, Rukstalis M, O'Brien CP, Volpicelli JR (2001) Predicting treatment response to naltrexone: the influence of craving and family history. Am J Addict 10(3):258–268. https://doi.org/10.1080/105504901750532148

Moos RH, Moos BS (2006) Rates and predictors of relapse after natural and treated remission from alcohol use disorders. Addiction 101:212–222. https://doi.org/10.1111/j.1360-0443.2006.01310.x

Mueller F, Lenz C, Dolder PC, Harder S, Schmid Y, Lang UE, Liechti ME, Borgwardt S (2017) Acute effects of LSD on amygdala activity during processing of fearful stimuli in healthy subjects. Transl Psychiatry 7:1–5. https://doi.org/10.1038/tp.2017.54

Nardou R, Lewis EM, Rothhaas R, Xu R, Yang A, Boyden E, Dölen G (2019) Oxytocin-dependent reopening of a social reward learning critical period with MDMA. Nature 569:116–120. https://doi.org/10.1038/s41586-019-1075-9

Nezlek JB (2002) Day-to-day relationships between self-awareness, daily events, and anxiety. J Pers 70:249–275. https://doi.org/10.1111/1467-6494.05005

Nichols DE (2016) Psychedelics. Pharmacol Rev 68:264–355. https://doi.org/10.1124/pr.115.011478

Noorani T, Garcia-Romeu A, Swift TC, Griffiths RR, Johnson MW (2018) Psychedelic therapy for smoking cessation: qualitative analysis of participant accounts. J Psychopharmacol (Oxf) 32: 756–769. https://doi.org/10.1177/0269881118780612

Nutt DJ, King LA, Phillips LD (2010) Drug harms in the UK: a multicriteria decision analysis. Lancet 376:1558–1565. https://doi.org/10.1016/S0140-6736(10)61462-6

Pasquini L, Palhano-Fontes F, Araujo DB (2020) Subacute effects of the psychedelic ayahuasca on the salience and default mode networks. J Psychopharmacol (Oxf) 34:623–635. https://doi.org/10.1177/0269881120909409

Petry NM, Petrakis I, Trevisan L, Wiredu G, Boutros NN, Martin B, Kosten TR (2001) Contingency management interventions: from research to practice. Am J Psychiatry 158:694–702. https://doi.org/10.1176/appi.ajp.158.5.694

Petry NM, Alessi SM, Rash CJ (2013) A randomized study of contingency management in cocaine-dependent patients with severe and persistent mental health disorders. Drug Alcohol Depend 130:234–237. https://doi.org/10.1016/j.drugalcdep.2012.10.017

Pettinati HM, Volpicelli JR, Pierce JD, O'Brien CP (2000) Improving naltrexone response: an intervention for medical practitioners to enhance medication compliance in alcohol dependent patients. J Addict Dis 19(1):71–83. https://doi.org/10.1300/J069v19n01_06

Pokorny T, Preller KH, Kometer M, Dziobek I, Vollenweider FX (2017) Effect of psilocybin on empathy and moral decision-making. Int J Neuropsychopharmacol 20:747–757. https://doi.org/10.1093/ijnp/pyx047

Pokorny T, Duerler P, Seifritz E, Vollenweider FX, Preller KH (2020) LSD acutely impairs working memory, executive functions, and cognitive flexibility, but not risk-based decision-making. Psychol Med 50:2255–2264. https://doi.org/10.1017/S0033291719002393

Preller KH, Herdener M, Schilbach L, Stämpfli P, Hulka LM, Vonmoos M, Ingold N, Vogeley K, Tobler PN, Seifritz E, Quednow BB (2014) Functional changes of the reward system underlie blunted response to social gaze in cocaine users. Proc Natl Acad Sci U S A 111:2842–2847. https://doi.org/10.1073/pnas.1317090111

Preller KH, Pokorny T, Hock A, Kraehenmann R, Stämpfli P, Seifritz E, Scheidegger M, Vollenweider FX (2016) Effects of serotonin 2A/1A receptor stimulation on social exclusion processing. Proc Natl Acad Sci U S A 113:5119–5124. https://doi.org/10.1073/pnas.1524187113

Preller KH, Burt JB, Ji JL, Schleifer CH, Adkinson BD, Stämpfli P, Seifritz E, Repovs G, Krystal JH, Murray JD, Vollenweider FX, Anticevic A, Hunt LT, Behrens TE (2018) Changes in global and thalamic brain connectivity in LSD-induced altered states of consciousness are attributable to the 5-HT2A receptor. elife 27:451–457. https://doi.org/10.7554/eLife.35082

Preller KH, Razi A, Zeidman P, Stämpfli P, Friston KJ, Vollenweider FX (2019) Effective connectivity changes in LSD-induced altered states of consciousness in humans. Proc Natl Acad Sci 116:2743–2748

Preller KH, Duerler P, Burt JB, Ji JL, Adkinson B, Stämpfli P, Seifritz E, Repovs G, Krystal JH, Murray JD (2020) Psilocybin induces time-dependent changes in global functional connectivity: psi-induced changes in brain connectivity. Biol Psychiatry 88:197–207. https://doi.org/10.1016/j.biopsych.2019.12.027

Prendergast M, Podus D, Finney J, Greenwell L, Roll J (2006) Contingency management for treatment of substance use disorders: a meta-analysis. Addiction 101:1546–1560. https://doi.org/10.1111/j.1360-0443.2006.01581.x

Quednow BB (2020) Social cognition in addiction. In: Verdejo-Garcia A (ed) Cognition and addiction. Academic Press, pp 63–78

Quednow BB, Kometer M, Geyer MA, Vollenweider FX (2012) Psilocybin-induced deficits in automatic and controlled inhibition are attenuated by ketanserin in healthy human volunteers. Neuropsychopharmacology 37:630–640. https://doi.org/10.1038/npp.2011.228

Raval NR, Johansen A, Donovan LL, Ros NF, Ozenne B, Hansen HD, Knudsen GM (2021) A single dose of psilocybin increases synaptic density and decreases 5-HT2A receptor density in the pig brain. Int J Mol Sci 22:1–14. https://doi.org/10.3390/ijms22020835

Riba J, Rodríguez-Fornells A, Barbanoj MJ (2002) Effects of ayahuasca on sensory and sensorimotor gating in humans as measured by P50 suppression and prepulse inhibition of the startle reflex, respectively. Psychopharmacology 165:18–28. https://doi.org/10.1007/s00213-002-1237-5

Rollnick S, Miller WR (1995) What is motivational interviewing? Behav Cogn Psychoth 23:325–334

Roseman L, Demetriou L, Wall MB, Nutt DJ, Carhart-Harris RL (2018) Increased amygdala responses to emotional faces after psilocybin for treatment-resistant depression. Neuropharmacology 142:263–269. https://doi.org/10.1016/j.neuropharm.2017.12.041

Ross S, Bossis A, Guss J, Agin-Liebes G, Malone T, Cohen B, Mennenga SE, Belser A, Kalliontzi K, Babb J, Su Z, Corby P, Schmidt BL (2016) Rapid and sustained symptom reduction following psilocybin treatment for anxiety and depression in patients with life-threatening cancer: a randomized controlled trial. J Psychopharmacol (Oxf) 30:1165–1180. https://doi.org/10.1177/0269881116675512

Salamone JD, Correa M (2012) The mysterious motivational functions of mesolimbic dopamine. Neuron 76:470–485. https://doi.org/10.1016/j.neuron.2012.10.021

Sampedro F, de La Fuente Revenga M, Valle M, Roberto N, Domínguez-Clavé E, Elices M, Luna LE, Crippa JAS, Hallak JEC, de Araujo DB, Friedlander P, Barker SA, Álvarez E, Soler J, Pascual JC, Feilding A, Riba J (2017) Assessing the psychedelic "after-glow" in Ayahuasca users: post-acute neurometabolic and functional connectivity changes are associated with enhanced mindfulness capacities. Int J Neuropsychopharmacol 20:698–711. https://doi.org/10.1093/ijnp/pyx036

Scarlata MJ, Keeley RJ, Stein EA (2021) Nicotine addiction: translational insights from circuit neuroscience. Pharmacol Biochem Behav 204:1–7. https://doi.org/10.1016/j.pbb.2021.173171

Schmid Y, Liechti ME (2018) Long-lasting subjective effects of LSD in normal subjects. Psychopharmacology 235:535–545. https://doi.org/10.1007/s00213-017-4733-3

Schmid Y, Enzler F, Gasser P, Grouzmann E, Preller KH, Vollenweider FX, Brenneisen R, Müller F, Borgwardt S, Liechti ME (2015) Acute effects of lysergic acid diethylamide in healthy subjects. Biol Psychiatry 78:544–553. https://doi.org/10.1016/j.biopsych.2014.11.015

Segawa T, Baudry T, Bourla A, Blanc J-V, Peretti C-S, Mouchabac S, Ferreri F (2019) Virtual reality (VR) in assessment and treatment of addictive disorders: a systematic review. Front Neurosci 13:1–14. https://doi.org/10.3389/fnins.2019.01409

Shao L-X, Liao C, Gregg I, Davoudian PA, Savalia NK, Delagarza K, Kwan AC (2021) Psilocybin induces rapid and persistent growth of dendritic spines in frontal cortex in vivo. Neuron 109:2535–2544. https://doi.org/10.1016/j.neuron.2021.06.008

Starosta AN, Leeman RF, Volpicelli JR (2006) The BRENDA model: integrating psychosocial treatment and pharmacotherapy for the treatment of alcohol use disorders. J Psychiatr Pract 12 (2):80–89. https://doi.org/10.1097/00131746-200603000-00003

Stein MD, Herman DS, Anderson BJ (2009) A motivational intervention trial to reduce cocaine use. J Subst Abus Treat 36:118–125. https://doi.org/10.1016/j.jsat.2008.05.003

Tobler PN, Preller KH, Campbell-Meiklejohn DK, Kirschner M, Kraehenmann R, Stämpfli P, Herdener M, Seifritz E, Quednow BB (2016) Shared neural basis of social and non-social reward deficits in chronic cocaine users. Soc Cogn Affect Neurosci 11:1017–1025. https://doi.org/10.1093/scan/nsw030

Vergara VM, Liu J, Claus ED, Hutchison K, Calhoun V (2017) Alterations of resting state functional network connectivity in the brain of nicotine and alcohol users. NeuroImage 151: 45–54. https://doi.org/10.1016/j.neuroimage.2016.11.012

Volkow ND, Fowler JS, Wang G-J (2003) The addicted human brain: insights from imaging studies. J Clin Invest 111:1444–1451. https://doi.org/10.1172/JCI18533

Volkow ND, Baler RD, Goldstein RZ (2011a) Addiction: pulling at the neural threads of social behaviors. Neuron 69:599–602. https://doi.org/10.1016/j.neuron.2011.01.027

Volkow ND, Tomasi D, Wang G-J, Fowler JS, Telang F, Goldstein RZ, Alia-Klein N, Wong C (2011b) Reduced metabolism in brain "control networks" following cocaine-cues exposure in female cocaine abusers. PLoS One 6:1–7. https://doi.org/10.1371/journal.pone.0016573

Volkow ND, Michaelides M, Baler R (2019) The neuroscience of drug reward and addiction. Physiol Rev 99:2115–2140. https://doi.org/10.1152/physrev.00014.2018

Vollenweider FX, Preller KH (2020) Psychedelic drugs: neurobiology and potential for treatment of psychiatric disorders. Nat Rev Neurosci 21:611–624. https://doi.org/10.1038/s41583-020-0367-2

Vollenweider FX, Csomor PA, Knappe B, Geyer MA, Quednow BB (2007) The effects of the preferential 5-HT2A agonist psilocybin on prepulse inhibition of startle in healthy human volunteers depend on interstimulus interval. Neuropsychopharmacology 32:1876–1887. https://doi.org/10.1038/sj.npp.1301324

Watts R, Luoma JB (2020) The use of the psychological flexibility model to support psychedelic assisted therapy. J Contextual Behav Sci 15:92–102. https://doi.org/10.1016/j.jcbs.2019.12.004

Watts R, Day C, Krzanowski J, Nutt D, Carhart-Harris R (2017) Patients' accounts of increased "connectedness" and "acceptance" after psilocybin for treatment-resistant depression. J Humanist Psychol 57:520–564. https://doi.org/10.1177/0022167817709585

Wise RA (2008) Dopamine and reward: the anhedonia hypothesis 30 years on. Neurotox Res 14: 169–183. https://doi.org/10.1007/BF03033808

Zhang G, Ásgeirsdóttir HN, Cohen SJ, Munchow AH, Barrera MP, Stackman RW (2013) Stimulation of serotonin 2A receptors facilitates consolidation and extinction of fear memory in C57BL/6J mice. Neuropharmacology 64:403–413. https://doi.org/10.1016/j.neuropharm.2012.06.007

Zhang R, Volkow ND (2019) Brain default-mode network dysfunction in addiction. NeuroImage 200:313–331. https://doi.org/10.1016/j.neuroimage.2019.06.036

Classic Psychedelics in Addiction Treatment: The Case for Psilocybin in Tobacco Smoking Cessation

Matthew W. Johnson

Contents

Abstract This manuscript reviews research suggesting that classic psychedelics (5-HT2A receptor agonists) are effective in treating addictions including tobacco use disorder. I review historical research from the 1950s to 1970s suggesting that classic psychedelics are associated with addiction recovery across pharmacologically distinct drugs of addiction. I then review anthropological reports about ceremonial use of classic psychedelics and epidemiological studies that are consistent with anti-addiction efficacy. I review modern research using psilocybin in the treatment of alcohol use disorder and tobacco use disorder. Both lines of research show high success rates in preliminary studies. General anti-addiction efficacy across a variety of classes of addictive drugs is consistent with the notion that the persisting positive behavior change prompted by psychedelic therapy is due to amplification of psychotherapeutic processes. Future research should examine classic psychedelic treatment of additional substance use disorders including for opioids, cocaine, methamphetamine, and cannabis, and other disorders broadly characterized as addictions (e.g., obesity, problem gambling, hypersexual disorder). Future research should also explore addiction treatments with other classic psychedelics including LSD, mescaline, DMT, 5-MeO-DMT, and yet-to-be-discovered

M. W. Johnson (✉)
Department of Psychiatry and Behavioral Sciences, Johns Hopkins University School of Medicine, Baltimore, MD, USA
e-mail: mwj@jhu.edu

Curr Topics Behav Neurosci (2022) 56: 213–228
https://doi.org/10.1007/7854_2022_327
Published Online: 16 June 2022

compounds. Experimental research is also needed to test different protocols for the delivery of classic psychedelic therapy for addictions. Given the staggering society costs of substance use disorders, including the mortality caused by tobacco smoking, it is critical that public funding be made available for scientists to follow up on promising early findings of classic psychedelics in addiction treatment. The costs and risks of not conducting such research are too great.

Keywords Addiction · Hallucinogens · Psilocybin · Psychedelics · Smoking cessation · Substance use disorders · Tobacco use disorder

Annual smoking-related mortalities are estimated at 480,000 in the USA (US Department of Health and Human Services 2014) and nine million annually worldwide (Murray et al. 2020). These numbers dwarf similar statistics for all other drugs of abuse. In the USA, smoking is responsible for approximately five to seven times more deaths than alcohol, and approximately 26 times more deaths than all illicit drugs combined (Danaei et al. 2009; Mokdad et al. 2004). Most behavioral interventions and pharmacotherapies for smoking cessation have only modest long-term success rates (e.g., typically <35% at 6 months post-quit attempt; Cahill et al. 2014; Mottillo et al. 2009). These statistics are particularly devastating because 68% of US smokers want to quit (Babb et al. 2017). The staggering death toll of smoking and unmet clinical need means that novel approaches are desperately needed.

This manuscript reviews research suggesting that classic psychedelics might be effective in treating substance use disorders including tobacco use disorder. The classic psychedelics are defined as drugs with agonist activity at 5-HT2A receptor (5-HT2AR) as a primary mechanism, with prototypes including psilocybin (which is a prodrug for psilocin), lysergic acid diethylamide (LSD), mescaline, and dimethyltryptamine (DMT). Within this context of classic psychedelics as potential anti-addiction medicines, a special focus will be placed on work examining psilocybin as a smoking cessation medication for tobacco use disorder. First, I will describe historical clinical research evidence from the 1950s to 1970s suggesting that classic psychedelics are associated with addiction recovery across pharmacologically distinct drugs of addiction. I will then review nonclinical lines of evidence suggesting that classic psychedelics may possess anti-addiction efficacy, including anthropological reports about ceremonial use of classic psychedelics by indigenous societies and syncretic religions, and epidemiological studies. Next, I will describe modern research examining psilocybin in the treatment of alcohol use disorder. Then I will review extensively the evidence related to psilocybin and other classic psychedelics and tobacco smoking cessation. I will conclude by recommending future directions.

1 Historical Research

Although anthropological evidence suggests that 5-HT2AR agonists have been used by various indigenous peoples as sacraments and healing agents before recorded history (Akers et al. 2011), in the mid-twentieth century they came to occupy a place at the cutting edge of psychiatric research (Johnson et al. 2008). Although studies generally did not meet modern scientific standards (Bonson 2018), over 1,000 papers were published during the mid-twentieth century that described the treatment of over 40,000 patients with 5-HT2AR agonists (Grinspoon 1981). In addition to basic research, therapeutic applications of 5-HT2AR agonists were investigated, with promising findings in the treatment of both addiction and cancer-related existential distress (Johnson and Griffiths 2017).

During the twentieth-century zenith of 5-HT2AR agonist research from the 1950s through the 1970s, addiction was a main therapeutic target. The primary form of addiction treated was alcoholism, and the most frequently studied classic psychedelic to treat alcoholism was LSD, although mescaline (e.g., Smith 1958, 1959) and dipropyltryptamine (Rhead et al. 1977) were also examined. Investigators in Saskatchewan, Canada hypothesized that classic psychedelics might mimic delirium tremens, which sometimes prompted sobriety (Chwelos et al. 1959; Dyck 2006; Smith 1958). However, undergoing delirium tremens was dangerous and sometimes fatal, so the rationale was to prompt a similar experience with a drug like LSD which was relatively safe physiologically. Instead of observing reactions resembling delirium tremens, researchers often observed "mind-manifesting" experiences that could be leveraged for psychotherapy and prompt sobriety. Some of the early studies in the literature showed nonsignificant positive trends for LSD, leading to conclusions that efficacy was undetermined (e.g., Abuzzahab and Anderson 1971; Mangini 1998). More recently, a meta-analysis examined all of the studies that randomized patients to an LSD or control condition (six studies; $N = 536$), and found in aggregate a significant and large decrease in alcohol misuse at the first follow-up ($\geq$1 month post-treatment) for the LSD groups compared to control groups (Krebs and Johansen 2012).

Early research hinted that the classic psychedelics might have broad applicability in treating substance use disorders beyond alcoholism. Savage and McCabe (1973) tested LSD in the treatment of opioid dependence in a study of 78 heroin-using individuals under judicial correctional supervision. The study randomized individuals to undergo either a single administration of LSD (300–450 microgram) during a six-week residential stay or undergo a control condition with outpatient clinic visits, daily urine drug testing, and weekly group therapy. At all timepoints examined, the group that received LSD showed significantly greater biologically confirmed drug abstinence. At this longest follow-up of 12 months post-treatment, continuous abstinence rates were 25% and 5% in the LSD group and control groups, respectively. Although the results were limited by the fact that the LSD group received a six-week residential stay not experienced by the control group, the large difference in biologically confirmed abstinence indicated the approach was worthy of more intensive follow-up.

2 Resurgence of Modern-Day Research

Despite initial excitement, research on 5-HT2AR agonists became increasingly marginalized due to their growing use outside of clinical research settings, and their association with the counter-culture movement in the late 1960s and early 1970s. In the 1990s a small number of investigators in Europe and the USA re-initiated human studies with 5-HT2AR agonists (e.g., Spitzer et al. 1996; Strassman and Qualls 1994; Vollenweider et al. 1997). Non-human research in the intervening decades had identified agonist activity at 5-HT2AR as a key mechanism underlying the effects of classic psychedelics (e.g., Glennon et al. 1984), which include LSD as well as psilocybin (present in many mushroom species), mescaline (present in peyote and some other cacti), and dimethyltryptamine (DMT; present in a variety of plants). Studies by researchers since the 1990s have followed safety guidelines for administering 5-HT2AR agonists (Johnson et al. 2008). These guidelines involve careful screening and preparation before drug administration sessions, close monitoring during sessions, and follow-up care involving both clinically supportive discussion of session experiences and assessment for any adverse effects resulting from the session. Therapeutic effects of 5-HT2AR agonists have been reported for depression and anxiety related to cancer and other life-threatening illness (Gasser et al. 2014; Griffiths et al. 2016; Grob et al. 2011; Ross et al. 2016), major depressive disorder (Carhart-Harris et al. 2016, 2021; Davis et al. 2021; Palhano-Fontes et al. 2019), tobacco use disorder (Johnson et al. 2014, 2017a), and alcohol use disorder (Bogenschutz et al. 2015). Some studies have been randomized trials, while others have been initial non-randomized pilot studies designed to establish safety in new populations and test the waters for future randomized trials. Some of these studies have reported rapid efficacy persisting for at least 6 months after one or a few 5-HT2AR agonist administrations (for review, see Johnson et al. 2019).

3 Anthropological Evidence

Various anthropological reports have suggested that 5-HT2A agonist use in the context of ritualized sacramental practices of indigenous cultures is associated with high rates of recovery from addiction, achievement of long-term abstinence, and lower rates of addictive substance use. Such reports include those of peyote (5-HT2AR agonist mescaline) ceremonies within the Native American Church in the USA (Albaugh and Anderson 1974; Bergman 1971; Blum et al. 1977; Calabrese 1997; Menninger 1971; Pascarosa and Futterman 1976; Roy 1973) and ayahuasca (5-HT2AR agonist dimethyltryptamine) ceremonies within certain Amazonian societies and syncretic religions (Dobkin de Rios et al. 2002; Doering-Silveira et al. 2005; Fábregas et al. 2010; Halpern et al. 2008). It is not possible to definitively differentiate the causal role of the drug from the larger context of religious guidance and community support provided in these settings. However, one large survey of over 8000 people in over 40 countries reported regression results suggesting that the association of ayahuasca use and lower use of addictive substances was apparent even after controlling for religious and social variables (Perkins et al. 2022).

4 Epidemiological Research

Using 6 years of data from the National Survey on Drug Use and Health, one analysis among people who had ever used illicit opioids ($N = 44{,}000$) found that a lifetime history of 5-HT2AR agonist use was associated with a significant 27% reduced risk of opioid dependence in the past year (Pisano et al. 2017). My laboratory recently conducted an online survey of individuals ($N = 343$) with prior AUD reporting cessation or reduction in alcohol use following 5-HT2AR agonist use in nonclinical settings (Garcia-Romeu et al. 2019). Participants reported 7 years of problematic alcohol use on average before the psychedelic experience to which they attributed reduced alcohol consumption. We also recently conducted an online survey of individuals reporting cessation or reduction in cannabis, opioid, or stimulant use following 5-HT2AR agonist use in nonclinical settings (Garcia-Romeu et al. 2020). Participants ($N = 444$) reported 4.5 years of problematic substance use on average before the psychedelic experience to which they attributed a reduction in drug consumption. In both Hopkins survey studies, greater 5-HT2AR agonist dose, insight, and personal meaning were associated with greater reductions in substance use. Like the anthropological studies, these surveys cannot definitively address the causal role of 5-HT2AR agonist use on addiction. However, combined with the clinical research described above, these survey data are suggestive of anti-addiction efficacy. It should be noted that the uncontrolled use of 5-HT2AR agonists is also associated with demonstrable harms (Carbonaro et al. 2016; Johnson et al. 2008). If these survey findings are suggesting anti-addictive effects can occur with uncontrolled use, carefully conducted clinical research (that includes screening, preparation, monitoring, and aftercare) is expected to further minimize the risks associated with uncontrolled use and to maximize the efficacy of 5-HT2AR agonists to increase abstinence.

5 Modern Research in Alcohol Use Disorder

Bogenschutz et al. (2015) conducted the first modern research using a 5-HT2AR agonist to treat alcohol use disorder. They administered one or two doses of psilocybin (0.3 and 0.4 mg/kg) and motivational enhancement therapy to ten treatment-seeking volunteers meeting DSM-IV criteria for alcohol dependence in an open-label trial. The percentage of drinking days and the percentage of heavy drinking days significantly decreased after the first psilocybin session. At 36 weeks after treatment, the mean percentage of drinking days dropped from ~32.5% in the 4 weeks of treatment preceding the psilocybin session to ~12.5% in the 4 weeks following the psilocybin session, and ~17.5% during the final follow-up period of 21–32 weeks after the psilocybin session. Mean percentage of heavy drinking days (i.e., ≥5 drinks for men, ≥4 drinks for women) dropped from ~26% in the 4 weeks of treatment preceding the psilocybin session to ~8% in the 4 weeks following the psilocybin session, and ~13% at the final follow-up period 21–32 weeks after the psilocybin session. Consistent with earlier studies of LSD (Kurland et al. 1967; Pahnke et al. 1970), and contemporary pilot research of psilocybin treatment for tobacco use disorder (Garcia-Romeu et al. 2014), subjective effects of the drug experience (e.g., mystical-type effects, intensity) appeared to be potential key factors facilitating subsequent behavior changes (Bogenschutz et al. 2015; Nielson et al. 2018). Additional controlled research on the psilocybin-facilitated treatment of alcohol use disorder is currently underway (US National Library of Medicine 2021a).

6 Psychedelics in Tobacco Smoking Cessation

Given the older laboratory studies and anthropological studies suggesting anti-addiction efficacy of 5-HT2AR agonists across distinct classes of addictive drugs, my laboratory embarked on a line of research examining 5-HT2AR agonist psilocybin therapy for a previously unstudied target: Tobacco use disorder. We conducted an open-label pilot study in 15 nicotine-dependent, treatment-resistant cigarette smokers (Johnson et al. 2014). Participants consisted of ten men and five women. They were on average 51 years old, had been smoking for an average of 31 years, had made six prior serious cessation attempts, and smoked an average of 19 cigarettes per day. In addition to up to three psilocybin sessions, the intervention involved a 15-week program of manualized cognitive behavioral therapy (CBT) for smoking cessation. There were four preparation sessions for a collective total of approximately 8 h. These preparation sessions included CBT material in preparation for quitting smoking as well as preparation for the psilocybin sessions (see Johnson et al. 2008). The dose delivered in the first session was 20 mg/70 kg. We included two additional sessions at 2 and 8 weeks after the target quit date. The timing of these was determined by examining typical survival curves for smoking cessation, and selecting times during the portion of the survival curve with greatest risk of lapse. For the second and third psilocybin sessions, we administered a higher dose of 30 mg/70 kg unless the participant judged the first session dose to be sufficiently high. Only of the 15 participants opted to not increase the dose after the first session. Intermixed with the psilocybin administrations, participants came to the laboratory once per week, during which CBT material was covered.

As shown in Fig. 1, at 6 months post-target quit date, 12 out of 15 participants (80%) showed biologically verified 7-day point-prevalence abstinence (via breath CO and urine cotinine analysis) (Johnson et al. 2014). At 12-month follow-up, and at

use compared to previous occasions in which they quit smoking (Johnson et al. 2017b). Although these data cannot test a causal role of 5-HT2AR agonist use on smoking behavior, they are consistent with laboratory research suggesting a causal effect. Moreover, these results may suggest that affective processing is a commonality in psychedelic therapy, considering the studies showing 5-HT2AR agonists to decrease depressive symptoms (Carhart-Harris et al. 2016, 2021; Davis et al. 2021; Gasser et al. 2014; Griffiths et al. 2016; Grob et al. 2011; Palhano-Fontes et al. 2019; Ross et al. 2016).

7 Future Directions

The reviewed data suggests that classic psychedelics hold promise as treatments for substance use disorders. Pilot research suggests promise in using psilocybin in the treatment of both tobacco use disorder and alcohol use disorder. Additional rigorous research should explore both indications, including double-blind and larger trials. The field also needs research examining dose effects in the treatment of disorders. If results continue to look promising, the conduct of Phase III trials will be appropriate, which may lead to the approval of psilocybin for these disorders by the Food and Drug Administration in the USA and other medical regulatory agencies internationally. In this respect the recent funding by the United States National Institute on Drug Abuse of a multi-site trials of psilocybin for tobacco smoking cessation may provide critical information for potential advancement along the medical pathway (Grant number U01DA052174).

Unlike most medications for substance use disorders, it may be that these therapies work against a wide variety of substance use disorders, regardless of the class of addictive drug being treated. This is suggested by early clinical evidence for alcohol use disorder (Bogenschutz et al. 2015; Krebs and Johansen 2012), opioid use disorder (Savage and McCabe 1973), and tobacco use disorder (Johnson et al. 2014). Supporting evidence also comes from anthropological and survey methods (Dobkin de Rios et al. 2002; Doering-Silveira et al. 2005; Fábregas et al. 2010; Garcia-Romeu et al. 2019, 2020; Halpern et al. 2008; Johnson et al. 2017b; Pisano et al. 2017). A general anti-addiction effect across classes of addictive drugs would be consistent with the notion that the persisting positive behavior change prompted by psychedelic therapy is due to amplification of psychotherapeutic processes such as the construction of meaning, corrective emotional experiences, and a change in self-narrative (Nayak and Johnson 2021). Such psychological mechanisms may be at play with the treatment of not only distinct substance use disorders, but also other disorders such as depression and cancer-related distress.

Given these potential general mechanisms, additional substance use disorders and other disorders which can be broadly characterized as addictions should be explored as treatment targets for classic psychedelics. As suggested by survey data (Garcia-Romeu et al. 2020), these include opioids, cocaine, methamphetamine, and cannabis. An ongoing trial is examining psilocybin in the treatment of cocaine use disorder

(US National Library of Medicine 2021c). Given ongoing concerns about the opioid crisis (Kolodny et al. 2015), we should test classic psychedelics in the treatment of opioid use disorder (e.g., clinicaltrials.gov: NCT04161066; NCT) and the treatment of chronic pain, an issue that partially drives the opioid crisis. In addition to substance use disorders, psychedelics should be examined as potential treatments for other disorders that could be considered forms of addiction. Examples include poor health behaviors such as obesity (Teixeira et al. 2021), problem gambling, and hypersexual disorder.

Another avenue for exploration is with other classic psychedelics compounds. While the large majority of therapeutic research in the last 20 years has been conducted with psilocybin, LSD was the classic psychedelic most frequently studied for addiction in the 1960s. Research should once again explore the efficacy and safety of LSD in substance use disorders. While LSD provides a much longer duration of action than psilocybin, which may possibly be beneficial in therapeutics, there should also be research with shorter acting classic psychedelics, including dimethyltryptamine (DMT) and 5-methoxy-dimethyltryptamine (5-MeO- DMT Ermakova et al. 2021) in the treatment of substance use disorders. If the longer duration is not always necessary for therapeutic effect, a shorter-duration treatment model would entail less cost and therefore increase dissemination. Beyond these, there are dozens of compounds that appear to be classic psychedelics awaiting both non-human and human research (e.g., Shulgin and Shulgin 1991, 1997).

Research should also examine how classic psychedelic therapy is delivered. One example is whether a particular psychotherapeutic backdrop (e.g., cognitive behavioral therapy, motivational enhancement therapy) or no formal therapy at all (but still including supportive preparation and follow-up sessions with treatment staff) is ideal for treating a particular disorder or type of patient with a disorder. The parameters of psychedelic therapy, including the use of eyeshades and encouragement toward introspection, are still yet to be experimentally tested. Indeed, a recent study experimentally manipulating the music played during sessions appears to be the only study to have randomized a "set and setting" variable in the context of the psychedelic treatment session (Strickland et al. 2021). Future research should manipulate a broader number of session variables (e.g., presence of music, extent of discussion, specific exercises such as examining a mirror or personal photos) to examine optimization of therapeutic response.

Given the staggering society costs of substance use disorders, including the mortality caused by tobacco smoking, it is critical that scientists conduct rigorous clinical trials to follow up on promising early findings to investigate with classic psychedelics in the treatment of substance use disorders and other addictions. To support this effort, it is time that public funding be provided to support such investigations. Twenty years of modern investigation indicates that such trials can be conducted with an acceptable level of safety when following appropriate research guidelines (Johnson et al. 2008). The costs and risks of not conducting research into classic psychedelics for addiction exceed those of conducting the science.

Acknowledgments Support was provided by Heffter Research Institute, William Harrison, Beckley Foundation, and the Johns Hopkins Center for Psychedelic and Consciousness Research which was funded by the Steven and Alexandra Cohen Foundation, Tim Ferriss, Matt Mullenweg, Blake Mycoskie, and Craig Nerenberg. I thank Hillary Jackson for assistance with manuscript preparation.

Disclosures The author has advisory relationships with the following organizations: AJNA Labs LLC, AWAKN Life Sciences Inc., Beckley Psytech Ltd., Entheon Biomedical Corp., Field Trip Psychedelics Inc., Mind Medicine Inc., and Otsuka Pharmaceutical Development & Commercialization Inc.

References

Abuzzahab FS, Anderson BJ (1971) A review of LSD treatment in alcoholism. Int Pharmacopsychiatry 6(4):223–235. https://doi.org/10.1159/000468273

Akers BP, Francisco J, Piper A, Ruck CAP (2011) A prehistoric mural in Spain depicting neurotropic *Psilocybe* mushrooms? Econ Bot 65(2):121–128. https://doi.org/10.1007/s12231-011-9152-5

Albaugh BJ, Anderson PO (1974) Peyote in the treatment of alcoholism among American Indians. Am J Psychiatry 131(11):1247–1250. https://doi.org/10.1176/ajp.131.11.1247

Babb S, Malarcher A, Schauer G, Asman K, Jamal A (2017) Quitting smoking among adults – United States, 2000–2015. MMWR Morb Mortal Wkly Rep 65(52):1457–1464. https://doi.org/10.15585/mmwr.mm6552a1

Bergman RL (1971) Navajo peyote use: its apparent safety. Am J Psychiatry 128(6):695–699. https://doi.org/10.1176/ajp.128.6.695

Blum K, Futterman SFL, Pascarosa P (1977) Peyote, a potential ethnopharmacologic agent for alcoholism and other drug dependencies: possible biochemical rationale. Clin Toxicol 11(4): 459–472. https://doi.org/10.3109/15563657708988210

Bogenschutz MP, Forcehimes AA, Pommy JA, Wilcox CE, Barbosa P, Strassman RJ (2015) Psilocybin-assisted treatment for alcohol dependence: a proof-of-concept study. J Psychopharmacol 29(3):289–299. https://doi.org/10.1177/0269881114565144

Bonson KR (2018) Regulation of human research with LSD in the United States (1949-1987). Psychopharmacology 235(2):591–604. https://doi.org/10.1007/s00213-017-4777-4

Cahill K, Stevens S, Lancaster T (2014) Pharmacological treatments for smoking cessation. J Am Med Assoc 311(2):193–194. https://doi.org/10.1001/jama.2013.283787

Calabrese JD (1997) Spiritual healing and human development in the native American church: toward a cultural psychiatry of peyote. Psychoanal Rev 84(2):237–255

Carbonaro TM, Bradstreet MP, Barrett FS, MacLean KA, Jesse R, Johnson MW et al (2016) Survey study of challenging experiences after ingesting psilocybin mushrooms: acute and enduring positive and negative consequences. J Psychopharmacol 30(12):1268–1278. https://doi.org/10.1177/0269881116662634

Carhart-Harris RL, Bolstridge M, Rucker J, Day CM, Erritzoe D, Kaelen M et al (2016) Psilocybin with psychological support for treatment-resistant depression: an open-label feasibility study. Lancet 3(7):619–627. https://doi.org/10.1016/S2215-0366(16)30065-7

Carhart-Harris R, Giribaldi B, Watts R, Baker-Jones M, Murphy-Beiner A, Murphy R et al (2021) Trial of psilocybin versus escitalopram for depression. N Engl J Med 384(15):1402–1411. https://doi.org/10.1056/NEJMoa2032994

Chwelos N, Blewett DB, Smith CM, Hoffer A (1959) Use of d-lysergic acid diethylamide in the treatment of alcoholism. Q J Stud Alcohol 20(3):577–590. https://doi.org/10.15288/qjsa.1959.20.577

Danaei G, Ding EL, Mozaffarian D, Taylor B, Rehm J, Murray CJL et al (2009) The preventable causes of death in the United States: comparative risk assessment of dietary, lifestyle, and metabolic risk factors. PLoS Med 6(4):e1000058. https://doi.org/10.1371/journal.pmed.1000058

Davis AK, Barrett FS, May DG, Cosimano MP, Sepeda ND, Johnson MW et al (2021) Effects of psilocybin-assisted therapy on major depressive disorder: a randomized clinical trial. JAMA Psychiat 78(5):481–489. https://doi.org/10.1001/jamapsychiatry.2020.3285

Dobkin de Rios M, Grob CS, Baker JR (2002) Hallucinogens and redemption. J Psychoactive Drugs 34(3):239–248. https://doi.org/10.1080/02791072.2002.10399960

Doering-Silveira E, Grob CS, Dobkin de Rios M, Lopez E, Alonso LK, Tacla C et al (2005) Report on psychoactive drug use among adolescents using ayahuasca within a religious context. J Psychoactive Drugs 37(2):141–144. https://doi.org/10.1080/02791072.2005.10399794

Dyck E (2006) 'Hitting highs at rock bottom': LSD treatment for alcoholism, 1950–1970. Soc Hist Med 19(2):313–329. https://doi.org/10.1093/shm/hkl039

Ermakova AO, Dunbar F, Rucker J, Johnson MW (2021) A narrative synthesis of research with 5-MeO-DMT. J Psychopharmacol. https://doi.org/10.1177/02698811211050543

Fábregas JM, González D, Fondevila S, Cutchet M, Fernández X, Barbosa PCR et al (2010) Assessment of addiction severity among ritual users of ayahuasca. Drug Alcohol Depend 111(3):257–261. https://doi.org/10.1016/j.drugalcdep.2010.03.024

Garcia-Romeu A, Griffiths R, Johnson M (2014) Psilocybin-occasioned mystical experiences in the treatment of tobacco addiction. Curr Drug Abuse Rev 7(3):157–164. https://doi.org/10.2174/1874473708666150107121331

Garcia-Romeu A, Davis AK, Erowid F, Erowid E, Griffiths RR, Johnson MW (2019) Cessation and reduction in alcohol consumption and misuse after psychedelic use. J Psychopharmacol 33(9):1088–1101. https://doi.org/10.1177/0269881119845793

Garcia-Romeu A, Davis AK, Erowid E, Erowid F, Griffiths RR, Johnson MW (2020) Persisting reductions in cannabis, opioid, and stimulant misuse after naturalistic psychedelic use: an online survey. Front Psych 10:955. https://doi.org/10.3389/fpsyt.2019.00955

Gasser P, Holstein D, Michel Y, Doblin R, Yazar-Klosinski B, Passie T et al (2014) Safety and efficacy of lysergic acid diethylamide-assisted psychotherapy for anxiety associated with life-threatening diseases. J Nerv Ment Dis 202(7):513–520. https://doi.org/10.1097/NMD.0000000000000113

Glennon RA, Titeler M, McKenney JD (1984) Evidence for 5-HT 2 involvement in the mechanism of action of hallucinogenic agents. Life Sci 35(25):2505–2511. https://doi.org/10.1016/0024-3205(84)90436-3

Griffiths RR, Johnson MW, Carducci MA, Umbricht A, Richards WA, Richards BD et al (2016) Psilocybin produces substantial and sustained decreases in depression and anxiety in patients with life-threatening cancer: a randomized double-blind trial. J Psychopharmacol 30(12):1181–1197. https://doi.org/10.1177/0269881116675513

Grinspoon L (1981) LSD reconsidered. Sciences 21(1):20–23. https://doi.org/10.1002/j.2326-1951.1981.tb01901.x

Grob CS, Danforth AL, Chopra GS, Hagerty M, McKay CR, Halberstadt AL et al (2011) Pilot study of psilocybin treatment for anxiety in patients with advanced-stage cancer. Arch Gen Psychiatry 68(1):71–78. https://doi.org/10.1001/archgenpsychiatry.2010.116

Halpern JH, Sherwood AR, Passie T, Blackwell KC, Ruttenber AJ (2008) Evidence of health and safety in American members of a religion who use a hallucinogenic sacrament. Med Sci Monit 14(8):SR15–SR22

Jackson SE, Kotz D, West R, Brown J (2019) Moderators of real-world effectiveness of smoking cessation aids: a population study. Addiction 114(9):1627–1638. https://doi.org/10.1111/add.14656

Johnson MW, Griffiths RR (2017) Potential therapeutic effects of psilocybin. Neurotherapeutics 14(3):734–740. https://doi.org/10.1007/s13311-017-0542-y

Johnson MW, Richards W, Griffiths RR (2008) Human hallucinogen research: guidelines for safety. J Psychopharmacol 22(6):603–620. https://doi.org/10.1177/0269881108093587

Johnson MW, Garcia-Romeu A, Cosimano MP, Griffiths RR (2014) Pilot study of the 5-HT2AR agonist psilocybin in the treatment of tobacco addiction. J Psychopharmacol 28(11):983–992. https://doi.org/10.1177/0269881114548296

Johnson MW, Garcia-Romeu A, Griffiths RR (2017a) Long-term follow-up of psilocybin-facilitated smoking cessation. Am J Drug Alcohol Abuse 43(1):55–60. https://doi.org/10.3109/00952990.2016.1170135

Johnson MW, Garcia-Romeu A, Johnson PS, Griffiths RR (2017b) An online survey of tobacco smoking cessation associated with naturalistic psychedelic use. J Psychopharmacol 31(7): 841–850. https://doi.org/10.1177/0269881116684335

Johnson MW, Hendricks PS, Barrett FS, Griffiths RR (2019) Classic psychedelics: an integrative review of epidemiology, therapeutics, mystical experience, and brain network function. Pharmacol Ther 197:83–102. https://doi.org/10.1016/j.pharmthera.2018.11.010

Kolodny A, Courtwright DT, Hwang CS, Kreiner P, Eadie JL, Clark TW et al (2015) The prescription opioid and heroin crisis: a public health approach to an epidemic of addiction. Annu Rev Public Health 36(1):559–574. https://doi.org/10.1146/annurev-publhealth-031914-122957

Krebs TS, Johansen P (2012) Lysergic acid diethylamide (LSD) for alcoholism: meta-analysis of randomized controlled trials. J Psychopharmacol 26(7):994–1002. https://doi.org/10.1177/0269881112439253

Kurland AA, Unger S, Shaffer JW, Savage C (1967) Psychedelic therapy utilizing LSD in the treatment of the alcoholic patient: a preliminary report. Am J Psychiatry 123(10):1202–1209. https://doi.org/10.1176/ajp.123.10.1202

MacLean KA, Johnson MW, Griffiths RR (2011) Mystical experiences occasioned by the hallucinogen psilocybin lead to increases in the personality domain of openness. J Psychopharmacol 25(11):1453–1461. https://doi.org/10.1177/0269881111420188

Mangini M (1998) Treatment of alcoholism using psychedelic drugs: a review of the program of research. J Psychoactive Drugs 30(4):381–418. https://doi.org/10.1080/02791072.1998.10399714

Menninger J (1971) Discussion of Bergman RL (1971): Navajo peyote use: its apparent safety. Am J Psychiatr 128:695–699. https://doi.org/10.1176/ajp.128.6.695

Mokdad AH, Marks JS, Stroup DF, Gerberding JL (2004) Actual causes of death in the United States, 2000. JAMA 291(10):1238–1245. https://doi.org/10.1001/jama.291.10.1238

Mottillo S, Filion KB, Bélisle P, Joseph L, Gervais A, O'Loughlin J et al (2009) Behavioural interventions for smoking cessation: a meta-analysis of randomized controlled trials. Eur Heart J 30(6):718–730. https://doi.org/10.1093/eurheartj/ehn552

Murray CJL, Aravkin AY, Zheng P, Abbafati C, Abbas KM, Abbasi-Kangevari M et al (2020) Global burden of 87 risk factors in 204 countries and territories, 1990–2019: a systematic analysis for the global burden of disease study 2019. Lancet 396(10258):1223–1249. https://doi.org/10.1016/S0140-6736(20)30752-2

Nayak S, Johnson MW (2021) Psychedelics and psychotherapy. Pharmacopsychiatry 54(4): 167–175. https://doi.org/10.1055/a-1312-7297

Nielson EM, May DG, Forcehimes AA, Bogenschutz MP (2018) The psychedelic debriefing in alcohol dependence treatment: illustrating key change phenomena through qualitative content analysis of clinical sessions. Front Pharmacol 9:132. https://doi.org/10.3389/fphar.2018.00132

Noorani T, Garcia-Romeu A, Swift TC, Griffiths RR, Johnson MW (2018) Psychedelic therapy for smoking cessation: qualitative analysis of participant accounts. J Psychopharmacol 32(7): 756–769. https://doi.org/10.1177/0269881118780612

Pahnke WN, Kurland AA, Unger S, Savage C, Grof S (1970) The experimental use of psychedelic (LSD) psychotherapy. J Am Med Assoc 212(11):1856–1863. https://doi.org/10.1001/jama.1970.03170240060010

Palhano-Fontes F, Barreto D, Onias H, Andrade KC, Novaes MM, Pessoa JA et al (2019) Rapid antidepressant effects of the psychedelic ayahuasca in treatment-resistant depression: a randomized placebo-controlled trial. Psychol Med 49(4):655–663. https://doi.org/10.1017/S0033291718001356

Pascarosa P, Futterman S (1976) Ethnopsychedelic therapy for alcoholics: observations in the peyote ritual of the native American church. J Psychedelic Drugs 8(3):215–221. https://doi.org/10.1080/02791072.1976.10472016

Perkins D, Opaleye ES, Simonova H, Bouso JC, Tófoli LF, GalvÃo-Coelho NL et al (2022) Associations between ayahuasca consumption in naturalistic settings and current alcohol and drug use: results of a large international cross-sectional survey. Drug Alcohol Rev 41(1): 265–274. https://doi.org/10.1111/dar.13348

Pisano VD, Putnam NP, Kramer HM, Franciotti KJ, Halpern JH, Holden SC (2017) The association of psychedelic use and opioid use disorders among illicit users in the United States. J Psychopharmacol 31(5):606–613. https://doi.org/10.1177/0269881117691453

Rhead JC, Soskin RA, Turek I, Richards WA, Yensen R, Kurland AA et al (1977) Psychedelic drug (DPT)-assisted psychotherapy with alcoholics: a controlled study. J Psychedelic Drugs 9(4): 287–300. https://doi.org/10.1080/02791072.1977.10472060

Roseman L, Nutt DJ, Carhart-Harris RL (2018) Quality of acute psychedelic experience predicts therapeutic efficacy of psilocybin for treatment-resistant depression. Front Pharmacol 8:974. https://doi.org/10.3389/fphar.2017.00974

Ross S, Bossis A, Guss J, Agin-Liebes G, Malone T, Cohen B et al (2016) Rapid and sustained symptom reduction following psilocybin treatment for anxiety and depression in patients with life-threatening cancer: a randomized controlled trial. J Psychopharmacol 30(12):1165–1180. https://doi.org/10.1177/0269881116675512

Roy C (1973) Indian peyotists and alcohol. Am J Psychiatry 130(3):329–330. https://doi.org/10.1176/ajp.130.3.329-b

Savage C, McCabe OL (1973) Residential psychedelic (LSD) therapy for the narcotic addict: a controlled study. Arch Gen Psychiatry 28(6):808–814. https://doi.org/10.1001/archpsyc.1973.01750360040005

Shulgin AT, Shulgin A (1991) Pihkal: a chemical love story. Transform Press, Berkeley

Shulgin AT, Shulgin A (1997) Tihkal: the continuation. Transform Press, Berkeley

Smith CM (1958) A new adjunct to the treatment of alcoholism: the hallucinogenic drugs. Q J Stud Alcohol 19(3):406–417. https://doi.org/10.15288/qjsa.1958.19.406

Smith CM (1959) Some reflections on the possible therapeutic effects of the hallucinogens; with special reference to alcoholism. Q J Stud Alcohol 20(2):292–301. https://doi.org/10.15288/qjsa.1959.20.292

Spitzer M, Thimm M, Hermle L, Holzmann P, Kovar K, Heimann H et al (1996) Increased activation of indirect semantic associations under psilocybin. Biol Psychiatry (1969) 39(12): 1055–1057. https://doi.org/10.1016/0006-3223(95)00418-1

Strassman RJ, Qualls CR (1994) Dose-response study of N,N-dimethyltryptamine in humans: I. neuroendocrine, autonomic, and cardiovascular effects. Arch Gen Psychiatry 51(2):85–97. https://doi.org/10.1001/archpsyc.1994.03950020009001

Strickland JC, Garcia-Romeu A, Johnson MW (2021) Set and setting: a randomized study of different musical genres in supporting psychedelic therapy. ACS Pharmacol Transl Sci 4(2): 472–478. https://doi.org/10.1021/acsptsci.0c00187

Teixeira PJ, Johnson MW, Timmermann C, Watts R, Erritzoe D, Douglass H, Kettner H, Carhart-Harris RL (2021) Psychedelics and health behaviour change. J Psychopharmacol 36:12–19

US Department of Health and Human Services (2014) The health consequences of smoking – 50 years of progress: a report of the surgeon general. U.S. Department of Health and Human Services, Centers for Disease Control and Prevention, National Center for Chronic Disease Prevention and Health Promotion, Office on Smoking and Health, Atlanta

US National Library of Medicine (2021a) A double-blind trial of psilocybin-assisted treatment of alcohol dependence. https://clinicaltrials.gov/ct2/show/NCT02061293. Accessed 10 Oct 2021

US National Library of Medicine (2021b) Psilocybin-facilitated smoking cessation treatment: a pilot study. https://clinicaltrials.gov/ct2/show/NCT01943994. Accessed 10 Oct 2021

US National Library of Medicine (2021c) Psilocybin-facilitated treatment for cocaine use. https://clinicaltrials.gov/ct2/show/NCT02037126. Accessed 10 Oct 2021

Vollenweider FX, Leenders KL, Scharfetter C, Maguire P, Stadelmann O, Angst J (1997) Positron emission tomography and fluorodeoxyglucose studies of metabolic hyperfrontality and psychopathology in the psilocybin model of psychosis. Neuropsychopharmacology 16(5):357–372. https://doi.org/10.1016/S0893-133X(96)00246-1

Wittchen H, Hoch E, Klotsche J, Muehlig S (2011) Smoking cessation in primary care - a randomized controlled trial of bupropione, nicotine replacements, CBT and a minimal intervention. Int J Methods Psychiatr Res 20(1):28–39. https://doi.org/10.1002/mpr.328

Psychedelics and Anti-inflammatory Activity in Animal Models

Thomas W. Flanagan and Charles D. Nichols

Contents

Abstract The serotonin (5-hydroxytryptamine, 5-HT) 2A receptor is most well known as the common target for classic psychedelic compounds. Interestingly, the 5-HT_{2A} receptor is the most widely expressed mammalian serotonin receptor and is found in nearly every examined tissue type including neural, endocrine, endothelial, immune, and muscle, suggesting it could be a novel and pharmacological target for several types of disorders. Despite this, the bulk of research on the 5-HT_{2A} receptor is focused on its role in the central nervous system (CNS). Recently, activation of 5-HT_{2A} receptors has emerged as a new anti-inflammatory strategy. This review will describe recent findings regarding psychedelics as anti-inflammatory compounds, as well as parse out differences in functional selectivity and immune regulation that exist between a number of well-known hallucinogenic compounds.

Keywords Anti-inflammatory · Asthma · Enhanced pause · IL-6 · Psilocybin · Psychedellic · (R)-DOI · Whole-body plethysmography

T. W. Flanagan and C. D. Nichols (✉)
Department of Pharmacology and Experimental Therapeutics, Louisiana State University Health Sciences Center, New Orleans, LA, USA
e-mail: cnich1@lsuhsc.edu

Curr Topics Behav Neurosci (2022) 56: 229–246
https://doi.org/10.1007/7854_2022_367
Published Online: 12 May 2022

1 Introduction

Current anti-inflammatory medications fall into three distinct categories (Nichols et al. 2017). The first and most commonly used group is comprised of nonsteroidal anti-inflammatory drugs (NSAIDs) (Vane and Botting 1998). NSAIDs exert their anti-inflammatory action through the inhibition of cyclooxygenase (COX), which prevents the production of proinflammatory lipid prostaglandins (PG). While highly effective and widely utilized, NSAID use is associated with gastric and renal toxicity. The second category, corticosteroids, is comprised of drugs such as prednisone, hydrocortisone, and dexamethasone, which are considered immunosuppressive agents (Barnes 2006). Corticosteroids control inflammation by suppressing the expression of genes that encode multiple inflammatory molecules like cytokines and chemokines via activation of glucocorticoid receptors, which are nuclear hormone receptors. Similar to NSAIDs, corticosteroids are widely used and are effective against a number of inflammatory disorders. Although corticosteroids are effective therapies for disorders such as asthma and inflammatory bowel disease, they have proven ineffective against other inflammatory maladies like chronic obstructive pulmonary disease (COPD) (Barnes 2010). Additionally, their overall suppression of the immune system disarms baseline immune homeostasis and increases the probability of opportunistic infection (Toruner et al. 2008). The third major class of anti-inflammatory therapeutic involves the use of recombinant proteins, or new biologic entities termed "biologics" (Rider et al. 2016), which neutralize inflammatory cytokines such as tumor necrosis factor α (TNFα (Li et al. 2017)), interleukins 12/23 (IL-12/23), and matrix metallopeptidase 9 (MMP9 (Moss 2018)) by binding and sequestering the proinflammatory agent itself or interacting with the agent's receptor to ultimately prevent the promotion of cellular inflammatory cascades. Biologics have proven very useful in their ability to inhibit specific, key inflammatory molecules involved in the pathogenesis of several inflammatory disorders such as rheumatoid arthritis and inflammatory bowel disease. Unfortunately, biologics are quite costly (Wu et al. 2020), must be administered by infusion in a clinical setting, and are associated with pain at the injection site (Curtis et al. 2011). Many new anti-inflammatory biologics target additional inflammatory agents like leukotrienes (Koeberle et al. 2016), or modify the structure of existing anti-inflammatory compounds (Kodela et al. 2012), but their overall mechanism of action is similar. We have previously proposed that activation of 5-HT_{2A} receptors with psychedelics represents a novel anti-inflammatory mechanism with broad clinical utility (Nichols et al. 2017). This review will describe recent findings in this emerging field and will highlight the benefits of psychedelic medication as therapeutics compared to other classes.

2 History of Psychedelics as Anti-Inflammatories

As mechanisms of inflammatory activation and psychedelic anti-inflammatory work performed prior to 2018 were described in a recent review article (Flanagan and Nichols 2018), these topics will not be repeated here. For historical perspective, a table containing work describing 5-HT$_2$-mediated anti-inflammatory and immunomodulatory effects between 2008 and the present is provided (Table 1). There have been some reports describing anti-inflammatory effects of N,N-DMT and 5-MeO-DMT (Szabo et al. 2014; Szabo 2015; Qi et al. 2018), however, the immunomodulatory effects of these compounds have been proposed to be mediated through the sigma-1 receptor. Therefore, as they are not 5-HT$_{2A}$ receptor dependent, they have not been included in this review, which is focused on the anti-inflammatory effect of psychedelics acting at 5-HT$_{2A}$ receptors. Anti-inflammatory effects have also been associated with 5-HT$_{2A}$ receptor antagonism/inverse agonism (Arzt et al. 1991; Maes et al. 1994; Marconi et al. 2003; Sugino et al. 2009; Akiyoshi et al. 2006; Hong et al. 2006; Sasaki et al. 2006; Hu et al. 2012; Todorović and Filipović 2017; Giridharan et al. 2020; Kaur and Krishan 2020); however, this is not discussed here because the likely mechanism of this class of drug is through blockade of the proinflammatory effects of serotonin itself at 5-HT$_{2A}$ receptors (Pierce et al. 1995; Harbuz et al. 1996, 1998; Afshar et al. 2019).

3 5-HT$_{2A}$ Receptor Activation in Allergic Asthma Models

A robust in vivo inflammatory model used to assess the putative therapeutic effects of 5-HT$_{2A}$ receptor activation is the chicken egg white ovalbumin (OVA)-induced allergic asthma model. As a research species mice are most commonly used in OVA asthma models due in part to their relative inexpensiveness and the availability of mouse-specific probes to study inflammatory biomarkers (Zosky and Sly 2007). OVA is a commonly used allergen as it induces intense pulmonary inflammation and can be produced inexpensively in mass quantities (Aun et al. 2017). Regardless of the allergen used (OVA, house dust mite [HDM], cockroach antigen, *Aspergillus fumigatus*, ragweed extracts, etc.) (Zosky and Sly 2007), the experimental procedure is fairly uniform (Bates et al. 2009). The foreign antigen is first injected intraperitoneally along with an adjuvant, most commonly aluminum hydroxide, to sensitize and "prime" the immune system (Joubert et al. 2020) with the adjuvant enhancing the allergen's immunogenicity (Bates et al. 2009). Two IP injections are spaced apart by a period of 7–14 days (Aun et al. 2017), during which time the host system mounts an immune response against the antigen (Bates et al. 2009). After 7–14 additional days, antigen is administered to challenge the sensitized host to induce an allergic immune response, typically once a day for three consecutive days (Aun et al. 2017). Challenge can be performed by either postnatal drip following nasal instillation or nose-only aerosol exposure, which serves to drive the allergen directly into

Table 1 In vivo inflammatory models which administer a 5-HT_2 agonist

Compound(s)	Inflammatory model	Key findings	Reference
(*R*)-DOI, lysergic acid 2,4-dimethylazetidine (LA-SS-Az), *2C-BCB* (4-Bromo-3,6-dimethoxybenzocyclobuten-1-yl) methylamine [2C-BCB], and lysergic acid diethylamide (LSD)	TNFα-induced in vitro model	5-HT_{2A} activation potently inhibits TNFα-induced inflammatory markers in rat aortic smooth muscle (RASM) cells	Yu et al. (2008)
PNU22394	3T3-L1 adipocytes	5-HT_{2A} agonism reduces adiponectin expression	Uchida-Kitajima et al. (2008)
(*R*)-DOI	CBA mice	5-HT_{2A} activation suppresses immune response and reduces spleen and peripheral blood CD8+ T-cell counts	Davydova et al. (2010)
(*R*)-DOI	ConA-induced in vivo model	5-HT_{2A} activation modulates T-helper 1 (Th1) and cytotoxic T-cell lymphocyte activation	Inoue et al. (2011)
(*R*)-DOI	TNFα-induced in vivo model	(*R*)-DOI blocks systemic effects of TNFα in a whole animal model	Nau et al. (2013)
(*R*)-DOI	Ovalbumin (OVA)-induced acute allergic asthma model	5-HT_{2A} activation reduces pulmonary inflammation and mucus hyperproduction in a murine model of acute asthma	Nau et al. (2015)
TCB-2	Streptozotocin-induced Alzheimer's disease/oxidative stress model	TCB-2 reduces oxidative stress and neuronal loss in a rodent model of Alzheimer's disease	Afshar et al. (2019)
(*R*)-DOI	High-fat diet induced inflammatory model	5-HT_{2A} activation reduces inflammatory markers, normalizes glucose homeostasis, and reduces circulating cholesterol levels in "Western" diet-fed apolipoprotein A knockout (Apoe−/−) mice	Flanagan et al. (2019a)
(*R*)-DOI	Ovalbumin (OVA)-induced chronic allergic asthma model	(*R*)-DOI reduces airways hyperresponsiveness (AHR), collagen deposition, and chronic pulmonary inflammation in a rescue model of allergic asthma	Flanagan et al. (2019b)
(*R*)-DOI, (*R*)-DOM, (*R*)-2,5-DMA, (*R*)-2,4-DMA, (*R*)-DOET, (*R*)-DOiBU, 2C-T-	Ovalbumin (OVA)-induced	The 2,5-dimethoxyphenethylamine structure represents the ideal	Flanagan et al. (2020)

(continued)

Table 1 (continued)

Compound(s)	Inflammatory model	Key findings	Reference
33, 5-iPro-2CE, 2C-B-Fly, TCB-2, 2C-B, 2C-I, 2C-H, TMA-2, LSD, ETH-LAD, lisuride, psilocin, 4-OH-DiPT	acute allergic asthma model	pharmacophore for 5-HT_2-mediated anti-inflammatory activity	

the airways. Aerosol challenge is less invasive than postnatal drip and does not require sensitization, making aerosol exposure with allergen the most popular challenge route. Typical symptoms of pulmonary allergic inflammation will peak at 48 h following the final allergen challenge, resulting in the escalation of a T-helper type 2 (Th-2) allergic response (Zosky and Sly 2007), characterized by elevated levels of antigen-specific immunoglobulin E (IgE), eosinophilia-driven airway inflammation, and a preponderance of Th2-associated cytokines (i.e., IL-4, IL-5, IL-9, IL-11, IL-13).

A thorough understanding of the OVA-induced allergic asthma model is critical for understanding the anti-inflammatory effects of 5-HT_{2A} receptor activation by psychedelics, because this model incorporates several aspects of immunity for study including both innate and adaptive that 5-HT_{2A} receptor activation modulates (Nau et al. 2015; Flanagan et al. 2019b, 2020). In 2015 we reported that nasally administered (*R*)-2,5-dimethoxy-4-iodoamphetamine ((*R*)-DOI), a 5-HT_2 receptor selective agonist and psychedelic drug, given prophylactically 30 min prior to allergen challenge, reduced or completely prevented pulmonary inflammation and mucus hyperproduction induced by repeated OVA exposure (Nau et al. 2015). Mechanistically, 5-HT_{2A} receptor activation does not act as an immunosuppressant, but serves to reduce the expression of only some proinflammatory markers (e.g., *Il-5*, *Il-13*, *monocyte chemoattractant protein-1* [*Mcp-1*], *granulocyte-macrophage colony-stimulating factor* [*gm-csf*]), but not others (*Il-4*). Interestingly, these data provided the first in vivo evidence that psychedelics act on only subsets of specific immune components, rather than simply blunting the entire immune system as a corticosteroid. It also suggests that 5-HT_{2A} therapeutic activity involves regulation/modulation of the pre-existing inflammatory environment, a hypothesis whose importance will be discussed below.

From a translational perspective, while acute mice models do reproduce several key features of atopic asthma (elevated IgE, airway inflammation, goblet cell hyperplasia, and elevated airways hyperresponsiveness [AHR]), the most commonly cited limitation is one of "chronicity" (Zosky and Sly 2007; Aun et al. 2017; Bates et al. 2009; Fulkerson et al. 2005). Essentially, in humans, allergic asthma is characterized by repeated exposure to an aeroallergen. This repeated exposure induces a negative feedback loop of inflammatory events which ultimately triggers constitutive goblet cell hyperplasia, subepithelial fibrosis, and thickening of airway smooth muscle to promote substantial airway remodeling. The consequence of this airway wall remodeling manifests itself on airflow and AHR, triggering labored breathing. A majority of the murine OVA models involve short-term administration of a high dose of aerosolized antigen, which not only results in a high degree of perivascular inflammation and eosinophilic infiltration not seen in human asthma (Fulkerson et al. 2005), but also fails to induce the extensive lung remodeling seen in adults as well. To address these concerns, we generated a murine chronic asthma model to evaluate the therapeutic effects of 5-HT_{2A} activation (Flanagan et al. 2019b).

One key consideration when attempting to establish a chronic asthma model is that mice repeatedly exposed to a particular allergen can display varying degrees of tolerance (Zosky and Sly 2007). Therefore, rather than the previously utilized single 3× challenge regimen (Nau et al. 2015), animals were exposed nose-only to low-dose OVA (1% wt/vol in sterile saline) once a week over a period of 12 weeks (Flanagan et al. 2019b). Following a one-week recovery period, respiratory parameters were assessed via enhanced pause (PenH) obtained from whole-body plethysmography (WBP), which will be discussed in greater detail below. Upon confirmation that AHR was significantly elevated in OVA-challenged animals, treatment groups were exposed to 1 mg/kg intranasal (*R*)-DOI for four consecutive days, 10 days after the final OVA exposure, followed by a final WBP analysis and sacrifice/organ harvesting. Of note, this treatment regimen allowed an exploration of 5-HT_{2A} activation in a *rescue* capacity in a model of established, persistent asthma. Similar to the acute model, a specific subset of inflammatory markers necessary for the maintenance of a chronic inflammatory state was upregulated with OVA-challenge and decreased upon 5-HT_{2A} receptor activation (e.g., *Il-9*, *Il-15*, *gm-csf*, *Muc5ac*). Interestingly, factors such as *Il-5* and *Il-13*, whose expression had been upregulated in the acute model, displayed blunted expression upon chronic OVA-challenge, a finding consistent with previous data regarding prolonged allergen challenge (Munroe et al. 2010). These data suggest that 5-HT_{2A} receptor activation by a psychedelic impacts the immune system differently in response to varying inflammatory conditions. Of note, 5-HT_{2A} receptor activation via psychedelics has been suggested as a novel therapy for numerous conditions involving misregulation of the immune system (major depressive disorder, Parkinson's disease, Alzheimer's disease, Amyotrophic Lateral Sclerosis, etc.) (Thompson and Szabo 2020; Inserra et al. 2021). Another possible area of therapy may be autoimmune diseases, which result from an imbalance between effector and regulatory immune responses where the immune system itself begins to attack healthy tissues (Rosenblum et al. 2015). The nature of the differential regulation of cytokines we have observed for acute versus chronic asthmatic states suggests that psychedelics may therefore be potent therapies for autoimmune disorders.

In addition to specific proinflammatory cytokines, (*R*)-DOI also downregulated *MMP-9* and *TGF*β, whose expression has been implicated in pulmonary collagen deposition and fibrosis (Ma et al. 2016; Hoshino et al. 1999; Kobayashi et al. 2014). Accordingly, histological analysis revealed that (*R*)-DOI not only reduced persistent mucus hyperproduction and peribronchial inflammation as in the previously used acute model, but also significantly reduced collagen deposition, airway fibrosis, and overall structural remodeling by about 70%. Surprisingly, few if any currently clinical therapies have shown promise at combating airway remodeling (Sumino et al. 2015) or reducing airway smooth muscle area (Girodet et al. 2015) associated with pulmonary disease. Thus, while many new asthma therapies target specific mediators of the inflammatory response, none have provided to be robust treatment alternatives for severe asthma associated with significant airway remodeling. 5-HT$_{2A}$ receptor activation with (*R*)-DOI dramatically reduces the degree of collagen deposition around the airways and the extent of peribronchial inflammation. The therapeutic implications for this reduction in airway smooth muscle area is evident, as a reduction of airway smooth muscle hypertrophy would likely decrease airway resistance and alleviate the persistent AHR seen in human cases of severe chronic asthma.

(*R*)-DOI is not the only 5-HT$_{2A}$ receptor agonist that has anti-asthma efficacy in vivo. When LSD, LA-SS-AZ, and 2C-BCB were utilized in an in vitro model (Yu et al. 2008) of TNFα-induced inflammation in rat aortic smooth muscle (RASM) cells, there were varying degrees of anti-inflammatory activity, with the more behaviorally potent LSD having the least anti-inflammatory activity. Unfortunately, not all psychedelics are predicted to be able to be tested in mice to determine structure-function relationships. One reason for this stems from the differing metabolic profiles of mice compared to other species. While the metabolism of phenethylamine compounds, like (*R*)-DOI, proceeds at a slower rate similar to humans, mice have a very active form of monoamine oxidase (Hope and Smith 1960) that rapidly deaminates tryptamine compounds like psilocybin and its active metabolite psilocin, as well as indoleamines like LSD and DMT, to an inactive form in a manner of minutes (Shen et al. 2010). Therefore, any structure-activity study looking at tryptamines or ergolines in addition to phenethylamines would need to incorporate a species such as rat, which has an enzymatic profile more similar to humans with a much slower metabolism of psychedelic tryptamine and ergoline compounds, to assess the anti-inflammatory potential of all three major psychedelic classes in an asthma model. To overcome this limitation of mice, we developed an acute OVA protocol for use in Brown Norway rats (Flanagan et al. 2020). This rat strain has a robust Th2-driven response, and easily develops the hallmark symptoms of atopic asthma (Du et al. 1996; Pini et al. 2006). Similar to the mouse protocol, each animal's respiratory parameters were measured via WBP 48 h following the final of three OVA challenges.

Utilizing this rat experimental system, we were able to perform structure-activity relationship with several psychedelics representing all three major classes. Substituents, functional groups, and structural requirements for anti-AHR activity (as a proxy for anti-inflammatory activity) were first evaluated for the phenethylamine class. The presence (DOx) or absence (2C-x) of an α-methyl does not affect potency or efficacy to reduce AHR. Tethering the side chain into a constrained conformation (TCB-2), however, significantly reduces anti-AHR efficacy, as does tethering the 5- and 2- methoxys (e.g., 2CB-fly). The nature of the 4-position substituent affects affinity and selectivity of compounds for 5-HT$_2$ receptors, and the size of the 4-moiety affects behavioral activity. For example, the addition of an aliphatic group larger than an ethyl- decreases behavioral potencies and/or efficacies (Nichols

2016). We found that a halogen at the 4-position (iodo or bromo)) allows for full anti-inflammatory activity as measured by prevention of AHR (Flanagan et al. 2020). Further, the lack of any substitution at the 4-position also affords full efficacy to reduce OVA-induced AHR (e.g. 2C-H). From our results, 2C-H appears to be the pharmacophore structure for anti-inflammatory activity of phenethylamines.

While phenethylamines offer the most easily modifiable chemical structure, it is the tryptamine class, in particular psilocybin, that has had the most widespread clinical use in human trials for psychiatric disease (Kyzar et al. 2017). For example, as a powerful agent in alleviating depression in certain patient populations, obsessive-compulsive disorder (OCD), and cluster headaches (Nichols 2016; Griffiths et al. 2016; Ross et al. 2016). In our rodent acute asthma model, psilocin, the active metabolite of psilocybin, displays a similar anti-AHR efficacy and potency to that of (*R*)-DOI. Surprisingly, other tryptamines with very similar structures like N,N-dimethyltryptamine (DMT) and 5-methoxy-N,N-dimethyltryptamine (5-MeO-DMT) show no efficacy to reduce OVA-induced AHR. Both DMT and 5-MeO-DMT have been shown by others to prevent the production of a variety of proinflammatory cytokines (IL-1β, IL-6, TFNα) in LPS and pathogen-stimulated human primary monocyte-derived dendritic cells through activation of the sigma-1 receptor (sigmar-1R) (Szabo et al. 2014; Szabo 2015). Likewise, ayahuasca (dos Santos 2014; Frecska et al. 2016; Galvão-Coelho et al. 2020; da Silva et al. 2020), whose main psychoactive compound is DMT, and other tryptamine derivatives like N,N-dimethylserotonin (bufotenine, 5-HO,DMT) (Qi et al. 2018; Zulfiker et al. 2016; Xu et al. 2021; Zhang et al. 2019) have been reported to display anti-inflammatory/immunomodulatory activity thought to be significantly mediated by the sigma-1 receptor rather than the 5-HT$_{2A}$ receptor. Our finding that DMT and 5-MeO-DMT is ineffective as an anti-inflammatory in an asthma model may have several explanations. The first is that the sigmar-1 receptor has varying expression depending on age (Ramakrishnan et al. 2016) (i.e., sigmar-1R has a higher expression in older rats to account for reduced receptor binding (Ishiwata et al. 2003). It therefore may be that in order to have a high enough degree of sigmar-1 expression, DMT and 5-MeO-DMT would only be effective in animals of advanced age, ages which exceed the experimental conditions necessary to reliably obtain PenH values. Another possibility may be an overall lack of sigmar-1R in the rodent lung. Sigmar-1R is predominantly expressed in the CNS, and while there is some expression in peripheral tissues, the amount present in the rodent lung is quite low (Mei and Pasternak 2001). In rats, sigmar-1R may simply not have enough receptor density to initiate a substantial anti-inflammatory response to allergen challenge. A third explanation may revolve around the metabolism of the drug itself. While DMT and 5-MeO-DMT are metabolized much more slowly in rats than mice, rats still maintain a terminal half-life of roughly 12 to 19 min (Shen et al. 2010). As such, it's quite possible that the drug has been completely excreted by the time the allergen challenge concludes, and any therapeutic prophylactic advantage has been overcome. The final explanation, which we believe to be the most plausible, may be that these particular tryptamines just are not anti-inflammatory at 5-HT$_{2A}$ receptors in the lung due to functionally selectivity.

While the tryptamines may be the most clinically researched psychedelic compounds (Nichols et al. 2017; Kyzar et al. 2017), LSD is arguably the most known psychedelic (Nichols 2016). LSD is one of the most potent psychoactive compounds ever discovered, but is only a partial agonist with respect to certain canonical signaling pathways at the 5-HT_{2A} receptor (Nichols 2016; Wacker et al. 2017). In vitro assays revealed LSD to have some anti-inflammatory activity, but orders of magnitude less than the potency of (*R*)-DOI (Yu et al. 2008). Likewise, in the acute rodent allergic asthma model, LSD demonstrated only partial efficacy to prevent OVA-induced AHR even at relatively high doses (Flanagan et al. 2020). As opposed to phenethylamines, ergolines like LSD maintain a complex structure where only a few modifications can be performed (Nichols 2012). One modification is alteration/extension of the N(6)-methyl to an ethyl. This drug, unlike LSD, is a full agonist at the 5-HT_{2A} receptor with respect to canonical signaling. However, it is still only partially efficacious in the rodent asthma model (Flanagan et al. 2020). Due to less efficacious and potent anti-inflammatory activity, and that repeated administration of LSD over several months has been shown to induce persistent abnormal behavior in rats (Martin et al. 2014), LSD does not represent the most promising psychedelic for development as an anti-inflammatory therapeutic.

4 5-HT_{2A} Activation by Psychedelics in Models of Cardiovascular Disease

Asthma is not the only inflammatory disease where 5-HT_{2A} receptor activation by psychedelics may have therapeutic potential. As mentioned above, 5-HT_{2A} receptors are expressed in most peripheral tissue, including those related to cardiovascular function (Nichols 2009). 5-HT_{2A} receptors are expressed both in vascular smooth muscle and endothelial cells, as well as cardiomyocytes, where the 5-HT_{2A} receptor activity is believed to mediate elements of cellular proliferation and vasoconstriction (Nagatomo et al. 2004; Brattelid et al. 2007; McKune and Watts 2001). Pharmacologically, most research has evaluated the effects of 5-HT_2 receptor antagonism on limiting cardiovascular disease and vascular inflammation. For example, sarpogrelate, a 5-HT_2 receptor antagonist, slows the progression of atherosclerosis in rabbits (Hayashi et al. 2003). Recently, co-treatment of sarpogrelate and an aromatic L-amino acid decarboxylase (AADC) inhibitor was shown to block 5-HT synthesis and improve hepatic steatosis, insulin resistance, and dyslipidemia in apolipoprotein E (ApoE) knockout mice fed a high-fat diet (Ma et al. 2021). The likely mechanism for an antagonist having anti-inflammatory activity is the above-mentioned blockade of the well-established proinflammatory effects of serotonin at the 5-HT_{2A} receptor (Flanagan and Nichols 2018; Arzt et al. 1991; Nau et al. 2013; Nishiyama 2009).

We evaluated the putative therapeutic anti-inflammatory properties associated with 5-HT_{2A} receptor activation in a high-fat "Western diet"-induced inflammatory model (Flanagan et al. 2019a). The standard Western diet elevates serum levels of cholesterol, triglycerides, and glucose to promote metabolic disease and cardiovascular dysfunction (Hawiger et al. 2018); additionally, prolonged Western dieting can induce systemic and immune cell dysregulation, as well as promote the expression of a number of proinflammation biomarkers (Manzel et al. 2014; Christ et al. 2018). In this experiment, we implanted subcutaneous osmotic minipumps do slowly deliver (*R*)-DOI during 16 weeks of feeding a high-fat high cholesterol diet to ApoE (−/−) knockout mice. The psychedelic-treated group fed a high-fat diet was found to have significantly less expression of proinflammatory genes in vascular tissue. Recent studies have shown that the selective 5-HT_{2A} receptor agonist 25CN-NBOH increases the production of connective tissue growth factor (CCN2) in a human chondrocyte cell line (Hori et al. 2017). Our data combined with this recent finding suggests that 5-HT_{2A} receptor activation at low sub-behavioral levels of drug may represent a new therapeutic strategy to treat metabolic and cardiovascular disease.

5 Conclusion

In order to fully gauge the importance of psychedelics as anti-inflammatory medications, a stringent, impartial analysis of the advantages and disadvantages of 5-HT_{2A}-mediated therapies must be described (Fig. 1). Psychedelics offer a clear departure from the traditional anti-inflammatory therapies listed above (NSAIDs, steroidal, biologics) and represent a new class of powerful anti-inflammatory medication based on preclinical data. Economically, psychedelics are predicted to be inexpensive to manufacture (Nichols et al. 2017), and methods are improving to allow for their cost-effective mass production (Londesbrough et al. 2020; Adams et al. 2019). Therefore, legalized, automated production of psychedelics would allow the manufacture and distribution of psychedelics to proceed in a cost-effective manner (Aldridge and Décary-Hétu 2016). As such, psychedelic medications could be purchased at prices far below those of biologics and corticosteroids, and at levels comparable to NSAIDs, which would also be more attractive for insurance reimbursement. One concern is the potential for abuse. Clinical trials involving psilocybin have been performed in rigid, standardized settings with tightly monitored patient evaluation, and have not been associated with harmful effects such as abuse or dependence (Griffiths et al. 2016; Ross et al. 2016; Carhart-Harris et al. 2016, 2017). Importantly, the predicted therapeutic dose in humans is predicted to be sub-behavioral. For example, the behavioral threshold for (*R*)-DOI in the sensitive 2-lever drug discrimination assay is ~0.2 mg/kg in rats (Glennon et al. 1982; Canal and Morgan 2012). The anti-inflammatory potency (IC_{50}) is ~0.005 mg/kg in our rat asthma model (Flanagan et al. 2020), with a potency 40-fold greater than that necessary for behavioral effects. The human behavioral threshold potency of (*R*)-DOI is likely ~0.5 mg (Shulgin 1991). This would suggest that a therapeutically relevant anti-inflammatory dose of (*R*)-DOI could be as low as 12 micrograms.

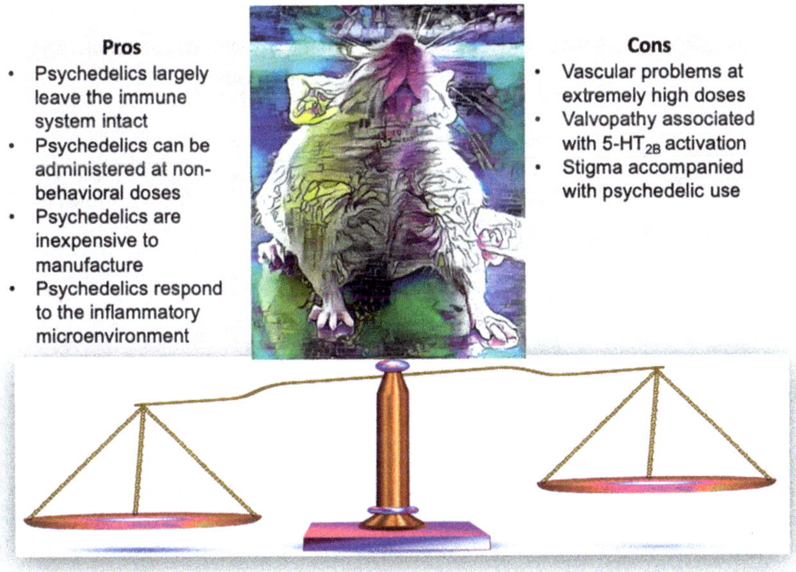

Fig. 1 Pros and cons of psychedelics as anti-inflammatory medications. Mass quantities of psychedelic medications can be produced at prices much lower than conventional anti-inflammatory therapies like corticosteroids and biologics. The potency of some 5-HT_{2A} agonists, like (*R*)-DOI, allows an administration at doses low enough to stave any potential of abuse. Key considerations before self-administration of any 5-HT_{2A} receptor agonist involve 5-HT_{2B}-mediated valvopathies, as well as coronary artery spasm at high dose usage. Unfounded stereotypes regarding psychedelics likewise encumber their widespread acceptance

The extraordinary potency of certain psychedelics belies another important drawback of psychedelic medication. High dose use of psychedelics can be associated with vascular muscle contraction, platelet aggregation, thrombus formation, and coronary artery spasms (Nagatomo et al. 2004). Fortunately, for several psychedelics, they are effective as an anti-inflammatory in preclinical models of human diseases at doses not only below those necessary to elicit a behavioral response (Flanagan and Nichols 2018; Yu et al. 2008; Nau et al. 2015; Flanagan et al. 2019a, 2020), but also greatly under those associated with cardiovascular problems, such as activation of 5-HT_{2B} receptors on heart valve leaflets which induces valvular heart disease (Hutcheson et al. 2011; Roth 2007; Rothman and Baumann 2009). Nevertheless, potential for negative cardiovascular effects must be considered (Kuypers et al. 2019) until more work is done to establish safety.

The final concern impeding psychedelic medication is their overall acceptance by the general populace. Psychedelics came to prominence in the late 1960s due to their widespread usage by elements of the counter-culture (Nichols 2016). While there is a renaissance and rehabilitation regarding the public's views on psychedelics (Nichols et al. 2017), there remains significant skepticism regarding their use as therapeutics (Forstmann and Sagioglou 2020). Similar to the paradigm shift regarding medical usage of cannabinoids (Hunt et al. 2021), continued unbiased, rigidly controlled, fully transparent laboratory and clinical research into psychedelic medication can mitigate these concerns and unlock the full therapeutic potential of 5-HT_{2A} receptor activation.

Acknowledgments Funding for this work was provided by Eleusis Therapeutics, and a generous contribution from the LSU Foundation. The authors would like to thank Brett C. Brown for figure design.

Declaration of Interest C. D. Nichols has a sponsored research contract with Eleusis Therapeutics and is a member of its Scientific Advisory Board.

References

Adams AM et al (2019) In vivo production of psilocybin in E. coli. Metab Eng 56:111–119

Afshar S, Shahidi S, Rohani AH, Soleimani Asl S, Komaki A (2019) Protective effects of 5-HT(1A) receptor antagonist and 5-HT(2A) receptor agonist on the biochemical and histological features in a rat model of Alzheimer's disease. J Chem Neuroanat 96:140–147. https://doi.org/10.1016/j.jchemneu.2019.01.008

Akiyoshi T et al (2006) Induction of indefinite survival of fully mismatched cardiac allografts and generation of regulatory cells by sarpogrelate hydrochloride. Transplantation 82:1051–1059. https://doi.org/10.1097/01.tp.0000233870.54297.9a

Aldridge J, Décary-Hétu D (2016) Hidden wholesale: the drug diffusing capacity of online drug cryptomarkets. Int J Drug Policy 35:7–15

Arzt E, Costas M, Finkielman S, Nahmod VE (1991) Serotonin inhibition of tumor necrosis factor-alpha synthesis by human monocytes. Life Sci 48:2557–2562. https://doi.org/10.1016/0024-3205(91)90612-f

Aun MV, Bonamichi-Santos R, Arantes-Costa FM, Kalil J, Giavina-Bianchi P (2017) Animal models of asthma: utility and limitations. J Asthma Allergy 10:293–301. https://doi.org/10.2147/jaa.s121092

Barnes PJ (2006) How corticosteroids control inflammation: quintiles prize lecture 2005. Br J Pharmacol 148:245–254. https://doi.org/10.1038/sj.bjp.0706736

Barnes PJ (2010) Inhaled corticosteroids in COPD: a controversy. Respiration 80:89–95. https://doi.org/10.1159/000315416

Bates JHT, Rincon M, Irvin CG (2009) Animal models of asthma. Am J Physiol Lung Cell Mol Physiol 297:L401–L410

Brattelid T et al (2007) Serotonin responsiveness through 5-HT2A and 5-HT4 receptors is differentially regulated in hypertrophic and failing rat cardiac ventricle. J Mol Cell Cardiol 43:767–779. https://doi.org/10.1016/j.yjmcc.2007.08.019

Canal CE, Morgan D (2012) Head-twitch response in rodents induced by the hallucinogen 2,5-dimethoxy-4-iodoamphetamine: a comprehensive history, a re-evaluation of mechanisms, and its utility as a model. Drug Test Anal 4:556–576. https://doi.org/10.1002/dta.1333

Carhart-Harris RL et al (2016) Psilocybin with psychological support for treatment-resistant depression: an open-label feasibility study. Lancet Psychiatry 3:619–627. https://doi.org/10.1016/s2215-0366(16)30065-7

Carhart-Harris RL et al (2017) Psilocybin with psychological support for treatment-resistant depression: six-month follow-up. Psychopharmacology (Berl). https://doi.org/10.1007/s00213-017-4771-x

Christ A et al (2018) Western diet triggers NLRP3-dependent innate immune reprogramming. Cell 172:162–175.e114. https://doi.org/10.1016/j.cell.2017.12.013

Curtis JR, Hobar C, Hansbrough K (2011) Injection-site burning and stinging in patients with rheumatoid arthritis using injectable biologics. Curr Med Res Opin 27:71–78. https://doi.org/10.1185/03007995.2010.534959

da Silva MG, Daros GC, de Bitencourt RM (2020) Anti-inflammatory activity of ayahuasca and its implications for the treatment of neurological and psychiatric diseases. Behav Brain Res 113003

Davydova SM, Cheido MA, Gevorgyan MM, Idova GV (2010) Effects of 5-HT2A receptor stimulation and blocking on immune response. Bull Exp Biol Med 150:219–221

dos Santos RG (2014) Immunological effects of ayahuasca in humans. J Psychoactive Drugs 46: 383–388

Du T et al (1996) Effect of nedocromil sodium on allergen-induced airway responses and changes in the quantity of airway smooth muscle in rats. J Allergy Clin Immunol 98:400–407

Flanagan TW, Nichols CD (2018) Psychedelics as anti-inflammatory agents. Int Rev Psychiatry:1–13. https://doi.org/10.1080/09540261.2018.1481827

Flanagan TW et al (2019a) Activation of 5-HT2 receptors reduces inflammation in vascular tissue and cholesterol levels in high-fat diet-fed apolipoprotein E knockout mice. Sci Rep 9:13444. https://doi.org/10.1038/s41598-019-49987-0

Flanagan TW et al (2019b) 5-HT2 receptor activation alleviates airway inflammation and structural remodeling in a chronic mouse asthma model. Life Sci 236:116790. https://doi.org/10.1016/j.lfs.2019.116790

Flanagan TW et al (2020) Structure–activity relationship analysis of psychedelics in a rat model of asthma reveals the anti-inflammatory pharmacophore. ACS Pharmacol Transl, Science

Forstmann M, Sagioglou C (2020) How psychedelic researchers' self-admitted substance use and their association with psychedelic culture affect people's perceptions of their scientific integrity and the quality of their research. Public Underst Sci:0963662520981728

Frecska E, Bokor P, Winkelman M (2016) The therapeutic potentials of ayahuasca: possible effects against various diseases of civilization. Front Pharmacol 7:35

Fulkerson PC, Rothenberg ME, Hogan SP (2005) Building a better mouse model: experimental models of chronic asthma. Clin Exp Allergy 35:1251–1253. https://doi.org/10.1111/j.1365-2222.2005.02354.x

Galvão-Coelho NL et al (2020) Changes in inflammatory biomarkers are related to the antidepressant effects of ayahuasca. J Psychopharmacol 34:1125–1133

Giridharan VV et al (2020) Clozapine prevents poly (I: C) induced inflammation by modulating NLRP3 pathway in microglial cells. Cells 9

Girodet PO et al (2015) Calcium channel blocker reduces airway remodeling in severe asthma. A proof-of-concept study. Am J Respir Crit Care Med 191:876–883. https://doi.org/10.1164/rccm.201410-1874OC

Glennon RA, Young R, Benington F, Morin RD (1982) Behavioral and serotonin receptor properties of 4-substituted derivatives of the hallucinogen 1-(2,5-dimethoxyphenyl)-2-aminopropane. J Med Chem 25:1163–1168. https://doi.org/10.1021/jm00352a013

Griffiths RR et al (2016) Psilocybin produces substantial and sustained decreases in depression and anxiety in patients with life-threatening cancer: a randomized double-blind trial. J Psychopharmacol 30:1181–1197. https://doi.org/10.1177/0269881116675513

Harbuz MS et al (1996) The role of endogenous serotonin in adjuvant-induced arthritis in the rat. Br J Rheumatol 35:112–116

Harbuz MS, Marti O, Lightman SL, Jessop DS (1998) Alteration of central serotonin modifies onset and severity of adjuvant-induced arthritis in the rat. Br J Rheumatol 37:1077–1083

Hawiger JJN, Veach RA, Zienkiewicz J, Liu Y, Collins RD, Major AS (2018) Compositions and methods for treating and preventing hyperlipidemia, fatty liver, atherosclerosis and other disorders associated with metabolic syndrome. United States patent

Hayashi T et al (2003) Sarpogrelate HCl, a selective 5-HT2A antagonist, retards the progression of atherosclerosis through a novel mechanism. Atherosclerosis 168:23–31

Hong Y, Ji H, Wei H (2006) Topical ketanserin attenuates hyperalgesia and inflammation in arthritis in rats. Pain 124:27–33

Hope DB, Smith AD (1960) Distribution and activity of monoamine oxidase in mouse tissues. Biochem J 74:101–107

Hori A, Nishida T, Takashiba S, Kubota S, Takigawa M (2017) Regulatory mechanism of CCN2 production by serotonin (5-HT) via 5-HT2A and 5-HT2B receptors in chondrocytes. PLoS One 12:e0188014. https://doi.org/10.1371/journal.pone.0188014

Hoshino M, Takahashi M, Takai Y, Sim J (1999) Inhaled corticosteroids decrease subepithelial collagen deposition by modulation of the balance between matrix metalloproteinase-9 and tissue inhibitor of metalloproteinase-1 expression in asthma. J Allergy Clin Immunol 104:356–363

Hu X et al (2012) Clozapine protects dopaminergic neurons from inflammation-induced damage by inhibiting microglial overactivation. J Neuroimmune Pharmacol 7:187–201

Hunt DA, Keefe J, Whitehead T (2021) Understanding cannabis: clinical considerations. J Nurse Pract 17:163–167

Hutcheson JD, Setola V, Roth BL, Merryman WD (2011) Serotonin receptors and heart valve disease – it was meant 2B. Pharmacol Ther 132:146–157. https://doi.org/10.1016/j.pharmthera.2011.03.008

Inoue M et al (2011) Regulation of antigen-specific CTL and Th1 cell activation through 5-hydroxytryptamine 2A receptor. Int Immunopharmacol 11:67–73. https://doi.org/10.1016/j.intimp.2010.10.007

Inserra A, De Gregorio D, Gobbi G (2021) Psychedelics in psychiatry: neuroplastic, immunomodulatory, and neurotransmitter mechanisms. Pharmacol Rev 73:202–277. https://doi.org/10.1124/pharmrev.120.000056

Ishiwata K, Kobayashi T, Kawamura K, Matsuno K (2003) Age-related changes of the binding of [3h]SA4503 to sigma1 receptors in the rat brain. Ann Nucl Med 17:73–77

Joubert AI et al (2020) Mechanisms of particles in sensitization, effector function and therapy of allergic disease. Front Immunol 11:1334

Kaur G, Krishan P (2020) Serotonin 5HT(2A) receptor antagonism mediated anti-inflammatory and anti-fibrotic effect in adriamycin-induced CKD in rats. Naunyn Schmiedebergs Arch Pharmacol 393:1269–1279. https://doi.org/10.1007/s00210-020-01826-0

Kobayashi T et al (2014) Matrix metalloproteinase-9 activates TGF-beta and stimulates fibroblast contraction of collagen gels. Am J Physiol Lung Cell Mol Physiol 306:L1006–L1015. https://doi.org/10.1152/ajplung.00015.2014

Kodela R, Chattopadhyay M, Kashfi K (2012) NOSH-aspirin: a novel nitric oxide-hydrogen Sulfide-releasing hybrid: a new class of anti-inflammatory pharmaceuticals. ACS Med Chem Lett 3:257–262. https://doi.org/10.1021/ml300002m

Koeberle A, Laufer SA, Werz O (2016) Design and development of microsomal prostaglandin E2 Synthase-1 inhibitors: challenges and future directions. J Med Chem 59:5970–5986. https://doi.org/10.1021/acs.jmedchem.5b01750

Kuypers KPC et al (2019) Microdosing psychedelics: more questions than answers? An overview and suggestions for future research. J Psychopharmacol 33:1039–1057

Kyzar EJ, Nichols CD, Gainetdinov RR, Nichols DE, Kalueff AV (2017) Psychedelic drugs in biomedicine. Trends Pharmacol Sci 38:992–1005. https://doi.org/10.1016/j.tips.2017.08.003

Li P, Zheng Y, Chen X (2017) Drugs for autoimmune inflammatory diseases: from small molecule compounds to anti-TNF biologics. Front Pharmacol 8:460. https://doi.org/10.3389/fphar.2017.00460

Londesbrough DJ et al (2020) Google Patents

Ma Y et al (2016) Immunization against TGF-beta1 reduces collagen deposition but increases sustained inflammation in a murine asthma model. Hum Vaccin Immunother 12:1876–1885. https://doi.org/10.1080/21645515.2016.1145849

Ma Y et al (2021) 5-HT(2A) receptor and 5-HT degradation play a crucial role in atherosclerosis by modulating macrophage foam cell formation, vascular endothelial cell inflammation, and hepatic steatosis. J Atheroscler Thromb. https://doi.org/10.5551/jat.58305

Maes M, Meltzer HY, Bosmans E (1994) Immune-inflammatory markers in schizophrenia: comparison to normal controls and effects of clozapine. Acta Psychiatr Scand 89:346–351

Manzel A et al (2014) Role of "Western diet" in inflammatory autoimmune diseases. Curr Allergy Asthma Rep 14:404. https://doi.org/10.1007/s11882-013-0404-6

Marconi A, Darquenne S, Boulmerka A, Mosnier M, D'Alessio P (2003) Naftidrofuryl-driven regulation of endothelial ICAM-1 involves nitric oxide. Free Radic Biol Med 34:616–625

Martin DA, Marona-Lewicka D, Nichols DE, Nichols CD (2014) Chronic LSD alters gene expression profiles in the mPFC relevant to schizophrenia. Neuropharmacology 83:1–8. https://doi.org/10.1016/j.neuropharm.2014.03.013

McKune CM, Watts SW (2001) Characterization of the serotonin receptor mediating contraction in the mouse thoracic aorta and signal pathway coupling. J Pharmacol Exp Ther 297:88–95

Mei J, Pasternak GW (2001) Molecular cloning and pharmacological characterization of the rat sigma1 receptor. Biochem Pharmacol 62:349–355. https://doi.org/10.1016/s0006-2952(01)00666-9

Moss AC (2018) Treatment of inflammatory bowel disease with biologics. Springer, pp 1–11

Munroe ME, Businga TR, Kline JN, Bishop GA (2010) Anti-inflammatory effects of the neurotransmitter agonist honokiol in a mouse model of allergic asthma. J Immunol 185:5586–5597. https://doi.org/10.4049/jimmunol.1000630

Nagatomo T, Rashid M, Abul Muntasir H, Komiyama T (2004) Functions of 5-HT2A receptor and its antagonists in the cardiovascular system. Pharmacol Ther 104:59–81. https://doi.org/10.1016/j.pharmthera.2004.08.005

Nau F Jr, Yu B, Martin D, Nichols CD (2013) Serotonin 5-HT2A receptor activation blocks TNF-alpha mediated inflammation in vivo. PLoS One 8:e75426. https://doi.org/10.1371/journal.pone.0075426

Nau F Jr et al (2015) Serotonin 5-HT(2) receptor activation prevents allergic asthma in a mouse model. Am J Physiol Lung Cell Mol Physiol 308:L191–L198. https://doi.org/10.1152/ajplung.00138.2013

Nichols CD (2009) Serotonin 5-HT(2A) receptor function as a contributing factor to both neuropsychiatric and cardiovascular diseases. Cardiovasc Psychiatry Neurol 2009:475108. https://doi.org/10.1155/2009/475108

Nichols DE (2012) Structure-activity relationships of serotonin 5-HT2A agonists. In: WIRES membrane transport and signaling, vol 1, pp 559–579

Nichols DE (2016) Psychedelics. Pharmacol Rev 68:264–355. https://doi.org/10.1124/pr.115.011478

Nichols DE, Johnson MW, Nichols CD (2017) Psychedelics as medicines: an emerging new paradigm. Clin Pharmacol Ther 101:209–219. https://doi.org/10.1002/cpt.557

Nishiyama T (2009) Acute effects of sarpogrelate, a 5-HT2A receptor antagonist on cytokine production in endotoxin shock model of rats. Eur J Pharmacol 614:122–127. https://doi.org/10.1016/j.ejphar.2009.03.041

Pierce PA, Xie GX, Peroutka SJ, Green PG, Levine JD (1995) 5-Hydroxytryptamine-induced synovial plasma extravasation is mediated via 5-hydroxytryptamine2A receptors on sympathetic efferent terminals. J Pharmacol Exp Ther 275:502–508

Pini L, Torregiani C, Martin JG, Hamid Q, Ludwig MS (2006) Airway remodeling in allergen-challenged Brown Norway rats: distribution of proteoglycans. Am J Physiol Lung Cell Mol Physiol 290:L1052–L1058. https://doi.org/10.1152/ajplung.00122.2005

Qi J, Zulfiker AHM, Li C, Good D, Wei MQ (2018) The development of toad toxins as potential therapeutic agents. Toxins 10:336

Ramakrishnan NK et al (2016) Sigma-1 agonist binding in the aging rat brain: a MicroPET study with [(11)C]SA4503. Mol Imaging Biol 18:588–597. https://doi.org/10.1007/s11307-015-0917-6

Rider P, Carmi Y, Cohen I (2016) Biologics for targeting inflammatory cytokines, clinical uses, and limitations. Int J Cell Biol 2016:9259646. https://doi.org/10.1155/2016/9259646

Rosenblum MD, Remedios KA, Abbas AK (2015) Mechanisms of human autoimmunity. J Clin Invest 125:2228–2233

Ross S et al (2016) Rapid and sustained symptom reduction following psilocybin treatment for anxiety and depression in patients with life-threatening cancer: a randomized controlled trial. J Psychopharmacol 30:1165–1180. https://doi.org/10.1177/0269881116675512

Roth BL (2007) Drugs and valvular heart disease. N Engl J Med 356:6–9. https://doi.org/10.1056/NEJMp068265

Rothman RB, Baumann MH (2009) Serotonergic drugs and valvular heart disease. Expert Opin Drug Saf 8:317–329. https://doi.org/10.1517/14740330902931524

Sasaki M, Obata H, Kawahara K, Saito S, Goto F (2006) Peripheral 5-HT2A receptor antagonism attenuates primary thermal hyperalgesia and secondary mechanical allodynia after thermal injury in rats. Pain 122:130–136. https://doi.org/10.1016/j.pain.2006.01.021

Shen HW, Jiang XL, Winter JC, Yu AM (2010) Psychedelic 5-methoxy-N,N-dimethyltryptamine: metabolism, pharmacokinetics, drug interactions, and pharmacological actions. Curr Drug Metab 11:659–666

Shulgin ATS (1991) PIHKAL: a chemical love story. Transform Press

Sugino H, Futamura T, Mitsumoto Y, Maeda K, Marunaka Y (2009) Atypical antipsychotics suppress production of proinflammatory cytokines and up-regulate interleukin-10 in lipopolysaccharide-treated mice. Prog Neuropsychopharmacol Biol Psychiatry 33:303–307

Sumino K, Sheshadri A, Castro M (2015) Calcium channel blocker reduces airway remodeling-or does it? Am J Respir Crit Care Med 191:863–864. https://doi.org/10.1164/rccm.201502-0322ED

Szabo A (2015) Psychedelics and immunomodulation: novel approaches and therapeutic opportunities. Front Immunol 6:358. https://doi.org/10.3389/fimmu.2015.00358

Szabo A, Kovacs A, Frecska E, Rajnavolgyi E (2014) Psychedelic N,N-dimethyltryptamine and 5-methoxy-N,N-dimethyltryptamine modulate innate and adaptive inflammatory responses through the sigma-1 receptor of human monocyte-derived dendritic cells. PLoS One 9: e106533. https://doi.org/10.1371/journal.pone.0106533

Thompson C, Szabo A (2020) Psychedelics as a novel approach to treating autoimmune conditions. Immunol Lett 228:45–54. https://doi.org/10.1016/j.imlet.2020.10.001

Todorović N, Filipović D (2017) The antidepressant-and anxiolytic-like effects of fluoxetine and clozapine in chronically isolated rats involve inhibition of hippocampal TNF-α. Pharmacol Biochem Behav 163:57–65

Toruner M et al (2008) Risk factors for opportunistic infections in patients with inflammatory bowel disease. Gastroenterology 134:929–936. https://doi.org/10.1053/j.gastro.2008.01.012

Uchida-Kitajima S et al (2008) 5-Hydroxytryptamine 2A receptor signaling cascade modulates adiponectin and plasminogen activator inhibitor 1 expression in adipose tissue. FEBS Lett 582: 3037–3044. https://doi.org/10.1016/j.febslet.2008.07.044

Vane JR, Botting RM (1998) Anti-inflammatory drugs and their mechanism of action. Inflamm Res 47(Suppl 2):S78–S87. https://doi.org/10.1007/s000110050284

Wacker D et al (2017) Crystal structure of an LSD-bound human serotonin receptor. Cell 168:377–389.e312. https://doi.org/10.1016/j.cell.2016.12.033

Wu AC, Fuhlbrigge AL, Robayo MA, Shaker M (2020) Cost-effectiveness of biologics for allergic diseases. J Allergy Clin Immunol Pract. https://doi.org/10.1016/j.jaip.2020.10.009

Xu D et al (2021) Evaluation of analgesic and anti-inflammatory actions of indolealkylamines from toad venom in mice using lipidomics and molecular docking. J Ethnopharmacol 269:113677

Yu B et al (2008) Serotonin 5-hydroxytryptamine(2A) receptor activation suppresses tumor necrosis factor-alpha-induced inflammation with extraordinary potency. J Pharmacol Exp Ther 327: 316–323. https://doi.org/10.1124/jpet.108.143461

Zhang Y et al (2019) Comparative analysis of hydrophilic ingredients in toad skin and toad venom using the UHPLC-HR-MS/MS and UPLC-QqQ-MS/MS methods together with the anti-inflammatory evaluation of indolealkylamines. Molecules 24:86

Zosky GR, Sly PD (2007) Animal models of asthma. Clin Exp Allergy 37:973–988. https://doi.org/10.1111/j.1365-2222.2007.02740.x

Zulfiker AH, Hashimi SM, Qi J, Grice ID, Wei MQ (2016) Aqueous and ethanol extracts of Australian cane toad skins suppress pro-inflammatory cytokine secretion in U937 cells via NF-kappaB Signaling pathway. J Cell Biochem 117:2769–2780. https://doi.org/10.1002/jcb.25577

Psilocybin for the Treatment of Obsessive-Compulsive Disorders

Katja Ehrmann, John J. B. Allen, and Francisco A. Moreno

Contents

Abstract Obsessive-compulsive disorder (OCD) is a highly prevalent and disabling condition for which currently available treatments are insufficiently effective and alternatives merit priority attention. Psilocybin may represent a safe and effective avenue for treatment of individuals affected by this condition. In this chapter we briefly introduce OCD symptoms, epidemiology, as well as relevant hypotheses on the mechanism of disease that may inform treatment interventions. We briefly describe currently available treatments, mechanisms of action, and efficacy limitations, as preamble to the potential use of psilocybin and perhaps similar compounds in the treatment of OCD and related conditions. Although much is reviewed throughout this book about the mechanisms of action of psychedelic agents, a

K. Ehrmann, J. J. B. Allen, and F. A. Moreno (✉)
University of Arizona, Tucson, AZ, USA
e-mail: fmoreno@email.arizona.edu

Curr Topics Behav Neurosci (2022) 56: 247–260
https://doi.org/10.1007/7854_2021_279
Published Online: 17 November 2021

focused discussion of psilocybin effects as they pertain to OCD is also included. Our experience with incidental observation, prospective research, and current explorations of psilocybin in OCD are also described.

Keywords Compulsions · Obsessions · Obsessive-compulsive disorder · OCD · Psilocybin · Psychedelics

1 OCD Defined

The American Psychiatric Association's Diagnostic and Statistical Manual of Mental Disorders 5th Edition (DSM-5) and the International Classification of Diseases 11th Edition (ICD-11) describe OCD as a mental condition characterized by the presence of either obsessions and/or compulsions that are usually recognized as excessive, unreasonable, and a product of the individual's own mind. These symptoms are time consuming (lasting at least 1 h per day and ranging to nearly constant) and cause marked distress and/or significant interference in the person's ability to function.

2 OCD Hypothesized Mechanisms of Disease

OCD is a heterogenous condition (Raines et al. 2018) with proposed clinical subtypes, some of which have been associated with biological, psychological, developmental, and environmental risk factors, and with distinct course of disease and treatment response. It is safe to assume that no single mechanism of disease explains OCD, and the complexity of the OCD phenotype may explain at least in part the variability of findings from pharmacological, psychotherapeutic, imaging, genetic, epidemiologic, and other study approaches.

2.1 Psychological Theories

Traditional psychoanalytic theories postulate that obsessional neurosis, a term that preceded the modern OCD diagnosis, results from unresolved fixations or regressions to primitive stages of psychosexual development, specifically the "Anal Phase." In the anal phase, aggressive and sexual impulses conflict with a rigid superego and certain ego defenses that attempt to keep such impulses out of consciousness, as they are unacceptable to personal imperatives as conscience and cultural development occur (Freud 1966). More recent formulations informed by *object relations theory* focus on the development of a fragmented or ambivalent self.

Patients feel threatened by thoughts that they are bad, imperfect, unreliable, uncontrollable, or immoral, and are unable to integrate these attributions into a coherent self-image. Cognitive Behavioral Theories have supported the development of current psychological best practices for OCD. Learning theory maintains that a neutral event stimulus (i.e., perception, thought) comes to elicit fear when it is repeatedly presented together with an event that causes pain/distress (obsession formation). Subsequently, escape or avoidance behaviors (i.e., compulsions) are developed and maintained to reduce the anxiety. OCD has been characterized by the following erroneous cognitions: (1) Assigning a high probability of danger to situations that are relatively safe; (2) Exaggerating the severity of anticipated negative outcomes; (3) Assuming dangerous qualities of an event or object in the absence of evidence of safety (Foa and Kozak 1985; Foa 2010).

Traditional analytical and modern cognitive-behavioral formulations overlap in that they emphasize contradictory and poorly integrated attributions of self and other (Chlebowski and Gregory 2009). Regardless of the particular psychodynamic formulation, obsessions and compulsions can serve the function to keep patients' attention away from longstanding conflicts, some of which may be brought to awareness during the psilocybin experience and which may be an important target for the support and integration that occurs in the context of psilocybin sessions.

2.2 *Biological Theories*

It is known that OCD is heritable, with estimates from twin studies that 27–65% of the variance in symptoms may be attributable to genetic variance. OCD also runs in families, and it is observed four times more frequently in relatives of people with OCD than in the general population. Certain subtypes of OCD are clearly more heritable, including early onset OCD (Mahjani et al. 2021), and the heterogeneity of ascertainment methods used in studies may explain the variability of results. Several neurotransmitters, neurohormones, and immunological alterations have been proposed as explanations for symptom generation, and treatment response, including specific elements of the serotoninergic, glutamatergic, GABAergic, and dopaminergic systems. Molecular genetic studies suggest a higher representation of certain genes related to serotonin function in patients with OCD, particularly the serotonin transporter and the serotonin 2a receptor gene (Taylor et al. 2016). Candidate gene association studies now available suggest the role of glutamatergic transmission, involving genes related to glutamatergic transport (SLC1A1), receptor regulation (DLGAP1), receptor and subunit synthesis (GRID2 and GRIN2B). The role of glutamate in OCD neurobiology and pathophysiology is also well supported, although treatments involving this neurochemical are in need of further research (Pittenger et al. 2011).

The serotonergic system, however, is among the most broadly supported given the seemingly selective pharmacological response to serotonin acting agents, the reports of serotonin related changes in the central nervous system, and other alterations in peripheral markers of serotonin function in OCD patients (Delgado and Moreno 1998). Intriguingly, laboratory-induced reductions in brain serotonin availability in patients recovered from OCD do not lead to a worsening of obsessive-compulsive symptoms. In contrast, specific blockade of serotonin subtype-2 receptors (5HT2) in patients recently improved and receiving treatment with serotonin enhancing medications causes an acute return of OCD symptoms. Furthermore, 5HT2 blockers used in practice such as atypical neuroleptics and mirtazapine appear to be beneficial in the treatment of OCD. These findings suggest that response to OCD is not only related to activity at a specific receptor, but rather the complex postsynaptic functional and structural changes that may involve the function of various neurochemicals and/or regions and circuits.

2.3 *Brain Basis of OCD*

Data from animal models as well as functional and structural human brain imaging studies point to the involvement of parallel and partially-segregated circuits that play a role in initiation and termination of thoughts and behaviors. The so-called Cortico-Striato-Thalamo-Cortical (CSTC) circuit refers to interconnectivity of the orbitofrontal cortex, the caudate nucleus, and the thalamus, however a larger number of regions frequently interacting with CSTC have been implicated in OCD (Tang et al. 2013). Interestingly, specific-symptom dimensions of OCD appear to be associated with findings in different brain regions, further supporting the heterogeneity of OCD.

The proposed OCD cortico-striatal hyperactivity leads to a persistently high error signal, ultimately resulting in its characteristic psychopathology, including irrational fears or obsessions, or that an action was not completed correctly according to a set of internal unattainable rules, triggering repetitive, compensatory behaviors (i.e., compulsions). An index of this error signal is larger in OCD patients, its magnitude relates to the continuum of OCD symptom severity (Zambrano-Vazquez and Allen 2014).

Other neural systems that play a role in a number of disorders characterized by persistent negative thinking are thought to play a role in OCD as well. Primary among these is the default mode network (DMN), a system of brain regions that coactivate during spontaneous self-generated deliberate thought, and that appear to be highly coactivated during times of persistent negative self-focused thought (Andrews-Hanna et al. 2014). In OCD, one of the main characteristics is an extreme focus on internally generated unpleasant, unwanted thoughts or images thought to be generated in the DMN. It has been shown that this network is hyperactive in patients with OCD during cognitive tasks, reflecting a failure in the normal process of DMN suppression to engage in externally-focused task-relevant activity. This lack of disengagement in OCD patients compared to healthy controls indicates a difference in communication between the DMN (posterior cingular cortex, medial frontal cortex, posterior inferior parietal lobule and parahippocampus) and frontal areas of the brain, leaving OCD patients less able to focus their attention on external cues, and thus the task at hand (Stern et al. 2012). This is even true for emotion provoking stimuli, especially pleasant stimuli. OCD patients, compared to healthy controls, have difficulties deactivating the DMN (Gonçalves et al. 2017) leaving them more engaged with internally generated negative thought content than with external information that has potential to elicit positive mood or thoughts.

3 OCD Treatments

Modern treatment guidelines support the use of Cognitive Behavioral Therapy (CBT) for OCD with emphasis on the value of Exposure and Response Prevention (ERP), and/or pharmacotherapy with serotonin reuptake inhibitor medications.

3.1 Pharmacotherapy

The U.S. Food and Drug Administration (FDA) has approved a small number of agents to treat this disorder. They are all potent serotonin reuptake inhibitors that lead to an increase in serotonin function and include: clomipramine (Anafranil® and others), fluoxetine (Prozac® and others), fluvoxamine (Luvox®), sertraline (Zoloft®), paroxetine (Paxil® and others). In spite of their established efficacy (Gosmann et al. 2021), a number of shortcomings limit their ability to improve a patient's function. For example, in spite of OCD medications being frequently prescribed at higher doses and for longer periods than for other disorders, only about half of the patients receiving adequate treatment-trials will reach a satisfactory response, while most patients that do improve only have a one third to one half decrease in severity ratings. Their residual symptoms continue to cause dysfunction and increase vulnerability to complications and exacerbations.

Drugs such as desipramine and bupropion, which may act primarily by enhancing function of norepinephrine and/or dopamine, have been found to be ineffective at treating OCD (Vulink et al. 2005). The apparent selectivity of treatment response to medication that acts on the serotonin system, and the finding that serotonin blockers cause a relapse of OCD symptoms further justifies pursuing options related to serotonin actions in this population.

3.2 Psychotherapy

CBT has been the mainstay of non-analytic therapy and a first-choice treatment for OCD. CBT comprises two components: cognitive reappraisal of distorted beliefs and a behavioral intervention to prevent symptom engagement. The treatment of choice for OCD is ERP, which involves gradual and prolonged exposure to fear-provoking stimuli combined with instructions to abstain from the compulsive behavior (Foa and Goldstein 1978; Öst et al. 2015). The integration of ERP with cognitive components, such as the discussion of feared consequences and other dysfunctional beliefs, can enhance the acceptability and effectiveness of ERP, particularly for patients with limited insight or those that may find exposure treatments difficult to tolerate (Stein et al. 2019).

3.3 *Treatment Strategies for Poor Responders*

Although both pharmacotherapy and psychotherapy can be useful alone, the most effective outcomes typically occur through a combination of these treatment forms (pharmacology and CBT) (March et al. 1997). When these treatments fail due to intolerance or lack of benefit after adequate trials of evidence supported first line interventions, an increasingly complex series of alternatives is possible, often with the goal of diminishing symptom severity rather than anticipating symptom remission. These include: alternative drug trials, augmentations and combination strategies, and the use of repeated transcranial magnetic stimulation (TMS), electroshock treatment, deep brain stimulation (DBS) in the anterior cingulate cortex and other targets, and surgical ablation such as cingulotomy and other lesion-based treatments. Surgical options may be tried only after other avenues have been exhausted in highly treatment-resistant cases. Given that a significant portion of individuals remain symptomatic and impaired in spite of escalation of treatment strategies with increasing associated morbidity and decreased likelihood of benefit, identifying new options for treatments of patients with severe and treatment resistant OCD has become a high priority.

4 Background of Psilocybin in OCD

4.1 *Incidental Findings*

A quarter century ago, we met a 34 year-old male patient in clinic who had suffered from severe OCD symptoms for nearly 30 years. His symptoms included upsetting obsessions of contamination and disgust with body secretions, and preoccupations with order, and his compulsions included excessive and ritualistic washing and cleaning, checking, arranging, and counting. This person reported having used multiple substances recreationally. Alcohol and marijuana relieved his anxiety but not his OCD, while stimulants made his OCD worse. He noticed that if he used 2 g of freeze-dried psilocybin mushrooms his symptoms would disappear during the time that he was intoxicated, and he would remain symptom free for several days before gradually returning. He then began chronic use of hallucinogens and found that the obsessions and compulsions actually went into remission for periods of several months after he stopped using them. The use of unspecified doses of peyote also improved his symptoms (Moreno and Delgado 1997). While reviewing the literature at the time we encountered previous similar reports, and a number of similar instances have been added to the literature documenting a suggested benefit of psilocybin use including lasting benefits with repeated use (Wilcox 2014), and in at least one case using psilocybin in combination with high dose SSRIs (Lugo-Radillo and Cortes-Lopez 2021), a practice currently discouraged based on concerns for increased risk of serotonin syndrome and concerns about potential decreased effects in people who take antidepressants chronically.

4.2 *Rationale for Psilocybin Use in OCD*

Psilocybin, LSD, and mescaline are known to bind to a large number of receptors, transporters, and other proteins; however, they are extremely potent agonists at 5-HT$_{1A}$, 5-HT$_{2A}$, and 5-HT$_{2C}$ receptors and their binding potency to these receptors is correlated with their human potency as hallucinogens (Pokorny et al. 2016). It is possible that involvement of other sites of action may explain the acute improvement in symptoms described in the published case reports, however it is likely that interactions with 5-HT$_{1A}$, 5-HT$_{2A}$, and 5-HT$_{2C}$ receptors may be an important component of anti-OCD drug action. SSRIs lead to increases in activation of postsynaptic 5-HT$_{1A}$, and chronic downregulation of 5-HT$_{2A}$. Additionally, the observations that administration of the non-selective 5-HT antagonists metergoline or ritanserin exacerbates OCD symptoms in patients recently remitted with SSRI treatment add complexity to the interpretation of a receptor specific effect. Knockout mice who lack 5-HT$_{2A}$ receptors lack the behavioral and cellular responses to psychedelics compared to wild type animals who have their natural 5-HT$_{2A}$ receptor expression.

Important questions arise from the reports above, including: why are the anti-obsessional effects so immediate with psilocybin, while potent serotonin enhancing agents like clomipramine and the SSRIs take weeks to months to show beneficial responses? Why do some 5HT2 blockers actually help treat OCD symptoms (atypical neuroleptic medications, mirtazapine), while experimental 5HT2 blockers ritanserin, ketanserin, and MDL-100,907 do not cause any effect? Delgado and Moreno (Delgado and Moreno 1998) published a mechanistic paper that reviews some of these topics offering an early theoretical foundation to support exploration of psilocybin and indole-based psychedelics in the treatment of OCD.

Advances in our understanding of OCD, psychedelic mechanisms, and brain circuitry in the last two decades suggest a more complex mechanism may be in play than initially proposed, involving a variety of postsynaptic effects, interactions with other neurochemical systems, and brain circuits in the anti-obsessional response. Considering specific receptors within the serotonergic system, Carhart-Harris and Nutt (Carhart-Harris and Nutt 2017) have proposed that serotonin neurotransmission enhances two distinct adaptive responses to adversity, in which postsynaptic 5-HT_{1A} signaling mediates passive coping characterized by stress moderation (SSRI treatments), whereas 5-HT_{2A} signaling facilitated by psychedelics mediates active coping and further enhances plasticity. In terms of interactions with other neurochemical systems, knockout mice who do not express the metabotropic glutamate receptor-2 (mGluR2) also fail to have the cellular and behavioral responses to LSD similar to 5-HT_{2A} knockouts, suggesting that the interaction of mGluR2 and 5-HT_{2A} is necessary for certain neuro-behavioral effects to take place (Moreno et al. 2011).

Successful treatments, both pharmacological and psychological therapies, show a normalization of CSTC circuits in patients with OCD (Van der Straten et al. 2017), indicating that a disruption of the CSTC circuits might be a good target for the treatment of OCD with psychedelics. Patients with OCD show higher functional connectivity and ventral striatal activity in sensorimotor and ventral cognitive circuits, while at the same time showing lower functional connectivity followed by disinhibition of dorsal striatal circuits (Stein et al. 2019).

The 5-HT_{2A} agonism of psychedelics is thought to disrupt the CSTC circuits. It has been proposed that this disruption lessens the sensory input filtering of the thalamus, thus allowing an increase of information in the cortex (Nichols 2004).

Providing another perspective on its potential therapeutic benefits, a single dose of psilocybin has been associated with a change in neocortical 5-HT_{2A} receptor binding followed by long-term increased mindfulness (Madsen et al. 2020). Furthermore, it has been shown that 5-HT_{2A} signaling can enhance neural plasticity (Lugo-Radillo and Cortes-Lopez 2021) and low-level learning as well as extinction learning (Vaidya et al. 1997), which is a key component of ERP and CBTs in OCD.

Psychedelics may also prompt neuroplasticity within the DMN (Carhart-Harris 2019). With the assumption that the DMN is controlling how information is integrated, it can be reasoned that incongruent or dissonant information deriving from perceptual anomalies and ambiguities will be dismissed by existing narratives, which in the case of OCD can be described as rigid and overlearned thought patterns and beliefs. Psychedelics are thought to interfere with this restrictive process. Excessive self-referential cognitive activity that is characteristic for OCD can be loosened by disrupting and "resetting" the DMN, allowing a healthier engagement with the patients' environment, by disengaging rigid top-down information processing patterns (Carhart-Harris and Friston 2019).

Thus multiple levels of analysis prompt optimism about the use of psychedelics to disrupt rigid patterns, both neural and psychological, in OCD. Independent of what level of analysis one prefers to describe the potential mechanism of action, there is sufficient support for the psychedelic class of drug as a tool to rapidly reduce OCD symptoms with the clinical benefit lasting significantly more than the subjective effects of the drug paving the way for important explorations of therapeutic potential.

5 Previous and Prospective Research

Near the turn of the century, we begun collaborations with the Multidisciplinary Association for Psychedelic Studies (MAPS) and the Heffter Research Institute to pursue the first clinical trial in psychiatric disorders in the USA in three decades. After overcoming multiple challenges, the University of Arizona, the Food and Drug Administration, and the Drug Enforcement Agency approved a proof-of-concept study to give psilocybin in a controlled research environment to individuals with OCD who had failed to improve with standard treatments.

5.1 Initial Study Design and Participants Characteristics

Seven men and two women, adults under the age of 65 years, with symptomatic OCD and previous experience using psychedelic drugs were treated with repeated escalating doses of psilocybin. They were free from antidepressants or other anti-OCD treatment for at least 2 weeks, and free of symptomatic relief medications such as anxiolytics or hypnotics, or any other prescription or over-the-counter medication, nutritional supplements or drugs of abuse, for at least 1 week as it is customary in all Phase I studies. OCD diagnosis was confirmed with the Structure Clinical Interview for DSM-IV.

Participants had failed to improve after an average of 3.4 adequate treatments with known anti-obsessional medications. Severity of symptoms was measured with the Yale-Brown Obsessive-Compulsive Scale (Y-BOCS, range 0–40) as is commonly done in clinical trials of OCD. The participants had an average baseline Y-BOCS score of 24.1 (severe) ranging from 18 (moderate) to 36 (extreme) prior to study drug ingestion in the first test day.

All participants signed written informed consent, which included a detailed discussion of potential physical and mental effects of psilocybin and risks of participation.

Psilocybin sessions were conducted in a specially adapted room (made to look as a home-like living room) in the outpatient offices of the Psychiatry Research Program, and subjects were subsequently housed overnight in the psychiatric unit to extend their monitoring. Special attention was given to the development of rapport and familiarization with the procedure and pertinent facilities. Psilocybin doses were selected to allow a range from non-hallucinogenic to frankly hallucinogenic. The low (100 μg/kg), medium (200 μg/kg), and high doses (300 μg/kg) were assigned in that order to secure escalating tolerability. A very low dose (25 μg/kg) thought to be almost inactive was assigned randomly as a control condition in a double-blind fashion during testing sessions 2, 3, or 4. The testing days were separated by at least 1 week. Trained sitters were present at all times. Subjects listened to a standardized set of music and wore eyeshades as they laid on a couch. They were advised to try to keep their experience focused internally, to avoid external distractions, and to engage with the theme of their experience (thoughts, emotions, perceptions, etc.) as they arose without avoidance or intentionally pursuing specific topics. Subjects were asked to follow these directions for 5–8 h. The duration of this phase coincided with the duration and intensity of their psychedelic experience, generally wearing off sooner for the very low dose and lasting as long as 8 h as the doses escalated. Within each dose session, participants were gradually allowed to modify the above-described routine, changing music to familiar or preferred content, take off their shades, engage in interesting activities like drawing, writing, and started debriefing with the principal investigator (FM) regarding aspects of their experience. At the end of 8 h, the participants went into our inpatient psychiatric unit for an overnight stay.

5.2 *Study Findings*

Psilocybin was clearly able to induce a psychedelic experience in a dose-dependent fashion. Some patients experienced altered states of consciousness similar to the classic descriptions of LSD effects. Subjects generally tolerated the procedure well, none of the participants experienced psychotic symptoms, or mood complications, and did not exhibit any threatening or dangerous behaviors. One individual experienced a transient asymptomatic elevation of blood pressure, which was not associated with nervousness, or the content of his psychedelic experience. Some of the subjects were uncomfortable with the overnight stay in the psychiatric unit and two decided not to continue primarily for this reason. Clinically, the most remarkable finding was the acute decrease in OCD symptoms of variable degree observed in every study participant during one or more of their test sessions. Improvement ranged from a modest reduction (23%) to a complete (100%) but temporary resolution of symptoms.

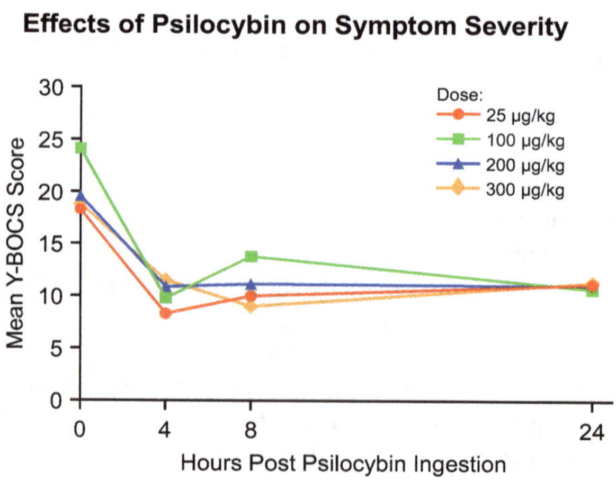

Interestingly as well, although the psychedelic effect had clearly worn off, the decreases in Y-BOCS scores generally lasted for at least 24 h.

Surprisingly, the very low dose (0.025 μg/kg) was found to have stronger psychedelic and anti-OCD effects than anticipated. For the group as a whole, psilocybin led to acute improvement of obsessive and compulsive symptoms comparable to higher doses (Moreno et al. 2006).

5.3 Important Implications

When administered in a supportive clinical environment, psilocybin is tolerated well and leads to a transient reduction of OCD symptoms. This protocol was developed in the year 2000, prior to the recent wave of psychedelic research experience, and the generalization of lay knowledge derived from self-experimentation that is currently available. Our very small dose was anticipated to be mostly inactive and serve as a placebo, however we now know that we were micro-dosing our subjects, which may have facilitated both a partial psychedelic-like experience and clinical symptom improvement. The draconian safety precaution to keep participants hospitalized in an inpatient psychiatric unit overnight significantly limited our ability to recruit and retain subjects in this protocol. Interestingly, although subjects were challenged by an overnight stay at a psychiatric facility that was generally perceived as an unpleasant setting, they reported continuation of OCD symptom relief overnight in ratings obtained at 24 h following study drug ingestion. Testing took place 1 week apart or a few days longer, and ratings of OCD severity were obtained prior to each session. Individuals consistently reported that their symptoms remain well controlled for several days after psilocybin ingestion. Unfortunately, a number of factors limit the interpretation of these preliminary data and are being addressed in ongoing studies, namely: (a) Prospective measurement after the last session is consistently obtained by collecting standardized data (Y-BOCS and various self-report symptom measures) for 6 months; (b) rather than using a psilocybin very low dose, the ongoing study includes an alternative active placebo (lorazepam 1 mg), which is anticipated to facilitate relaxation but not to lead to significant or lasting OCD symptom reduction; (c) our new protocol explores the effects of blinded and repeated psilocybin administration on lasting symptom reductions.

Furthermore, in this ongoing study, we are including psychedelic-naïve as well as experienced study participants. Moreover, we are gathering information that may inform our understanding of the potential psilocybin mechanisms of anti-OCD action by exploring brain imaging, and electrophysiological data, and their relations to the subjective experience.

6 Conclusion

As explained in detail in other chapters in this book. Psychedelic substances like psilocybin hold promise as a future treatment option for a broad spectrum of mental disorders, and in particular for individuals with disorders that include the transdiagnostic symptoms of rigid and repetitive negative thinking, and the inability to inhibit unhelpful and unwanted behaviors.

Preliminary data suggest that psilocybin may be an important agent to pursue as we explore alternative treatment options for OCD. Moreover, the current research base concerning brain chemistry and network function in OCD and how patients respond to the administration of psychedelics further motivates the promise of using psychedelics as a part of an intervention package for those with treatment-resistant OCD.

References

Andrews-Hanna JR, Smallwood J, Spreng RN (2014) The default network and self-generated thought: component processes, dynamic control, and clinical relevance. Ann N Y Acad Sci 1316(1):29

Carhart-Harris RL (2019) How do psychedelics work? Curr Opin Psychiatry 32(1):16–21

Carhart-Harris RL, Friston KJ (2019) REBUS and the anarchic brain: toward a unified model of the brain action of psychedelics. Pharmacol Rev 71(3):316–344

Carhart-Harris RL, Nutt DJ (2017) Serotonin and brain function: a tale of two receptors. J Psychopharmacol 31(9):1091–1120

Chlebowski S, Gregory RJ (2009) Is a psychodynamic perspective relevant to the clinical management of obsessive–compulsive disorder? Am J Psychother 63(3):245–256

Delgado PL, Moreno FA (1998) Hallucinogens, serotonin and obsessive-compulsive disorder. J Psychoactive Drugs 30(4):359–366

Foa EB (2010) Cognitive behavioral therapy of obsessive-compulsive disorder. Dialogues Clin Neurosci 12(2):199–207

Foa EB, Goldstein A (1978) Continuous exposure and complete response prevention in the treatment of obsessive-compulsive neurosis. Behav Ther 9(5):821–829

Foa EB, Kozak MJ (1985) Treatment of anxiety disorders: implications for psychopathology. In: Tuma AH, Maser JD (eds) Anxiety and the anxiety disorders. Lawrence Erlbaum Associates, Hillsdale, pp 421–452

Freud A (1966) Obsessional neurosis: a summary of psycho-analytic views as presented at the congress. Int J Psychoanal 47:116–122

Gonçalves ÓF, Soares JM, Carvalho S, Leite J, Ganho-Ávila A, Fernandes-Gonçalves A, Pocinho F, Sampaio A (2017) Patterns of default mode network deactivation in obsessive compulsive disorder. Sci Rep 7(1):1–7

Gosmann NP, Costa MDA, Jaeger MDB, Motta LS, Frozi J, Spanemberg L, Manfro GG, Cuijpers P, Pine DS, Salum GA (2021) Selective serotonin reuptake inhibitors, and serotonin and norepinephrine reuptake inhibitors for anxiety, obsessive-compulsive, and stress disorders: a 3-level network meta-analysis. PLoS Med 18(6):e1003664

Lugo-Radillo A, Cortes-Lopez JL (2021) Long-term amelioration of OCD symptoms in a patient with chronic consumption of psilocybin-containing mushrooms. J Psychoactive Drugs 53 (2):146–148

Madsen MK, Fisher PM, Stenbæk DS, Kristiansen S, Burmester D, Lehel S, Páleníček T, Kuchař M, Svarer C, Ozenne B, Knudsen GM (2020) A single psilocybin dose is associated with long-term increased mindfulness, preceded by a proportional change in neocortical 5-HT2A receptor binding. Eur Neuropsychopharmacol 33:71–80

Mahjani B, Bey K, Boberg J, Burton C (2021) Genetics of obsessive-compulsive disorder. Psychol Med 51(13):2247–2259

March JS, Liebowitz M, Carpenter D, Kahn DA, Francis A (1997) The expert consensus guideline series: treatment of obsessive-compulsive disorder. J Clin Psychiatry 58(4):2–72

Moreno FA, Delgado PL (1997) Hallucinogen-induced relief of obsessions and compulsions. Am J Psychiatry 154(7):1037–1038

Moreno FA, Wiegand CB, Taitano EK, Delgado PL (2006) Safety, tolerability, and efficacy of psilocybin in 9 patients with obsessive-compulsive disorder. J Clin Psychiatry 67 (11):1735–1740

Moreno JL, Holloway T, Albizu L, Sealfon SC, González-Maeso J (2011) Metabotropic glutamate mGlu2 receptor is necessary for the pharmacological and behavioral effects induced by hallucinogenic 5-HT2A receptor agonists. Neurosci Lett 493(3):76–79

Nichols DE (2004) Hallucinogens. Pharmacol Ther 101(2):131–181

Öst LG, Havnen A, Hansen B, Kvale G (2015) Cognitive behavioral treatments of obsessive–compulsive disorder. A systematic review and meta-analysis of studies published 1993–2014. Clin Psychol Rev 40:156–169

Pittenger C, Bloch MH, Williams K (2011) Glutamate abnormalities in obsessive compulsive disorder: neurobiology, pathophysiology, and treatment. Pharmacol Ther 132(3):314–332

Pokorny T, Preller KH, Kraehenmann R, Vollenweider FX (2016) Modulatory effect of the 5-HT1A agonist buspirone and the mixed non-hallucinogenic 5-HT1A/2A agonist ergotamine on psilocybin-induced psychedelic experience. Eur Neuropsychopharmacol 26(4):756–766

Raines AM, Oglesby ME, Allan NP, Mathes BM, Sutton CA, Schmidt NB (2018) Examining the role of sex differences in obsessive-compulsive symptom dimensions. Psychiatry Res 259:265–269

Stein DJ, Costa D, Lochner C, Miguel EC, Reddy Y, Shavitt RG, van den Heuvel OA, Simpson HB (2019) Obsessive-compulsive disorder. Nat Rev Dis Primers 5(1):52

Stern ER, Fitzgerald KD, Welsh RC, Abelson JL, Taylor SF (2012) Resting-state functional connectivity between fronto-parietal and default mode networks in obsessive-compulsive disorder. PLoS One 7(5):e36356

Tang W, Zhu Q, Gong X, Zhu C, Wang Y, Chen S (2013) Cortico-striato-thalamo-cortical circuit abnormalities in obsessive-compulsive disorder: a voxel-based morphometric and fMRI study of the whole brain. Behav Brain Res 313:17–22

Taylor S, Asmundson GJ, Jang KL (2016) Etiology of obsessions and compulsions: general and specific genetic and environmental factors. Psychiatry Res 237:17–21

Vaidya VA, Marek GJ, Aghajanian GK, Duman RS (1997) 5-HT2A receptor-mediated regulation of brain-derived neurotrophic factor mRNA in the hippocampus and the neocortex. J Neurosci 17(8):2785–2795

Van der Straten AL, Denys D, Van Wingen GA (2017) Impact of treatment on resting cerebral blood flow and metabolism in obsessive compulsive disorder: a meta-analysis. Sci Rep 7(1):1–8

Vulink NC, Denys D, Westenberg HG (2005) Bupropion for patients with obsessive-compulsive disorder: an open-label, fixed-dose study. J Clin Psychiatry 66(2):228–230

Wilcox JA (2014) Psilocybin and obsessive compulsive disorder. J Psychoactive Drugs 46 (5):393–395

Zambrano-Vazquez L, Allen JJB (2014) Response monitoring in obsessive-compulsive, worrying, and anxious individuals. Neuropsychologia 61:197–209

Psychedelics in the Treatment of Headache and Chronic Pain Disorders

Emmanuelle A. D. Schindler

Contents

Abstract The therapeutic potential of psychedelics in headache and chronic pain disorders is documented over decades of anecdotal and early investigational reports, which have paved the way for the first controlled studies of psilocybin and lysergic acid diethylamide (LSD) in these disorders. The reported lasting clinical effects after limited dosing with psychedelics present a novel means for disease management, but considerable further study will be required to address disease-specific treatments, uncover mechanism(s) of action, and verify safety. In this chapter, these topics are reviewed with particular attention to the neurobiological systems that offer potential sources of psychedelics' unique clinical effects in headache and pain.

E. A. D. Schindler (✉)
Headache Center of Excellence, Veterans Affairs Connecticut Healthcare System, West Haven, CT, USA

Yale School of Medicine, New Haven, CT, USA
e-mail: emmanuelle.schindler@yale.edu

Curr Topics Behav Neurosci (2022) 56: 261–286
https://doi.org/10.1007/7854_2022_365
Published Online: 12 May 2022

Keywords Chronic Pain · Cluster Headache · Headache · LSD · Migraine · Psilocybin · Psychedelics

1 Introduction

Due to the severe nature of the pain and the lack of available therapies, in the 1990s, patients with cluster headache started using classic psychedelic compounds, defined as 5-hydroxytryptamine (5-HT)2A receptor agonists with particular acute psychotropic effects, to manage their disease, fine-tuning their treatment regimens and sharing this information in early online forums. Decades before this, the therapeutic effects of psychedelics in headache disorders were appreciated in various reports, though the first controlled study did not begin until 2017 (Schindler et al. 2021a). The recognition and study of psychedelics on cancer pain also has a long history (Kast 1967; Kast and Collins 1964). As has been demonstrated in the treatment of psychiatric conditions, psychedelics effect lasting reductions in disease burden in headache and pain disorders after limited dosing (i.e., one or a few doses). In this chapter, these effects and others are reviewed, and potential mechanisms of action discussed. The safety of different methods of treatment in pain management is also discussed.

2 Background

2.1 Pain and Chronic Pain

Pain is an evolutionarily beneficial biological function, warning the sensor of danger (i.e., hot stove) or pathology (i.e., appendicitis). It is defined by the International Association for the Study of Pain (IASP) as "an unpleasant sensory and emotional experience associated with, or resembling that associated with, actual or potential tissue damage" (Raja et al. 2020). Pain is a complex biological signal involving multiple levels of sensation and perception that are modulated by experiential, biological, psychological, and societal factors (Raja et al. 2020; Bushnell et al. 2013; Greenspan et al. 2007). While such components as the somatosensory tracts and thalamic nuclei translate incoming stimuli to neuronal signals, other regions including the periaqueductal gray (PAG) and anterior cingulate cortex (ACC) are integral to the conscious appreciation and reaction to those signals (Bushnell et al. 2013; Chen et al. 2018; Wager et al. 2013). If pain is severe enough in intensity, disruptive enough to basic functions, or results in prolonged suffering, treatment is warranted, though the approaches vary based on individual, sex, and societal factors (Greenspan et al. 2007; Urits et al. 2019).

Chronic pain involves "pain in one or more anatomic regions that persists or recurs for longer than three months and is associated with significant emotional distress or significant functional disability" (Treede et al. 2015). Chronic pain disorders are further categorized into primary and secondary disorders. Primary chronic pain is seen in such disorders as fibromyalgia, low back pain, and primary headache disorders. Secondary chronic pain stems from another underlying disease and includes such conditions as cancer-related pain, post-traumatic pain, and secondary headache disorders. In some cases, pain may remain despite treatment of the underlying disease in these secondary chronic pain disorders (Treede et al. 2019). The development of chronic pain involves peripheral and central plasticity, long-term potentiation at spinal and central levels, and sensitization of both ascending and descending pain pathways (Bushnell et al. 2013; Moayedi and Davis 2013; Fenton et al. 2015). Pain disorders may furthermore be inherited or acquired. Those with inherited erythromelalgia, for example, endure pain attacks in their limbs due to a mutation in voltage-gated sodium channel functioning (McDonnell et al. 2016). Thalamic pain syndrome arises in approximately 10% of patients after thalamic stroke or other injury, leaving them with allodynia and persisting pain in the corresponding body part (Levi et al. 2019). Similarly, in phantom limb syndrome, which occurs in three quarters or more of patients after limb amputation, the reorganization and sensitization of peripheral and central systems result in persistent pain represented in the location of the missing limb (Collins et al. 2018; Ramachandran et al. 2018).

2.2 *Techniques in Pain Management*

The various neurobiological systems involved with chronic pain disorders present several potential therapeutic targets. Given the complexity and great diversity of pain disorders, no single target is expected to address all pain equally. In considering pain management, the distinction between persisting pain (i.e., phantom limb pain, thalamic pain syndrome) and acute or paroxysmal pain (i.e., inherited erythromelalgia, headache disorders) is vital. Furthermore, depending on the disorder, the treatment of acute pain or exacerbations may be distinct from pain prophylaxis or treatment of the underlying disorder itself. For example, treatment of an individual migraine attack with a non-steroidal anti-inflammatory drug (NSAID) does not treat the underlying migraine disorder. In fact, frequent use of NSAIDs and other acute medications can transform episodic migraine ($<$15 headache days per month) to the chronic form ($\geq$15 headache days per month) (Bigal and Lipton 2011). Examples of specific forms of treatment in headache and chronic pain disorders are discussed in the respective sections below.

Pain management often fails to target the pain disorder as a whole and instead, focuses on the symptom of pain. The use of opioids, in particular, leads to sensitization and chronification of pain disorders, and furthermore risks dependence and addiction (Bigal and Lipton 2011; Lee et al. 2011). Opioids are so ingrained in the traditional concept of pain management, however, that they continue to be used routinely even when not indicated. For example, opioids are neither recognized nor recommended under the prevailing guidelines for the management of migraine or cluster headache, though they continue to be prescribed in these disorders (Robbins et al. 2016; Loder et al. 2012; Mazer-Amirshahi et al. 2014). The most widely used drug in the world, acetaminophen, can be a safe analgesic when dosed appropriately and used infrequently, but the misconception that it is completely safe (at any dose and frequency) and the focus on treating acute pain (as opposed to the disorder) have made it one of the more toxic substances on (or off) the market (Brune et al. 2015).

Targeting higher-level pain processing and perception is another technique in pain management. Deep brain stimulation targeting the ACC has been employed in such pain conditions as thalamic pain syndrome (Levi et al. 2019). Nerivio™ (Theranica Bio-electronics Ltd.) is a remote electrical neuromodulation device that employs conditioned pain modulation (CPM, aka "pain inhibiting pain") for the acute treatment of migraine (Yarnitsky et al. 2019; Rapoport and Lin 2019). CPM recruits the endogenous analgesic system through descending inhibitory pathways from the periaqueductal gray and rostral ventromedial medulla (Rapoport and Lin 2019). Other methods of pain control include psychological treatments, such as cognitive behavioral therapy (CBT), Acceptance and Commitment Therapy (ACT), Mindfulness-to-Meaning Theory, and meditation which harness higher-level systems to produce analgesic effects (Bushnell et al. 2013; Hughes et al. 2017; Garland and Fredrickson 2019).

3 Headache Disorders

The International Classification of Headache Disorders (3rd edition) contains over 100 headache and facial pain disorders (more when counting subdivisions) (Headache Classification Committee of the International Headache Society (IHS) 2018). The most commonly known disorders are tension-type headache and migraine headache, which occur in approximately 40% and 15% of the population, respectively (Burch 2019). Despite widespread familiarity and/or personal experience with headache (or perhaps because of it), the burden of headache disorders is often minimized and misunderstood. Headache disorders involve pathological function of numerous peripheral and central systems and include other neurological symptoms (e.g., lacrimation, nausea) characteristic to the specific disorder (Headache Classification Committee of the International Headache Society (IHS) 2018; Akerman et al. 2011). Being largely paroxysmal pain disorders, they manifest with spontaneous or triggered attacks of pain accompanied by other neurological symptoms, followed by pain-free periods. Some headache disorders may involve persistent (24 h a day/7 days a week) pain, but these are less common and are usually secondary or emerge from a primary headache disorder. Diagnosis of headache disorders is largely made clinically and is based on the characteristics of the attacks and of the patient. Table 1 outlines the basic characteristics of migraine and cluster headache, two distinct disorders that are discussed in this chapter.

Table 1 Comparative characteristics of migraine and cluster headache

Feature	Migraine	Cluster headache
Disorder characteristics (Horton 1952; Burch et al. 2018; Burch et al. 2019; GBD Disease and Injury Incidence and Prevalence Collaborators 2017; Robbins 2013)		
Prevalence	15%	0.05–0.1%
Sex ratio	Female 3:1	Male 3:1
Comorbidities	Cardiovascular disease, cerebrovascular disease, depression, anxiety	Sleep apnea, tobacco use, depression, anxiety, suicidality
Impact	Top cause of disability worldwide under age 50	Also known as "suicide headache"
Attack characteristics (Headache Classification Committee of the International Headache Society (IHS) 2018)		
Location	Unilateral or bilateral, variable locations	Unilateral, retro-orbital
Quality	Throbbing, pulsing	Stabbing, searing
Intensity	Moderate to severe	Very severe
Duration	3–72 h	15–180 min
Associated symptoms	Light sensitivity, sound sensitivity, nausea/vomiting	Ptosis, lacrimation, sinus congestion, rhinorrhea
Restlessness	Absent (activity worsens attack)	Present (pacing, rocking, etc. alleviates attack)
Frequency	Usually no more than once daily	Up to 8 times daily
Subtypes (Headache Classification Committee of the International Headache Society (IHS) 2018)		
Episodic	14 or fewer headache days per month	Attacks occur regularly for weeks to months on an annual (or otherwise specified) cycle
Chronic	15 or more headache days per month	Attacks occur year-round with no remission >3 months
Treatment[a] (Robbins et al. 2016; Loder et al. 2012; Rapoport and Lin 2019; Dodick et al. 2020; Goadsby et al. 2019; Schindler and Gottschalk 2019)		
Acute (treats the attacks)	NSAID, APAP, triptans, antidopaminergics, TENS, REN	High-flow oxygen, triptans (subcutaneous or intranasal only), nVNS
Preventive (treats the disorder)	Topiramate, valproic acid, beta-blockers, tricyclic antidepressants, anti-CGRP (or receptor) monoclonal antibodies	Verapamil, lithium, galcanezumab, nVNS

NSAID non-steroidal anti-inflammatory drug, *APAP* acetaminophen/paracetamol, *TENS* transcutaneous electrical nerve stimulation, *REN* remote electrical neuromodulation, *nVNS* non-invasive vagus nerve stimulation, *CGRP* calcitonin gene-related peptide

[a] Non-inclusive list of more commonly used treatments or drug classes; off-label treatments included

3.1 Migraine Headache

Migraine headache, preferentially referred to as migraine or migraine disease (Young et al. 2012), is at once one of the best-known and most misunderstood headache disorders. The paroxysms that identify the disease are migraine attacks, also colloquially referred to as migraines. These attacks encompass much more than just head pain, however. Light sensitivity and nausea are common accompanying symptoms, and some may have more complex neurological features, such as aphasia or hemiplegia. There is even a form of migraine in which there is no head pain, only the classic visual aura – a scintillating scotoma within a homonymous visual field, which reflects cortical spreading depression across the contralateral occipital lobe. These features of migraine attacks reflect underlying dysfunction in several brain regions, from cortical to subcortical to brainstem, and also involving contribution from peripheral nervous system input (Akerman et al. 2011; Olesen et al. 2009).

Migraine headache (the disease) is also comorbid with cardiovascular and cerebrovascular disease, as well as affective and mood disorders (Burch et al. 2019), reflecting an underlying pathology affecting the entire body not only the head. Much progress has been made in uncovering the pathophysiology of migraine (both the disease and the attacks), which over the years has led to specific treatments (i.e., triptans, anti-calcitonin gene-related peptide [CGRP] agents). While there are numerous treatments for migraine (Table 1), good control of the disease remains a challenge. The heterogeneity of migraine ensures that no single treatment will be effective in all patients. Furthermore, as with all headache disorders, attacks can be aborted but the underlying disorder remains. The goal of preventive treatment (which treats the disease) is to reduce the frequency and intensity of attacks and raise the threshold for triggering attacks (i.e., from alcohol, stress let-down, drops in barometric pressure). In rare cases, remission might be achieved, but this is more likely to occur naturally with age.

Based on the relationship between serotonin and migraine, methysergide, a semisynthetic ergot alkaloid derived from lysergic acid diethylamide (LSD), was tested as a preventive agent starting in 1959 (Koehler and Tfelt-Hansen 2008; Rapoport 2012). Methysergide has antagonist activity at several serotonin receptors, as well as adrenergic receptors. Other chemically related drugs used in migraine include ergotamine and dihydroergotamine (Rapoport 2012). It should come to no surprise, then, that LSD appears to have therapeutic effects in migraine as well, with reports extending back to the early days of LSD research. These reports include preventive effects of the drug (Sicuteri 1963) (Table 2). The acute and preventive use of other indoleamine psychedelics, including psilocybin (mushroom), *N,N*-dimethyltryptamine (DMT), 4-acteoxy-dimethyltryptamine (4-ACO-DMT), and 4-acetoxy-*N*-methyl-*N*-ethyltryptamine (4-ACO-MET) in migraine is also anecdotally reported (Sewell and Gottschalk 2011; Andersson et al. 2017).

The first controlled study of psilocybin in migraine was recently completed (Schindler et al. 2021a) (Table 2). In this exploratory double-blind, placebo-controlled, cross-over study, subjects with migraine headache, as defined by ICHD (excluding variants) with a baseline frequency of two attacks per week or more, received a single oral administration of placebo (microcrystalline cellulose) followed at least 2 weeks later by a single oral administration of psilocybin (0.143 mg/kg). This weight-based dose produced a final average amount of psilocybin of 9.4 mg (range 6.8–13.7 mg), a dose less than half the commonly used in mental health studies (approximately 25 mg). Subjects and staff were under a standard enhanced

Table 2 Reports on therapeutic effects of psychedelics in headache disorders

Disorder	Study type	*N*	Drug, oral dose, and regimen	Findings	Adverse events	Reference
"Vascular headache" (likely migraine)	Case series	390	LSD 50–100 μg BOL 2–4 mg (unclear regimens)	*Acute*: NR *Preventive*: LSD had moderate effect and BOL a mild effect compared to methysergide	NR	Sicuteri (1963)
Migraine	Randomized controlled trial	10	Psilocybin 0.143 mg/kg vs. placebo, (cross-over design, once each)	*Acute*: NR *Preventive*: In the 2 weeks after a single administration, migraine days/week, attacks per week, pain intensity, attack-related functional disability, abortive medication consumption were reduced	– Lightheaded – Nausea – Anxiety – Sore muscles – Headache (acute and subacute) No events were sig. different in incidence from placebo	Schindler et al. (2021a)
Cluster headache	Survey	53	LSD, psilocybin (mushroom) (unclear regimens)	*Acute*: psilocybin more effective than oxygen, as effective as triptans; LSD as effective as oxygen and triptans *Preventive*: Psilocybin as effective as prednisone; LSD more effective than prednisone	NR	Sewell et al. (2006)
Cluster headache	Case report	1	Psilocybin 1 g dried mushroom every 1–2 months	*Acute*: NR *Preventive*: Pain free for 2–6 weeks	NR	Matharu et al. (2005)
Cluster headache	Case series	5	BOL 30 μg/kg, three doses 5 days apart each	*Acute*: NR *preventive*: Out to 12 weeks after completion of three doses, attack frequency was reduced	– Slightly tipsy – Flabby feeling – Funny feeing	Karst et al. (2010)

(continued)

Table 2 (continued)

Disorder	Study type	*N*	Drug, oral dose, and regimen	Findings	Adverse events	Reference
Cluster headache	Case series	2	5-MeO-DALT 15 mg, 30 mg	*Acute*: NR *Preventive*: Single administration prevents attacks out to 2 weeks, regular use (15 mg every 5 days) prevented attacks entirely	– Drowsiness – Lightheadedness – Spatial disorientation – Closed-eye visuals – Temperature drop (30 mg dose)	Post (2014)
Cluster headache	Survey	496	Psilocybin (mushroom), LSD, LSA (seeds), BOL, DMT Every few weeks to annually	*Acute*: Psilocybin equal to high-flow oxygen and intranasal sumatriptan, but less effective than injectable sumatriptan *Preventive*: Psilocybin, LSD, and LSA superior to verapamil and prednisone; DMT and BOL as effective as verapamil and prednisone	– Acute headache – Wooziness – Abdominal cramping (in an IBS patient)	Schindler et al. (2015)
Cluster headache	Survey	54	Psilocybin, LSD, LSA (sources unspecified) Some used only 1–3 times per year, sub-hallucinogenic doses	*Acute*: NR *Preventive*: ~75% who used any of the drugs had a response (“fully or in part”)	NR	DiLorenzo et al. (2016)
Cluster headache	Case report	1	LSA (single ingestion of 6 HBWR seeds)	*Acute*: “Abated completely” *Preventive*: relief for 2 weeks	– Stomach cramping – Hospitalized	Johnson and Black (2020)
Cluster headache	Survey	400	Psilocybin, LSA (seeds)	*Acute*: use reported in one subject each for psilocybin and LSA; efficacy NR *Preventive*: NR	NR	Petersen et al. (2021)

NR not reported, *LSD* lysergic acid diethylamide, *BOL* 2-bromo-LSD, *DMT* dimethyltryptamine, *5-MeO-DALT N,N*-dially-5-methoxytryptamine, *IBS* irritable bowel syndrome, *LSA* lysergic acid amide, *HWBR* Hawaiian baby woodrose blinding procedure such that the drug, dose, and order of drug administration were unknown. On each of the two test days, subjects received the blinded drug capsule in the morning and remained in the outpatient research unit for at least 6 h after capsule ingestion. Vital signs and subject-reported physiological and neuropsychological effects were obtained throughout the test session. Follow-up calls with subjects took place the day after each test day, weekly for 2 weeks after each test day, and again at 2 and 3 months after the second test day. Subjects kept a headache diary starting 2 weeks before the first drug session, until 2 weeks after the second drug session. This short duration of diary was chosen given the primary goal of identifying a "signal" of therapeutic effect, which would then be more fully explored in larger, more definitive studies. Data from this study showed that psilocybin reduced several measures of migraine burden over the two-week period measured, while placebo did not effect significant change. This single administration of psilocybin reduced migraine days per week by half (−49.3%), pain intensity by over a third (−39.1%), and attack-related functional impairment by over two thirds (−68.9%). Psilocybin also delayed the time to the next migraine attacks, some subjects enjoying complete remission during the 2 weeks measured.

While this study provides proof-of-concept that psilocybin has lasting effects in migraine after limited dosing and that a single low oral dose is safe in migraine patients under experimental conditions, very little can be extrapolated with regard to how psilocybin might be utilized in the management of migraine. The full duration of effects of this dose, as well as others, needs to be investigated. The safety of repeated psilocybin administration is also required before this drug can be considered as a viable preventive agent in migraine. The abortive effects of psilocybin on an individual migraine attack were neither a primary outcome of this initial trial nor could it be extrapolated from available data, but acute therapy remains plausible, so long as safety and efficacy can be demonstrated in a controlled trial.

3.2 Cluster Headache

Cluster headache is the most common type of trigeminal autonomic cephalalgia (TAC), a grouping of headache disorders that also includes paroxysmal hemicrania and short-lasting unilateral neuralgiform headache attacks (SUNHA). TAC attacks are severely painful and characteristically include ipsilateral activation of autonomic systems. Cluster headache patients have ranked the pain of their attacks at least 30% higher than that of childbirth, pancreatitis, or nephrolithiasis (Burish et al. 2021). The condition is also known as "suicide headache," a term given by patients over a half a century ago (Horton 1952). Suicidal ideation and behavior are indeed increased among cluster headache patients, with particular association to demoralization (Ji Lee et al. 2019; Koo et al. 2021).

Despite the advent of new therapies for cluster headache, including the monoclonal antibody against CGRP, galcanezumab (Eli Lilly), and the non-invasive vagus nerve simulator (nVNS; electroCore), successful and accessible treatment eludes many. Both these treatments have clinical efficacy in episodic patients (between one third and two thirds reduction in weekly attacks); chronic patients have minimal response (Dodick et al. 2020; Goadsby et al. 2019; Marin et al. 2018; Gaul et al. 2016). These new treatments also have monthly costs of several hundred US dollars and may or may not be reimbursed by patients' health insurance carrier. Additional treatment challenges remain. For instance, high-flow oxygen is a safe and effective abortive therapy in cluster headache and yet, patients struggle to obtain prescriptions, insurance coverage, and proper equipment (Robbins et al. 2016; Pearson et al. 2019; Schindler et al. 2018a). Despite its use for decades in cluster headache and its diagnostic value in the disorder, oxygen is not formally approved by the US Food and Drug Administration and was only recognized as a valid therapy by the Centers for Medicare and Medicaid Services in 2021 (May et al. 1998).

In addition to lack of effective treatments, or lack of access to effective treatments, cluster headache patients often suffer diagnostic delays, some on the order of decades (Schindler et al. 2015; Rozen and Fishman 2012; Bahra and Goadsby 2004). Out of desperation and self-preservation, patients have uncovered several disease management techniques outside of traditional medicine. Such methods include megadose vitamin D supplementation, the use of licorice and ginger to abort attacks, the use of energy drinks and coffee to abort attacks, and the use of psychedelics (Schindler et al. 2021b). In the 1990s, a Scottish cluster headache patient serendipitously discovered the therapeutic effects of LSD after recreational use. He began to use LSD and psilocybin mushrooms medicinally and shared this information with other cluster headache patients online.

Given the relatively low prevalence of cluster headache, online communication among patients is a vital source for information, support, and frankly, it has been a live saving tool. The group that most commonly discusses psychedelics in cluster headache is Clusterbusters, Inc., a patient advocacy group comprised of patients, physicians, researchers, and donors striving to increase access to available treatment and support research into new therapies. Over the past quarter century, Clusterbusters has fine-tuned treatment found to be most effective in treating cluster headache (the disease). The classic method of "cluster busting" involves a short course, or pulse regimen, of low dose (mildly psychedelic) oral psilocybin (in the form of psychedelic mushroom) used to terminate a cluster cycle or induce a period of remission in chronic cluster headache (Andersson et al. 2017; Schindler et al. 2015). In addition to psilocybin, LSD and seeds containing lysergic acid amide (LSA) have been reported to have similar effects (Schindler et al. 2015; Johnson and Black 2020; Sewell et al. 2006; Di Lorenzo et al. 2016). Psilocybin and other psychedelics may be used to abort cluster attacks as well (Schindler et al. 2015; Johnson and Black 2020; Petersen et al. 2021), but given their relatively short duration (100 min on average) and high frequency (2–4 daily on average) (Rozen and Fishman 2012; Bahra et al. 2002; Lund et al. 2017), this drug class is neither practical for acute treatment nor is it typically used by patients in this way.

In a survey executed by Clusterbusters, patients ranked the preventive effectiveness of psilocybin significantly above that for verapamil and prednisone, two commonly used conventional preventive treatments (Schindler et al. 2015). LSD and LSA were also rated above verapamil and prednisone. Importantly, psilocybin and other psychedelics were reported to be used infrequently for the purposes of prevention, on intervals of every few weeks to once annually (Schindler et al. 2015). In this same survey, patients remarked on the effectiveness of psychedelics after failing so many conventional therapies, though several also mentioned reservations or concerns about taking the drug (Schindler et al. 2021b). One patient in particular described a dislike of the acute psychedelic effects and therefore he only used very low doses (Schindler et al. 2021b). In fact, among cluster headache patients who self-medicate with psychedelics, sub-psychedelic and minimally psychedelic doses are commonly used (Schindler et al. 2015; Sewell et al. 2006; Di Lorenzo et al. 2016). A congener of LSD, 2-bromo-LSD (BOL or BOL-148), which has greatly reduced acute psychotropic effects, was shown in a case series of five cluster headache patients to reduce weekly attacks out to 16 weeks after a pulse of three oral administrations (Karst et al. 2010). Use of another indoleamine drug with reduced psychotropic effects, *N,N*-dially-5-methoxytryptamine (5-MeO-DALT), was reported in two cluster headache patients to terminate attacks for approximately 2 weeks after single administration (Post 2014). It is notable that the psychedelics reported to produce lasting therapeutic effects in cluster headache are in the indoleamine class. One patient reported via personal communication that mescaline affords him preventive effects but at this time, there is not an existing body of evidence for therapeutic effects of phenethylamine psychedelics in cluster (or other) headache.

To date, the therapeutic effects of psychedelics and related compounds in cluster headache are sourced from anecdotal and uncontrolled investigations (Table 2). Controlled studies are under way, however, and currently there are randomized controlled studies investigating psilocybin (NCT02981173) and LSD (NCT03781128) in cluster headache and an open label study of psilocybin in SUNHA (NCT04905121). An open-label neuroimaging study of psilocybin in cluster headache is also ongoing (NCT04280055). Through these studies, the therapeutic effects of psychedelics in cluster headache will be partially characterized but many questions will remain. Importantly, in these short-term studies, the question of safety and efficacy of long-term cluster headache management (i.e., years, decades) will not be answered. A single pulse treatment may be sufficient for termination of an annual cluster cycle, but for those with long cycles or chronic cluster headache, repeated treatments might be required (Schindler et al. 2015). Therefore, just as in migraine, the understanding of psychedelics in cluster headache will be incomplete without examining the safety and efficacy of repeated treatment.

3.3 Source of Psychedelics' Effects in Headache Disorders

Psychedelics have pharmacological overlap with existing headache medications and thus, that this re-emerging class of drugs would have therapeutic effects in headache disorders is not wholly unexpected. Of those headache preventives with 5-HT2A receptor activity (e.g., tricyclic antidepressants, DHE), none can effect lasting change after a single dose, however. DHE is also used in a pulse regimen to transition patients out of a high burden of migraine or cluster headache, but a regimen of thrice daily administration for 5 days (or similar) is required (Nagy et al. 2011; Mather et al. 1991; Magnoux and Zlotnik 2004). Therefore, while targeting 5-HT2A receptors or the serotonergic system may be relevant, this does not alone explain psychedelics' unique ability to suppress headache burden over a prolonged duration after limited dosing (Schindler et al. 2021a). Similarly, while psychedelics may serve as acute anti-inflammatories (Flanagan and Nichols 2018; Whelan and Johnson 2018), a one-time reduction in inflammation would not be expected to cause prolonged change in underlying disease. Theoretically, means of effecting long-lasting change after limited exposure, relevant to psychedelics and headache, include genetic and hormonal effects (Schindler et al. 2018b), as well as neuroplasticity (Ly et al. 2018). An operational "switch" in cluster headache may lie in the hypothalamus, which demonstrates heightened activity in cycle and during attacks (May et al. 1998; Buture et al. 2019). The posterior hypothalamus is the target for deep brain stimulation in cluster headache (Bartsch et al. 2009; Schoenen et al. 2005), and tumors in the region of the hypothalamus or pituitary gland can lead to secondary cluster headache (Favier et al. 2007).

Unlike the relationship between acute psychedelic effects and outcomes in mental health conditions (Griffiths et al. 2016; Garcia-Romeu et al. 2014), such a relationship is not yet evident in headache disorders. Several pieces of evidence suggest that acute psychedelics effects may not be required for therapeutic effects: (1) over 25 years of self-management by cluster headache patients using low and sub-hallucinogenic doses of psychedelics; (2) the case series of the clinical effects of BOL (LSD analog with greatly reduced psychedelic effects) in cluster headache (Karst et al. 2010); (3) the psilocybin migraine pilot described above, wherein psychedelic effects during drug administration, which were mild, did not correlate with the reduction in migraine days over the two-week period measured (Schindler et al. 2021a). Furthermore, unlike studies in mental health, subjects in this migraine study, as well as the ongoing cluster headache study, are not guided through drug sessions and there is not a formal integration session upon follow-up. Preparation, monitoring during sessions, and close follow-up are carried out, but these procedures are done from the standpoint of safety and tolerability and not with the intent to emphasize or capitalize on the acute psychedelic effects. This is not to say that standard psychedelic-assisted therapy procedures (i.e., therapy, guided sessions, integration) would not affect outcomes in headache, particularly as psychological and behavioral techniques, such as CBT and ACT, can be effective in headache management (Vasiliou et al. 2021; Grazzi et al. 2019; Perez-Munoz et al. 2019). The potential added benefit of standard psychedelic-assisted therapy in addition to drug-alone administration in headache disorders would need to be studied systematically. Of note, in these migraine and cluster studies employing drug-alone administration, there have been no serious or unexpected adverse events as determined by close follow-up with subjects out to 6 months after drug administration (Schindler et al. 2021a; Schindler 2020). The relatively low dose of psilocybin used and the lack of clinical depression or other significant mental health conditions in these headache patients may allow for this technique of drug administration, but this method may not be appropriate for all psychedelic drugs, doses, and populations.

Psychedelics can serve as enhancers of suggestibility, set, and setting, all of which play into the placebo effect (Hartogsohn 2016; Carhart-Harris et al. 2015). The magnitude of the placebo effect for conventional migraine and cluster headache preventive medications can surpass 50% for certain agents (Goadsby et al. 2017, 2019). There was very little effect of placebo in the primary outcome of the exploratory migraine study (0.04% reduction of weekly migraine days), though some secondary outcome measures were more noticeably affected (e.g., attack duration, −16%; nausea/vomiting severity, −43%) (Schindler et al. 2021a). While not formally assessed in this study, the success of subject blinding is another important factor and a notorious challenge for psychedelic studies (Muthukumaraswamy et al. 2021). Selecting a compound with similar acute physiologic and psychological effects without anticipated therapeutic effects can be difficult, but necessary in order to avoid overestimating the therapeutic abilities of psychedelics. In the follow-up migraine study where repeated drug dosing is being investigated (NCT04218539), the control agent for 10 mg psilocybin is diphenhydramine 25 mg, which is anticipated to share some of the acute vigilance reduction and general drug effects of low dose psilocybin without the lasting therapeutic effects.

4 Chronic Pain Disorders (Non-headache)

As discussed above, chronic pain disorders involve various etiologies and numerous sources for pain. Table 3 outlines characteristics and management strategies for some chronic pain disorders that are of interest in psychedelic therapy.

Much of the literature on the effects of psychedelics in non-headache pain is in patients with terminal cancer undergoing LSD-assisted psychotherapy. Some of these studies and others included patients with other types of pain, such as gangrene, herpes zoster, phantom limb pain, and low back pain. Given the variety of sources of pain presented and the general absence of placebo or control agents in these studies, at this time it is not possible to precisely quantify the pain relief offered by psychedelics nor is it possible to predict what effects a psychedelic might have in any one pain disorder. Among these studies, however, both acute and preventive effects were reported, offering anecdotal evidence that psychedelics have lasting therapeutic effects in pain disorders. Table 4 summarizes reports on psychedelics' effects in pain disorders in humans.

Table 3 Characteristics and management of chronic pain disorders of interest for psychedelic treatment

Type	Cancer pain (Fallon et al. 2018; Aman et al. 2021; Syrjala et al. 2014)	Phantom limb syndrome (Collins et al. 2018; Ramachandran et al. 2018)	Low back pain (Urits et al. 2019; GBD Disease and Injury Incidence and Prevalence Collaborators 2017)
Prevalence	– 1.8 million new cancer diagnoses in the USA in 2020 – Moderate-severe pain in 35% all cancer patients, 80% in advanced stage cancer	– About two million amputees currently in the USA – 75% or more amputees develop phantom limb syndrome – Phantom pain can emerge after removal of other body parts (e.g., breast)	– Annual prevalence up to 30% – Lifetime prevalence up to 80% – #1 cause of disability worldwide
Source(s) of pain	– Tissue invasion – Inflammation – Other depending on type and comorbidity – Surgical intervention – Radiation therapy – Chemotherapy – Hormonal therapy	– Cortical reorganization – Subcortical (e.g., thalamus) reorganization – Proprioceptive memory – Peripheral reorganization	– Radiculopathy ("pinched nerve") – Myelopathy (spinal stenosis) – Arthritis (vertebral, sacroiliac) – Muscle spasm – Myofascial pain
Treatment options	– APAP, NSAIDs – Opioids (oral, epidural, intrathecal) – Ketamine (IV) – Steroids – Nerve block, neurolysis, cordotomy – Spinal cord stimulation – Other depending on type (e.g., gabapentin for neuropathy, baclofen for spasticity) – CBT, relaxation, meditation, hypnosis	– Gabapentin, pregabalin – Memantine – Opioids – Lidocaine (injected into dorsal root ganglion) – Mirror therapy – Virtual reality therapy	– Gabapentin, pregabalin – Duloxetine, venlafaxine – Amitriptyline, nortriptyline – APAP, NSAIDs – Opioids – Muscle relaxers, botulinum toxin – Acupuncture – CBT, progressive relaxation, biofeedback – Physical therapy, chiropractic treatment – Steroids (systemic, intra-articular, epidural) – Nerve block, radiofrequency neurotomy – Spinal cord stimulation
Additional disease management	– Cancer treatment (i.e., chemotherapy, radiation, surgery)	–	– Surgical intervention

APAP acetaminophen/paracetamol, *NSAIDs* non-steroidal anti-inflammatory drugs, *CBT* cognitive behavioral therapy

Table 4 Reports on therapeutic effects of psychedelics in pain disorders

Disorder	Study type	N	Drug, oral dose, and regimen	Findings	Adverse events	Reference
Cancer, gangrene, herpes zoster	Case series	50	LSD 100 μg, dihydromorphinone (D) 2 mg, meperidine (M) 100 mg, once each	*Acute*: Between the first and third hours, LSD reduced pain by ~80%, D and M each reduced pain by ~30% *Preventive*: NR	– Nausea – Vomiting – Psychosis 8 subjects refused, 12 wished to repeat treatment	Kast and Collins (1964)
Terminal cancer	Case series	128	LSD 100 μg once (subjects told "potent medicine")	*Acute*: Reduced pain score by ~80% for 12 h *Preventive*: Pain remained overall reduced by ~30% for 3 weeks	– Anxiety – Panic 30% subjects unwilling to repeat treatment	Kast (1967)
Terminal cancer	Case series	22	LSD 300 μg IM (in one case)	*Acute*: NR in case described *Preventive*: Better able to tolerate pain	NR	Pahnke et al. (1969)
Terminal cancer	Case series	31	LSD 200–500 μg once	*Acute*: NR *Preventive*: ~25% reduction in pain (scored by staff and family) over "weeks to months"	– Headache – Tremor – Nausea – Palpitations – Dyspnea – Fatigue	Grof et al. (1973)
Terminal cancer	Case series	4	LSD 200–600 μg once (some doses IM or IV; sessions repeated up to 4 times) (overlap with Pahnke et al. (1969))	*Acute*: 2 subjects had relief, 1 subject initially had pain increase but after 50 μg IV had relief *Preventive*: 3/4 had some degree of lasting pain relief or increased ability to bear the pain 1 subject had no acute or preventive response	– Headache – Tremor – Nausea – Palpitations – Dyspnea – Fatigue – Worsening pain – Worsening GI distress	Kurland (1985)

(continued)

Table 4 (continued)

Disorder	Study type	N	Drug, oral dose, and regimen	Findings	Adverse events	Reference
Phantom limb syndrome	Case series	7	LSD 25 μg daily × 1 week, then 50 μg daily × 2 weeks	*Acute*: NR *Preventive*: Pain and analgesic consumption were reduced in 5 subjects; residual pain reduction for several weeks after stopping LSD in 1 subject	NR	Fanciullacci et al. (1977)
Phantom limb syndrome	Case report	1	Psilocybin (Psi; dried mushroom) 0.2–3 g and mirror therapy (MT)	*Acute*: Psi alone = pain relief; Psi + MT = longer lasting acute relief *Preventive*: Psi + MT = weeks to months of pain relief	NR	Ramachandran et al. (2018)
Low back pain	Case report	1	LSA (single ingestion of 6 HBWR seeds)	*Acute*: "abated completely" *Preventive*: relief for 2 weeks Same subject as Table 2	– Stomach cramping – Hospitalized	Johnson and Black (2020)

LSD lysergic acid diethylamide, *LSA* lysergic acid amide, *HBWR* Hawaiian baby woodrose

4.1 Cancer Pain

The aforementioned cancer studies largely involve sessions with single dosing of LSD, which may have then been repeated after a certain interval (from 1 week to 6 months in one study (Kurland 1985)). Acute pain relief is reported during drug administration with oral doses ranging from 100 to 600 μg (Kast and Collins 1964; Collins et al. 2018; Kurland 1985). In one patient with pain and depression in the context of pancreatic adenocarcinoma, her pain level increased after ingesting 250 μg of LSD, but this was temporarily relived after 50 μg of intravenous LSD (Kurland 1985). One study with control agents compared the acute effects of 100 μg LSD, 2 mg dihydromorphinone, and 100 mg meperidine in a design wherein subjects received all three drugs at least 6 h apart (Kast and Collins 1964). The first two drugs given were always dihydromorphinone or meperidine, given in random order, and the third drug was always LSD. In these study subjects, LSD's acute pain reduction after 3 h was greater than that of the other two drugs (Kast and Collins 1964).

In an early study with 128 terminally ill cancer patients, pain scores were reduced by ~30% out to 3 weeks after the single administration of 100 μg LSD (Kast 1967). In another study, significant reductions in pain (scored by observers) and a trend toward reduced narcotic consumption were also reported to last "weeks to months" (Grof et al. 1973). In a qualitative report, lasting effects on the order of weeks (at minimum) could be deduced from patient and family reports (Kurland 1985; Pahnke et al. 1969). These lasting effects were described as "now adequately controlled with the aid of narcotics" and "much better. . .able to bear the pain than previously" and "she could push the pain out of her mind by remembering her out-of-body LSD experience" (Kurland 1985). Currently there are two active open label clinical trials investing psilocybin in cancer patients that include pain outcomes (NCT04593563, NCT04950608).

4.2 Phantom Limb Syndrome

Lasting effects of LSD were also reported in phantom limb syndrome, a condition wherein pain and other sensory distortions are perceived in the represented area of a missing limb after amputation. Both central and peripheral neural reorganization is involved in the development of phantom limb pain (Collins et al. 2018). In a 1977 study, daily oral LSD over 3 weeks (25 μg increased to 50 μg) reduced phantom limb pain and analgesic medication use in five of seven subjects (Fanciullacci et al. 1977). In one subject, pain severity remained reduced to "light" in the 2 weeks after LSD was stopped, and then returned to "intense" by 4 weeks (Fanciullacci et al. 1977). In a recent case report of phantom limb syndrome treatment, lasting pain reduction with psilocybin (dried mushroom) alone was not appreciated, but the combination of psilocybin mushroom with mirror therapy had a synergistic effect reducing pain for weeks or months (Ramachandran et al. 2018). In particular, the combination "dosed" three times led to "a striking and long-term reduction of pain" (Ramachandran et al. 2018). A controlled trial of psilocybin in phantom limb syndrome is currently under way (NCT 05224336).

4.3 Source of Psychedelics' Effects in Pain Disorders

The anti-inflammatory capacity of psychedelics through their activity at 5-HT2A receptors (Flanagan and Nichols 2018; Whelan and Johnson 2018) may be relevant for acute analgesia. Additional pharmacological effects via 5-HT2A and 5-HT1A receptors on such regions implicated in pain processing as the spinal cord, rostral ventromedial medulla, and dorsal raphe, (Chen et al. 2018; Whelan and Johnson 2018; De Gregorio et al. 2016; Castellanos et al. 2020). The ACC, a known functional target of psychedelics, is another relevant player in descending pain modulation (Chen et al. 2018; Carhart-Harris et al. 2012). In a recent study with healthy human volunteers, 20 μg LSD increased the tolerance of the cold pressor test (right hand submerged longer in 3°C water) and reduced the subjective measure of pain in this challenge (Ramaekers et al. 2021). Both serotonergic receptor-mediated effects on pain processing and a hypertension-associated analgesia effect were discussed as potentially relevant to these findings. Given the very mild psychological effects induced by the low dose in this study, it was not felt to be sufficient to produce dissociation and ego-dissolution, which are also possible sources for acute analgesia (Kast 1967; Ramaekers et al. 2021). In those reporting pain relief out a few weeks or months after the single administration of a psychedelic, the acute psychological effects are long gone. While in this extended period of pain relief, the residual effects of reset functional connectivity, as well as psychological factors related to the perception of pain, may remain (Grof et al. 1973; Castellanos et al. 2020). Indeed, cancer pain studies were typically carried out with the model of drug-assisted therapy currently used in mental health. Given the existing psychological techniques used in pain management, including cancer-related pain (Syrjala et al. 2014), employing the psychedelic-assisted therapy model may provide additional benefit than a drug-alone model, though this will require further systematic study. As discussed in headache disorders, the role of the placebo effect in pain management is not to be overlooked. This effect is a complex phenomenon involving opioid, cannabinoid, oxytocinergic, and dopaminergic systems in such brain regions as prefrontal cortex, ACC, PAG, insula, and nucleus accumbens (Fenton et al. 2015; Klosterhalfen and Enck 2008; Colloca 2019). The magnitude of the placebo effect in conventional pain treatment will vary among different conditions, but can actually be harnessed through pharmacological and psychological means to improve outcomes (Colloca 2019).

In 1977, Fanciullacci et al. remarked on the similarity between treatment targets in migraine and phantom limb syndrome, proposing the role of central pain processes (Fanciullacci et al. 1977). Indeed, studies in phantom limb syndrome have shown that cortical reorganization and pain are jointly reduced after such treatments as mirror therapy (Foell et al. 2014) and oral morphine sulfate (Huse et al. 2001). While those studies used daily treatments, as opposed to single administration, whether a single or few doses of a psychedelic can similarly reduce cortical reorganization and pain in phantom limb syndrome would speak to the ability of this treatment method to have effects at the disease level. In the report of lasting clinical effects in phantom limb syndrome after combined treatment with mirror therapy + psilocybin mushroom, Ramachandran et al. suggested that psilocybin might make the brain more receptive to the effects of mirror therapy through the cross-model sensory experience and neural plasticity (Ramachandran et al. 2018). While only a single case report, it might suggest that psilocybin in addition to standard treatment, be it pharmaceutical, rehabilitative, psychological, etc., may have additive or synergistic effects in pain management in phantom limb syndrome, as well as other chronic pain conditions. Understanding the specific effects would again require further comparative study.

5 Safety

The advent of a new method of pain management is exciting and deserving of continued rigorous study. The long history of human consumption of psychedelic compounds may be used as an argument for their immediate or fast-track course through regulatory (or de-regulatory) processes. However, safety remains a significant concern, particularly as the doses and regimens sometimes used in self-medication are distinct from those that have long been used in the ceremonial, recreational, and investigational contexts (i.e., moderate to high doses, taken infrequently). As some psychedelic compounds can be obtained rather easily outside of a controlled clinical study, it is vital that their safety in headache and pain disorders, which are very common conditions, not be overlooked.

5.1 *Safety of Infrequent Use*

As mentioned above, while cluster headache patients report on the acute effects of psilocybin and LSD on their attacks (Schindler et al. 2015; Sewell et al. 2006), the drug is not commonly used for acute use. Not only is it not practical to take psilocybin numerous times daily, as attacks can occur up to eight times a day, but this method would almost certainly lead to tolerance and loss of efficacy. In conditions where infrequent use of psilocybin or another psychedelic might be done for relief from exacerbations of pain (i.e., episodic migraine, sciatica), the drug class might theoretically have a role as an acute therapy.

As demonstrated in the first controlled study of the preventive effects of a psychedelic in a pain condition (migraine), the single administration of a low dose of psilocybin was safe when patients were carefully screened, prepared, monitored, and followed up (Schindler et al. 2021a). This study excluded patients with physical or psychiatric conditions for which psychedelic use may not be safe, including those comorbid with migraine (e.g., cardiovascular disease, cerebrovascular disease). The numerous other human clinical trials of the preventive effects of psilocybin in affective disorders and addiction also employ infrequent dosing, have similar exclusion criteria, study procedures, and safety outcomes (Johnson et al. 2008). Together these studies demonstrate the safety of a single or few doses of psychedelics for disease prevention, but they do not address the safety of lifetime disease management. For instance, chronic cluster headache patients report using pulses of psychedelics as frequently as every few weeks (Schindler et al. 2015); the safety and reliability of such repeated ingestion in relatively close succession has not been confirmed under controlled conditions. Currently, one cluster headache study is looking at a second psilocybin pulse administration after at least 6 months (NCT02981173).

5.2 *Safety of Chronic (Daily or Frequent) Use*

Conventional preventive pain medications, such as amitriptyline or gabapentin, are taken on a daily basis to effect changes in pain processing. Even the headache preventive therapies that are administered monthly (galcanezumab) or quarterly (onabotulinum toxin) remain in the body for the duration. The term "microdosing" is frustratingly non-specific on dose and regimen, but it is commonly used and typically regards the frequent use of small doses of psychedelics. Firstly, this technique is similar to taking daily preventive medications, so unless it has demonstrable superiority over conventional drugs, such treatment is not novel and would not seem worth the unknown risk. Secondly, while "micro" might presume safety, there is no conclusive evidence that the frequent use of psychedelics at any dose is safe over the long term (i.e., months, years, decades). A very clear warning comes from methysergide, the very effective migraine and cluster headache preventive (taken daily) that was removed from the market after some patients developed retroperitoneal fibrosis and cardiac valve pathology (Koehler and Tfelt-Hansen 2008; Rapoport 2012). The source of this damage is understood to stem from activity at the 5-HT2B receptor, which is also a target, albeit with low affinity, of psychedelics (Koehler and Tfelt-Hansen 2008; Horvath et al. 2004; Rickli et al. 2016). In order to avoid unintentional harm through common use (e.g., acetaminophen (Brune et al. 2015)), a clear understanding of the effects of different doses and regimens of psychedelic compounds is still required.

6 Conclusions

Recognition of the therapeutic effects of psychedelics in headache and chronic pain disorders has a long history, though rigorous controlled studies are required to fully characterize and understand these effects. The potential for this drug class to serve as a tool in pain management requires a keen understanding of the disorder in which they are used (i.e., pathophysiology and source of pain), the purpose for which they are used (i.e., acute vs. preventive), and the safety of their dosing regimen (i.e., infrequent vs. frequent). Varied mechanisms of pain control may be involved, including effects on sensory and perceptual aspects of pain processing, as well as effects on underlying disease itself. The capacity of psychedelics to produce sustained reductions in pain after limited dosing (i.e., one or a few administrations) further supports the truly unique nature of this drug class and may in turn help improve our understanding of the headache and chronic pain disorders themselves.

Acknowledgments Dr. Schindler is an employee of the U.S. Department of Veterans Affairs. Any opinions, findings, and conclusions or recommendations expressed in this material are those of the author and do not necessarily reflect the views of the U.S. Department of Veterans Affairs.

References

Akerman S, Holland PR, Goadsby PJ (2011) Diencephalic and brainstem mechanisms in migraine. Nat Rev Neurosci 12:570–584

Aman MM, Mahmoud A, Deer T et al (2021) The American Society of Pain and Neuroscience (ASPN) best practices and guidelines for the interventional management of cancer-associated pain. J Pain Res 14:2139–2164

Andersson M, Persson M, Kjellgren A (2017) Psychoactive substances as a last resort-a qualitative study of self-treatment of migraine and cluster headaches. Harm Reduct J 14:60

Bahra A, Goadsby PJ (2004) Diagnostic delays and mis-management in cluster headache. Acta Neurol Scand 109:175–179

Bahra A, May A, Goadsby PJ (2002) Cluster headache: a prospective clinical study with diagnostic implications. Neurology 58:354–361

Bartsch T, Paemeleire K, Goadsby PJ (2009) Neurostimulation approaches to primary headache disorders. Curr Opin Neurol 22:262–268

Bigal ME, Lipton RB (2011) Migraine chronification. Curr Neurol Neurosci Rep 11:139–148

Brune K, Renner B, Tiegs G (2015) Acetaminophen/paracetamol: a history of errors, failures and false decisions. Eur J Pain 19:953–965

Burch R (2019) Migraine and tension-type headache: diagnosis and treatment. Med Clin North Am 103:215–233

Burch R, Rizzoli P, Loder E (2018) The prevalence and impact of migraine and severe headache in the United States: figures and trends from government health studies. Headache 58:496–505

Burch RC, Buse DC, Lipton RB (2019) Migraine: epidemiology, burden, and comorbidity. Neurol Clin 37:631–649

Burish MJ, Pearson SM, Shapiro RE, Zhang W, Schor LI (2021) Cluster headache is one of the most intensely painful human conditions: results from the international cluster headache questionnaire. Headache 61:117–124

Bushnell MC, Ceko M, Low LA (2013) Cognitive and emotional control of pain and its disruption in chronic pain. Nat Rev Neurosci 14:502–511

Buture A, Boland JW, Dikomitis L, Ahmed F (2019) Update on the pathophysiology of cluster headache: imaging and neuropeptide studies. J Pain Res 12:269–281

Carhart-Harris RL, Erritzoe D, Williams T et al (2012) Neural correlates of the psychedelic state as determined by fMRI studies with psilocybin. Proc Natl Acad Sci U S A 109:2138–2143

Carhart-Harris RL, Kaelen M, Whalley MG, Bolstridge M, Feilding A, Nutt DJ (2015) LSD enhances suggestibility in healthy volunteers. Psychopharmacology (Berl) 232:785–794

Castellanos JP, Woolley C, Bruno KA, Zeidan F, Halberstadt A, Furnish T (2020) Chronic pain and psychedelics: a review and proposed mechanism of action. Reg Anesth Pain Med 45:486–494

Chen T, Taniguchi W, Chen QY et al (2018) Top-down descending facilitation of spinal sensory excitatory transmission from the anterior cingulate cortex. Nat Commun 9:1886

Collins KL, Russell HG, Schumacher PJ et al (2018) A review of current theories and treatments for phantom limb pain. J Clin Invest 128:2168–2176

Colloca L (2019) The placebo effect in pain therapies. Annu Rev Pharmacol Toxicol 59:191–211

De Gregorio D, Comai S, Posa L, Gobbi G (2016) d-Lysergic acid diethylamide (LSD) as a model of psychosis: mechanism of action and pharmacology. Int J Mol Sci 17

Di Lorenzo C, Coppola G, Di Lorenzo G, Bracaglia M, Rossi P, Pierelli F (2016) The use of illicit drugs as self-medication in the treatment of cluster headache: results from an Italian online survey. Cephalalgia 36:194–198

Dodick DW, Goadsby PJ, Lucas C et al (2020) Phase 3 randomized, placebo-controlled study of galcanezumab in patients with chronic cluster headache: results from 3-month double-blind treatment. Cephalalgia 40:935–948

Fallon M, Giusti R, Aielli F et al (2018) Management of cancer pain in adult patients: ESMO clinical practice guidelines. Ann Oncol 29:iv166–iiv91

Fanciullacci M, Bene ED, Franchi G, Sicuteri F (1977) Brief report: phantom limp pain: sub-hallucinogenic treatment with lysergic acid diethylamide (LSD-25). Headache 17:118–119

Favier I, Haan J, van Duinen SG, Ferrari MD (2007) Typical cluster headache caused by granulomatous pituitary involvement. Cephalalgia 27:173–176

Fenton BW, Shih E, Zolton J (2015) The neurobiology of pain perception in normal and persistent pain. Pain Manag 5:297–317

Flanagan TW, Nichols CD (2018) Psychedelics as anti-inflammatory agents. Int Rev Psychiatry 30: 363–375

Foell J, Bekrater-Bodmann R, Diers M, Flor H (2014) Mirror therapy for phantom limb pain: brain changes and the role of body representation. Eur J Pain 18:729–739

Garcia-Romeu A, Griffiths RR, Johnson MW (2014) Psilocybin-occasioned mystical experiences in the treatment of tobacco addiction. Curr Drug Abuse Rev 7:157–164

Garland EL, Fredrickson BL (2019) Positive psychological states in the arc from mindfulness to self-transcendence: extensions of the mindfulness-to-meaning theory and applications to addiction and chronic pain treatment. Curr Opin Psychol 28:184–191

Gaul C, Diener HC, Silver N et al (2016) Non-invasive vagus nerve stimulation for PREVention and acute treatment of chronic cluster headache (PREVA): a randomised controlled study. Cephalalgia 36:534–546

GBD Disease and Injury Incidence and Prevalence Collaborators (2017) Global, regional, and national incidence, prevalence, and years lived with disability for 328 diseases and injuries for 195 countries, 1990–2016: a systematic analysis for the Global Burden of Disease Study 2016. Lancet 390:1211–1259

Goadsby PJ, Reuter U, Hallstrom Y et al (2017) A controlled trial of erenumab for episodic migraine. N Engl J Med 377:2123–2132

Goadsby PJ, Dodick DW, Leone M et al (2019) Trial of galcanezumab in prevention of episodic cluster headache. N Engl J Med 381:132–141

Grazzi L, Bernstein C, Raggi A et al (2019) ACT for migraine: effect of acceptance and commitment therapy (ACT) for high-frequency episodic migraine without aura: preliminary data of a phase-II, multicentric, randomized, open-label study. Neurol Sci 40:191–192

Greenspan JD, Craft RM, LeResche L et al (2007) Studying sex and gender differences in pain and analgesia: a consensus report. Pain 132(Suppl 1):S26–S45

Griffiths RR, Johnson MW, Carducci MA et al (2016) Psilocybin produces substantial and sustained decreases in depression and anxiety in patients with life-threatening cancer: a randomized double-blind trial. J Psychopharmacol 30:1181–1197

Grof S, Goodman LE, Richards WA, Kurland AA (1973) LSD-assisted psychotherapy in patients with terminal cancer. Int Pharmacopsychiatry 8:129–144

Hartogsohn I (2016) Set and setting, psychedelics and the placebo response: an extra-pharmacological perspective on psychopharmacology. J Psychopharmacol 30:1259–1267

Headache Classification Committee of the International Headache Society (IHS) (2018) The international classification of headache disorders, 3rd edition. Cephalalgia 38:1–211

Horton BT (1952) Histaminic cephalgia. J Lancet 72:92–98

Horvath J, Fross RD, Kleiner-Fisman G et al (2004) Severe multivalvular heart disease: a new complication of the ergot derivative dopamine agonists. Mov Disord 19:656–662

Hughes LS, Clark J, Colclough JA, Dale E, McMillan D (2017) Acceptance and commitment therapy (ACT) for chronic pain: a systematic review and meta-analyses. Clin J Pain 33:552–568

Huse E, Larbig W, Flor H, Birbaumer N (2001) The effect of opioids on phantom limb pain and cortical reorganization. Pain 90:47–55

Ji Lee M, Cho SJ, Wook Park J et al (2019) Increased suicidality in patients with cluster headache. Cephalalgia 39:1249–1256

Johnson S, Black QC (2020) Can psychedelics alleviate symptoms of cluster headache and accompanying mental health problems? A case report involving Hawaiian baby Woodrose. J Psychoactive Drugs:1–5

Johnson M, Richards W, Griffiths R (2008) Human hallucinogen research: guidelines for safety. J Psychopharmacol 22:603–620

Karst M, Halpern JH, Bernateck M, Passie T (2010) The non-hallucinogen 2-bromo-lysergic acid diethylamide as preventative treatment for cluster headache: an open, non-randomized case series. Cephalalgia 30:1140–1144

Kast E (1967) Attenuation of anticipation: a therapeutic use of lysergic acid diethylamide. Psychiatry Q 41:646–657

Kast EC, Collins VJ (1964) Study of lysergic acid diethylamide as an analgesic agent. Anesth Analg 43:285–291

Klosterhalfen S, Enck P (2008) Neurophysiology and psychobiology of the placebo response. Curr Opin Psychiatry 21:189–195

Koehler PJ, Tfelt-Hansen PC (2008) History of methysergide in migraine. Cephalalgia 28:1126–1135

Koo BB, Bayoumi A, Albanna A et al (2021) Demoralization predicts suicidality in patients with cluster headache. J Headache Pain 22:28

Kurland AA (1985) LSD in the supportive care of the terminally ill cancer patient. J Psychoactive Drugs 17:279–290

Lee M, Silverman SM, Hansen H, Patel VB, Manchikanti L (2011) A comprehensive review of opioid-induced hyperalgesia. Pain Physician 14:145–161

Levi V, Cordella R, D'Ammando A et al (2019) Dorsal anterior cingulate cortex (ACC) deep brain stimulation (DBS): a promising surgical option for the treatment of refractory thalamic pain syndrome (TPS). Acta Neurochir 161:1579–1588

Loder E, Burch R, Rizzoli P (2012) The 2012 AHS/AAN guidelines for prevention of episodic migraine: a summary and comparison with other recent clinical practice guidelines. Headache 52:930–945

Lund N, Barloese M, Petersen A, Haddock B, Jensen R (2017) Chronobiology differs between men and women with cluster headache, clinical phenotype does not. Neurology 88:1069–1076

Ly C, Greb AC, Cameron LP et al (2018) Psychedelics promote structural and functional neural plasticity. Cell Rep 23:3170–3182

Magnoux E, Zlotnik G (2004) Outpatient intravenous dihydroergotamine for refractory cluster headache. Headache 44:249–255

Marin J, Giffin N, Consiglio E, McClure C, Liebler E, Davies B (2018) Non-invasive vagus nerve stimulation for treatment of cluster headache: early UK clinical experience. J Headache Pain 19: 114

Matharu MS, van Vliet JA, Ferrari MD, Goadsby PJ (2005) Verapamil induced gingival enlargement in cluster headache. J Neurol Neurosurg Psychiatry 76:124–127

Mather PJ, Silberstein SD, Schulman EA, Hopkins MM (1991) The treatment of cluster headache with repetitive intravenous dihydroergotamine. Headache 31:525–532

May A, Bahra A, Buchel C, Frackowiak RS, Goadsby PJ (1998) Hypothalamic activation in cluster headache attacks. Lancet 352:275–278

Mazer-Amirshahi M, Dewey K, Mullins PM et al (2014) Trends in opioid analgesic use for headaches in US emergency departments. Am J Emerg Med 32:1068–1073

McDonnell A, Schulman B, Ali Z et al (2016) Inherited erythromelalgia due to mutations in SCN9A: natural history, clinical phenotype and somatosensory profile. Brain 139:1052–1065

Moayedi M, Davis KD (2013) Theories of pain: from specificity to gate control. J Neurophysiol 109:5–12

Muthukumaraswamy SD, Forsyth A, Lumley T (2021) Blinding and expectancy confounds in psychedelic randomized controlled trials. Expert Rev Clin Pharmacol:1–20

Nagy AJ, Gandhi S, Bhola R, Goadsby PJ (2011) Intravenous dihydroergotamine for inpatient management of refractory primary headaches. Neurology 77:1827–1832

Olesen J, Burstein R, Ashina M, Tfelt-Hansen P (2009) Origin of pain in migraine: evidence for peripheral sensitisation. Lancet Neurol 8:679–690

Pahnke WN, Kurland AA, Goodman LE, Richards WA (1969) LSD-assisted psychotherapy with terminal cancer patients. Curr Psychiatr Ther 9:144–152

Pearson SM, Burish MJ, Shapiro RE, Yan Y, Schor LI (2019) Effectiveness of oxygen and other acute treatments for cluster headache: results from the cluster headache questionnaire, an international survey. Headache 59:235–249

Perez-Munoz A, Buse DC, Andrasik F (2019) Behavioral interventions for migraine. Neurol Clin 37:789–813

Petersen AS, Lund N, Jensen RH, Barloese M (2021) Real-life treatment of cluster headache in a tertiary headache center – results from the Danish cluster headache survey. Cephalalgia 41:525–534

Post MD (2014) Treatment of cluster headache symptoms using synthetic tryptamine N,N-diallyl-5 methoxytryptamine. J Contrib 20:12

Raja SN, Carr DB, Cohen M et al (2020) The revised international association for the study of pain definition of pain: concepts, challenges, and compromises. Pain 161:1976–1982

Ramachandran V, Chunharas C, Marcus Z, Furnish T, Lin A (2018) Relief from intractable phantom pain by combining psilocybin and mirror visual-feedback (MVF). Neurocase 24: 105–110

Ramaekers JG, Hutten N, Mason NL et al (2021) A low dose of lysergic acid diethylamide decreases pain perception in healthy volunteers. J Psychopharmacol 35:398–405

Rapoport AM (2012) What happens to the old headache medicines? Headache 52:701–706

Rapoport AM, Lin T (2019) Device profile of the Nerivio for acute migraine treatment: overview of its efficacy and safety. Expert Rev Med Devices 16:1017–1023

Rickli A, Moning OD, Hoener MC, Liechti ME (2016) Receptor interaction profiles of novel psychoactive tryptamines compared with classic hallucinogens. Eur Neuropsychopharmacol 26: 1327–1337

Robbins MS (2013) The psychiatric comorbidities of cluster headache. Curr Pain Headache Rep 17: 313

Robbins MS, Starling AJ, Pringsheim TM, Becker WJ, Schwedt TJ (2016) Treatment of cluster headache: the American headache society evidence-based guidelines. Headache 56:1093–1106

Rozen TD, Fishman RS (2012) Cluster headache in the United States of America: demographics, clinical characteristics, triggers, suicidality, and personal burden. Headache 52:99–113

Schindler EAD (2020) Sustained reductions in headache burden after the limited administration of low dose psilocybin in migraine and cluster headache: results from two preliminary studies. Neuropsychopharmacology:57–58

Schindler EAD, Gottschalk CH (2019) Cluster headache preventive therapies. Bryn Mawr Communications, Wayne

Schindler EA, Gottschalk CH, Weil MJ, Shapiro RE, Wright DA, Sewell RA (2015) Indoleamine hallucinogens in cluster headache: results of the clusterbusters medication use survey. J Psychoactive Drugs 47:372–381

Schindler EAD, Wright DA, Weil MJ, Gottschalk CH, Pittman BP, Sico JJ (2018a) Survey analysis of the use, effectiveness, and patient-reported tolerability of inhaled oxygen compared with injectable Sumatriptan for the acute treatment of cluster headache. Headache 58:1568–1578

Schindler EAD, Wallace RM, Sloshower JA, D'Souza DC (2018b) Neuroendocrine associations underlying the persistent therapeutic effects of classic serotonergic psychedelics. Front Pharmacol 9:177

Schindler EAD, Sewell RA, Gottschalk CH et al (2021a) Exploratory controlled study of the migraine-suppressing effects of psilocybin. Neurotherapeutics 18:534–543

Schindler EAD, Cooper V, Quine DB et al (2021b) "You will eat shoe polish if you think it would help"-familiar and lesser-known themes identified from mixed-methods analysis of a cluster headache survey. Headache 61:318–328

Schoenen J, Di Clemente L, Vandenheede M et al (2005) Hypothalamic stimulation in chronic cluster headache: a pilot study of efficacy and mode of action. Brain 128:940–947

Sewell RA, Gottschalk CH (2011) Problem child is no headache. Headache 51:306

Sewell RA, Halpern JH, Pope HG Jr (2006) Response of cluster headache to psilocybin and LSD. Neurology 66:1920–1922

Sicuteri F (1963) Prophylactic treatment of migraine by means of lysergic acid derivatives. Triangle 6:116–125

Syrjala KL, Jensen MP, Mendoza ME, Yi JC, Fisher HM, Keefe FJ (2014) Psychological and behavioral approaches to cancer pain management. J Clin Oncol 32:1703–1711

Treede RD, Rief W, Barke A et al (2015) A classification of chronic pain for ICD-11. Pain 156: 1003–1007

Treede RD, Rief W, Barke A et al (2019) Chronic pain as a symptom or a disease: the IASP classification of chronic pain for the international classification of diseases (ICD-11). Pain 160: 19–27

Urits I, Burshtein A, Sharma M et al (2019) Low back pain, a comprehensive review: pathophysiology, diagnosis, and treatment. Curr Pain Headache Rep 23:23

Vasiliou VS, Karademas EC, Christou Y, Papacostas S, Karekla M (2021) Acceptance and commitment therapy for primary headache sufferers: a randomized controlled trial of efficacy. J Pain 22:143–160

Wager TD, Atlas LY, Lindquist MA, Roy M, Woo CW, Kross E (2013) An fMRI-based neurologic signature of physical pain. N Engl J Med 368:1388–1397

Whelan A, Johnson MI (2018) Lysergic acid diethylamide and psilocybin for the management of patients with persistent pain: a potential role? Pain Manag 8:217–229

Yarnitsky D, Dodick DW, Grosberg BM et al (2019) Remote electrical neuromodulation (REN) relieves acute migraine: a randomized, double-blind, placebo-controlled, multicenter trial. Headache 59:1240–1252

Young WB, Kempner J, Loder EW et al (2012) Naming migraine and those who have it. Headache 52:283–291

Psychedelics as Novel Therapeutics in Alzheimer's Disease: Rationale and Potential Mechanisms

Albert Garcia-Romeu, Sean Darcy, Hillary Jackson, Toni White, and Paul Rosenberg

Contents

Abstract Serotonin 2A receptor (5-HT_{2A}R) agonist "classic psychedelics" are drawing increasing interest as potential mental health treatments. Recent work suggests psychedelics can exert persisting anxiolytic and antidepressant effects

A. Garcia-Romeu (✉), S. Darcy, and H. Jackson
Department of Psychiatry and Behavioral Sciences, Johns Hopkins University School of Medicine, Baltimore, MD, USA

Center for Psychedelic and Consciousness Research, Johns Hopkins University School of Medicine, Baltimore, MD, USA
e-mail: agarci33@jhmi.edu; sdarcy2@jhu.edu; hjacks18@jhmi.edu

T. White and P. Rosenberg
Department of Psychiatry and Behavioral Sciences, Johns Hopkins University School of Medicine, Baltimore, MD, USA

Memory and Alzheimer's Treatment Center, Johns Hopkins University School of Medicine, Baltimore, MD, USA
e-mail: toni.white@som.umaryland.edu; prosenb9@jhmi.edu

Curr Topics Behav Neurosci (2022) 56: 287–318
https://doi.org/10.1007/7854_2021_267
Published Online: 4 November 2021

lasting up to several months after a single administration. Data indicate acute subjective drug effects as important psychological factors involved in observed therapeutic benefits. Additionally, animal models have shown an important role for 5-$HT_{2A}R$ agonists in modulating learning and memory function with relevance for Alzheimer's Disease (AD) and related dementias. A number of biological mechanisms of action are under investigation to elucidate 5-$HT_{2A}R$ agonists' therapeutic potential, including enhanced neuroplasticity, anti-inflammatory effects, and alterations in brain functional connectivity. These diverse lines of research are reviewed here along with a discussion of AD pathophysiology and neuropsychiatric symptoms to highlight classic psychedelics as potential novel pharmacotherapies for patients with AD. Human clinical research suggests a possible role for high-dose psychedelic administration in symptomatic treatment of depressed mood and anxiety in early-stage AD. Preclinical data indicate a potential for low- or high-dose psychedelic treatment regimens to slow or reverse brain atrophy, enhance cognitive function, and slow progression of AD. In conclusion, rationale and potential approaches for preliminary research with psychedelics in patients with AD are presented, and ramifications of this line of investigation for development of novel AD treatments are discussed.

Keywords Alzheimer's disease · Dementia · Hallucinogen · Mild cognitive impairment (MCI) · Psilocybin · Psychedelic

1 Introduction

Alzheimer's Disease (AD) is a growing concern amid a rapidly increasing population aged 65 and older worldwide, and projected rising global life expectancy (He et al. 2016). Currently, more than five million adults in the USA and 36 million worldwide are living with AD, and this number is expected to triple by 2050 (Alzheimer's Association 2021). However, there has been little success in development of strategies for AD pharmacotherapy. Symptomatic treatment of AD with acetylcholinesterase inhibitors such as donepezil, rivastigmine, and galantamine has been available since the 1990s with modest benefits for some patients (Tayeb et al. 2012). The *N*-Methyl-*D*-aspartate (NMDA) antagonist memantine was approved for treating moderate to severe AD by the US Food and Drug Administration (FDA) in 2003, but to date no cure or well-established disease modifying treatment for AD is available, despite extensive research and drug development efforts involving over 240 failed candidate drugs (Dubois et al. 2014; Wimo et al. 2014). Although recent progress has been made toward developing novel antibody-based pharmacotherapies such as aducanumab (Sevigny et al. 2016) and donanemab (Mintun et al. 2021), controversy remains whether these will prove safe, accessible, and substantially effective treatments for patients with AD (Ayton 2021; Doggrell 2021; Knopman et al. 2021). Given the enormous morbidity and mortality associated with AD, it is clear that novel approaches to AD treatment are urgently needed.

The past two decades have seen a resurgence in research involving hallucinogenic serotonin 2A receptor (5-HT_{2A}R) agonists, known as "classic psychedelics," as potential treatments across a range of medical and mental health conditions. Preliminary studies in animals and humans suggest that classic psychedelics such as psilocybin, lysergic acid diethylamide (LSD), and the dimethyltryptamine (DMT) containing decoction ayahuasca may have promising antidepressant, anxiolytic, and antiaddictive properties (Garcia-Romeu et al. 2016). So much so, that the FDA has granted psilocybin "breakthrough therapy" designation as a potential treatment for major depressive disorder, with clinical trials of therapeutic safety and efficacy currently underway (Nichols 2020). To date, psychedelics' psychological mechanisms of action appear related to acute subjective drug effects associated with positive therapeutic outcomes (Bogenschutz et al. 2015; Garcia-Romeu et al. 2014; Griffiths et al. 2016; Roseman et al. 2018; Ross et al. 2016). Additionally, preclinical and neuroimaging research indicate a number of compelling biological mechanisms of psychedelics related to stimulation of 5-HT_{2A}R and downstream signaling pathways relevant to AD. These mechanisms include promotion of structural and functional neuroplasticity (Catlow et al. 2013; Lima da Cruz et al. 2018; Ly et al. 2018), post-acute changes in key signaling pathways such as brain-derived neurotrophic factor (BDNF) (Hutten et al. 2021; Ly et al. 2018), anti-inflammatory effects (Flanagan and Nichols 2018), as well as acute and post-acute changes in brain functional connectivity (Barrett et al. 2020a, b; Carhart-Harris et al. 2012; Carhart-Harris et al. 2017; Preller et al., 2020). This review provides a detailed examination of potential mechanisms of classic psychedelics as possible treatments for patients with AD and describes the rationale for targeted investigation of psychedelics in patients with early AD (e.g., ClinicalTrials.gov NCT04123314).

2 Pathophysiology and Etiology of Alzheimer's Disease

Both normal aging and Alzheimer's Disease (AD) have been associated with decreased functional brain activity and connectivity (Dennis and Thompson 2014; Tomasi and Volkow 2012). Network hypersynchrony and abnormalities such as impaired default mode network (DMN) deactivation have been linked to cognitive dysfunction and implicated as potential targets for therapeutic intervention in AD (Palop and Mucke 2016). The neuropathological hallmarks that typically define AD are amyloid-β (Aβ) plaques, neurofibrillary tangles, and neuronal and synaptic loss (Serrano-Pozo et al. 2011a; Shoghi-Jadid et al. 2002). This neurodegeneration is associated with cognitive and functional decline typically starting with loss of episodic memory and progressing to include aphasia, apraxia, and agnosia (Butzlaff and Ponimaskin 2016; Weintraub et al. 2012). While amyloid is thought to be the "prime mover" in AD pathobiology, we are still ascertaining the mechanisms of progressive neurodegeneration, which likely include tau deposition as the next phase, leading on to neuronal loss. Aβ accumulation has been suggested to facilitate formation of pathological tau, and together these seem to trigger additive neurotoxic effects functioning as a systemic feedback loop resulting in acute neuron death and synaptic dysfunction (Bloom 2014).

AD is thought to begin up to 20 years prior to symptoms with a lengthy preclinical, "prodromal" phase during which cleavage of Amyloid Precursor Protein (APP) by Beta-secretase 1 (BACE-1) and Gamma-secretase results in the aggregation of Aβ protein and Aβ plaques (Sperling et al. 2011). This accrual results in neurodegeneration in characteristic brain regions (including the hippocampus, posterior cingulate cortex, and precuneus) and impaired synaptic function over time (Bateman et al. 2012; Dubois et al. 2014). Post-mortem data suggest a temporal pattern of neurofibrillary tangle formation from the transentorhinal layer in early stages of AD proceeding to the entorhinal cortex before subsequent degeneration in the isocortical association areas in later stages of disease progression (Braak and Braak 1991). This focus has led to exploration in clinical trials of anti-Aβ therapies for AD treatment, which have thus far garnered little success (Karran et al. 2011; Karran and Hardy 2014). AD patients present with heterogeneous symptoms that may be conceptualized as distinct clinical syndromes with relatively greater disturbances in language, visuospatial functions, apraxia, or behavioral manifestations (Stopford et al. 2008). These variations have also been associated with particular clinical biomarkers. For instance, visual perception problems and/or spatial difficulties are often accompanied by posterior cortical atrophy including hypometabolism in these areas as observed by magnetic resonance imaging (MRI) and positron emission tomography (PET) (Graff-Radford et al. 2021; Jack Jr et al. 2019). Furthermore, while genetic variants such as apolipoprotein E4 (ApoE4) have long been known to play a role in development of AD, which is highly heritable (Tanzi 2012), contemporary research is shedding new light on genetic and environmental factors related to AD, such as amyloid precursor protein metabolism (Kunkle et al. 2019) and pesticide exposure (Killin et al. 2016). Below, we review selected aspects of AD biological mechanisms which are potentially relevant to psychedelics' mechanisms of action.

2.1 *Decreased Serotonergic Neurotransmission in AD*

Evidence indicates reduced serotonergic neurotransmission in AD may be associated with psychiatric symptoms (Butzlaff and Ponimaskin 2016). Animal models of AD suggest selective neurodegeneration of serotonin pathways and reduced serotonergic neurotransmission (Liu et al. 2008). Preclinical research has shown β-amyloid accumulation leads to a decline in 5-HT_{2A}R levels in the cortex of mice (Holm et al. 2010). In prodromal AD, PET imaging reveals a reduced density of serotonin transporter which is associated with early cognitive changes (Smith et al. 2017). Several studies report decreased 5-HT_{2A}R levels in widespread areas of the brain in AD (Marner et al. 2012; Mecca 2019). These changes are associated with neuropsychiatric symptoms including agitation, depression, and psychosis in AD (Chakraborty et al. 2019). Relevant to classic psychedelics, $5\text{-HT}_{2A}R$ density declines in healthy aging throughout the brain and specifically in the hippocampus, and the degree of temporal lobe $5\text{-HT}_{2A}R$ decrease is associated with cognitive decline in AD (Marner et al. 2012; Versijpt et al. 2003). Some studies point to possible genetic influences of the serotonin system in AD, such as the $5\text{-HT}_{2A}R$ T102C polymorphism, where the CC genotype has been associated with risk of psychotic symptoms in AD (Tang et al. 2017). In human PET studies, neocortical regions including the orbitofrontal cortex (OFC) showed reduced $5\text{-HT}_{2A}R$ binding in both Mild cognitive impairment (MCI) and AD patients (Hasselbalch et al. 2008; Lai et al. 2005; Versijpt et al. 2003). In addition to serotonin, other neurotransmitter systems such as norepinephrine (Theofilas et al. 2017) and acetylcholine (Grothe et al. 2012) have been implicated in AD pathology and identified as targets for AD pharmacotherapies (Marucci et al. 2021). However, the current review focuses primarily on serotonergic neurotransmission due to its key role in psychedelics' biological mechanisms (Nichols 2016),

2.2 *Loss of Synaptic Function in AD*

Data suggest the loss of synaptic function in AD prior to neuronal loss (Selkoe 2002). For example, synaptophysin (a characteristic marker of synaptic integrity) is decreased in prodromal AD (Masliah et al. 2001; Sze et al. 1997; Yuki et al. 2014). One well-validated marker of synaptic density is synaptic vesicle glycoprotein 2 (SV2) which is expressed in virtually all synapses and is located in synaptic vesicles at presynaptic terminals (Mecca 2019). A PET tracer for imaging SV2 density in vivo is now available ([11C]UCB-J) and has demonstrated decreased SV2 density in the hippocampus of patients with AD compared to cognitively unimpaired older adults (Chen et al. 2018).

2.3 *Key Signaling Pathways in AD*

Brain-derived neurotrophic factor (BDNF), a protein critical to neuronal growth and survival, is affected by the accumulation of Aβ protein in prodromal AD (Peng et al. 2005). This accumulation interferes with the conversion of proBDNF to BDNF such that parietal cortex levels of both proBDNF and BDNF are reduced in prodromal AD and MCI, correlating with cognitive decline (Peng et al. 2005; Tanila 2017). In addition, decreased Tropomyosin receptor kinase B (TrkB) BDNF receptor expression and increased expression of TrkB.T1 (a primary inhibitor of TrkB) result in further BDNF inhibition as well as the prevention of long-term potentiation (LTP) and long-term depression (Eide et al. 1996; Michaelsen et al. 2010) – processes essential to memory formation (Minichiello 2009). The steady decrease of BDNF and its action, which may occur as a direct result of Aβ deposition, could be a critical link between the prodromal phase of AD and the beginning of neurodegeneration and cognitive decline, eventually culminating in dementia (Arancibia et al. 2008; Ciaramella et al. 2013).

The Mammalian Target of Rapamycin (mTOR) signaling pathway also has a role in LTP and memory formation (Cammalleri et al. 2003; Hoeffer and Klann 2010). Hyperactivation of mTOR complex I leads to downstream inhibition of cell autophagy, which could result in further Aβ deposition and tau hyperphosphorylation (Caccamo et al. 2013). Conversely, the activated mTOR signaling pathway has also been shown to induce structural plasticity and neuritogenesis by regulating the behavior of axonal growth cones, dendrite arborization, and dendritic spine morphology via control of local protein synthesis (Jaworski and Sheng 2006). Targeting mTOR signaling in key cognitive brain regions could help delay or even prevent cognitive decline during the neurodegeneration phase of AD (Tramutola et al. 2017).

2.4 Inflammation in AD

Accumulating evidence has implicated neuroinflammation in the progression of AD (Kinney et al. 2018; Zotova et al. 2010). In post-mortem studies, brain tissue of patients with AD exhibits signs of persisting inflammatory activity, such as activated microglia and astrocyte clusters (Serrano-Pozo et al. 2011b). These in turn release proinflammatory cytokines and interleukins (e.g., interleukin [IL]-1β, IL-6, and tumor necrosis factor-α [TNF-α]), which cause tissue damage with prolonged exposure, and interact with accumulating Aβ and NFT to contribute to neuronal loss (Garwood et al. 2011; Wang et al. 2015). Early patient data suggested use of medications such as Non-Steroidal Anti-Inflammatory Drugs (NSAIDs) may be associated with reduced severity of AD symptoms (Rich et al. 1995) and reduced risk of developing AD (Stewart et al. 1997). To date, clinical trials of NSAIDs as a treatment for AD have not shown positive results (Miguel-Álvarez et al. 2015). However, targeted strategies for modulating neuroinflammation remain a viable pathway for developing novel AD therapeutics that continue to be explored (Ozben and Ozben 2019).

2.5 Changes in Brain Metabolism and Functional Connectivity in AD

In addition to cellular and molecular mechanisms, human neuroimaging and post-mortem studies provide insight into the neurobiology of AD. PET is a critical tool for understanding and diagnosing AD, allowing Aβ and tau deposition to be evaluated in vivo (Brier et al. 2016), and using ligands such as fluorodeoxyglucose (FDG) to assess functional brain metabolism (Rice and Bisdas 2017). These methods have demonstrated differential patterns of atrophy, hypometabolism, and Aβ and tau aggregation across the course of disease progression, informing the neurodegenerative processes underlying AD (Joie et al. 2012; Ossenkoppele et al. 2016). Current imaging data suggest early-stage AD is marked by Aβ deposition, atrophy, and metabolic dysfunction in posterior cortical regions, which are active during memory retrieval in healthy individuals (Buckner et al. 2005). As cognitive function declines, concurrent increases in tau deposition are observed in the temporal lobe (Brier et al. 2016) and other key domain-specific regions (e.g., occipital cortex for individuals with visual impairment) (Ossenkoppele et al. 2016).

Functional MRI data also show notable decrease in connectivity in normal aging across several brain networks, while AD mainly shows alterations in the default mode network (DMN), dorsal attention network (DAN), and the precuneus (Hafkemeijer et al. 2012; Klaassens et al. 2017; Tomasi and Volkow 2012). The DMN is primarily composed of the medial prefrontal cortex (mPFC), posterior cingulate cortex (PCC), precuneus, and angular gyrus, which are involved in episodic memory retrieval (Sestieri et al. 2011), a function known to deteriorate in older adults with AD (Mevel et al. 2011; Weintraub et al. 2012). In some cases of AD, DMN desynchronization has been posited to contribute to cognitive decline, consistent with evidence that differences in DMN activation in the precuneus and PCC predicted lower Mini-Mental State Exam (MMSE) scores, typically indicative of more severe dementia (Schwindt et al. 2013). Amyloid accumulation may also affect DMN function, for instance leading to hypoconnectivity within the DMN in early AD (Buckner et al. 2005; Palmqvist et al. 2017). The salience network (SN) has also shown reduced gray matter volume and altered functional connectivity in AD that were associated with neuropsychiatric symptoms (Balthazar et al. 2014), as well as cognitive impairment (He et al. 2014). It has been hypothesized that increased SN connectivity in AD is associated with greater "emotionality" which might contribute to the expression of affective and other neuropsychiatric symptoms (Zhou and Seeley 2014). Lasting alterations in brain network connectivity have been observed after a single dose of psilocybin (Barrett et al. 2020a) and are correlated with psilocybin's antidepressant effects (Carhart-Harris et al. 2017), indicating a potential biological mechanism by which psychedelics could affect AD progression and related symptoms. Notably, the medial temporal lobes have been found to show increases in activation during MCI and early-stage AD, which may be associated with local tau formation and subsequent neurodegeneration and hypoactivation in these regions that then spreads as AD progresses (Pasquini et al. 2019; Putcha et al. 2011).

2.6 Neuropsychiatric Comorbidities in AD

Patients with AD have a high prevalence of comorbid neuropsychiatric symptoms, with more than 40% exhibiting clinically significant symptoms of depression (Lyketsos et al. 2002; Zhao et al. 2016). Beyond other challenges posed by AD, depression adversely impacts both patient and caregiver quality of life (Karttunen et al. 2011; Shin et al. 2005). Moreover, depression is known to mediate the progression of AD, with more pronounced symptoms being associated with greater risk of cognitive decline (Dotson et al. 2010; Herbert and Lucassen 2016; Ruthirakuhan et al. 2019). Typical antidepressant medications have not shown clear evidence of efficacy in patients with dementia, indicating a need for novel treatments (Banerjee et al. 2013; Nelson and Devanand 2011; Rosenberg et al. 2010). Neuropsychiatric symptoms are frequently the first symptoms of prodromal dementia (Leoutsakos et al. 2015) and associated with early manifestations of AD biomarkers (Banning et al. 2021), which has led to the concept of Mild Behavioral Impairment defining five domains of such symptoms which are predictive of incident MCI and dementia (Ismail et al. 2017).

The symptoms of AD also extend beyond cognitive complaints to include highly prevalent, comorbid neuropsychiatric symptoms such as agitation, apathy, sleep disturbances, and anxiety (Lyketsos et al. 2002; Steinberg et al. 2008). These symptoms contribute to disability, worse life quality, impaired activities of daily living, caregiver burden, institutionalization, and accelerated mortality (Lanctôt et al. 2017; Lyketsos et al. 2011; Peters et al. 2015; Soto et al. 2015). While practice guidelines consistently refer to managing such symptoms as central to treating AD (Lyketsos et al. 2006), there are no established effective treatments, highlighting this as an important area for further research into novel therapies. In particular, depressed mood, anxiety, apathy, and reduced quality of life represent compelling targets for brief interventions involving moderate to high-dose psychedelic administration based on existing clinical research described in more detail below. Conversely, other neuropsychiatric symptoms related to AD such as delusions and hallucinations are generally considered contraindications for high-dose psychedelic treatments (Johnson et al. 2008), and it remains unclear how symptoms such as motor disturbances may be influenced by psychedelics.

3 Neurobiology of Psychedelics

A growing body of research supports the administration of 5-HT_{2A}R agonist classic psychedelics such as psilocybin and LSD as a potential treatment for various conditions, including anxiety, mood, and substance use disorders (Garcia-Romeu et al. 2016; Reiff et al. 2020). These compounds represent a novel frontier in the field of psychiatry as possibly transdiagnostic pharmacotherapies with low toxicity and addiction risk, and the potential for long-lasting benefits (Johnson et al. 2018). The underlying neurobiological mechanisms responsible for these effects are now being explored in basic translational and clinical research, indicating additional potential for these substances as possible novel treatment options for patients with AD.

3.1 Data on the Role of 5-HT$_{2A}$R in Learning and Memory

Evidence indicates that serotonin has a key modulatory role in learning alongside other neurotransmitters such as dopamine (Aznar and Hervig 2016; Frick et al. 2015; Harvey et al. 2004). In particular, 5-HT$_{2A}$R agonists like psilocybin have long been studied as potential modulators of learning and memory with early experiments identifying pretreatment with 25ug/kg LSD as a facilitator of reversal learning in rats compared to placebo (King et al. 1972). More recently, administration of a selective 5-HT$_{2A}$R antagonist was shown to dose-dependently impair spatial reversal learning and increase perseverative errors in rats, further implicating 5-HT$_{2A}$R signaling in cognitive flexibility processes (Boulougouris et al. 2008). Similarly, reversal learning deficits in chronically stressed rats can be alleviated with chronic SSRI treatment, and this improvement is blocked by injection of a 5-HT$_{2A}$R antagonist (Furr et al. 2012).

Additionally, LSD injections in the hippocampus have been shown to accelerate classical conditioning of the eyeblink response in rabbits, with chronic LSD injections desensitizing 5-HT$_{2A}$R but not 5-HT$_{2C}$R-mediated behavioral responses (Romano et al. 2010). Related work found that chronic treatment of rabbits with an inverse 5-HT$_{2A}$R agonist increased 5-HT$_{2A}$R density in the frontal cortex and produced similar acceleration in classical conditioning (Harvey et al. 2004). The extinction of fear memories in rats can be accelerated by administration of 5-HT$_{2A}$R agonist (4-Bromo-3,6-dimethoxybenzocyclobuten-1-yl)methylamine hydrobromide (TCB-2) and delayed by administration of a 5-HT$_{2A}$R antagonist (Zhang et al. 2013). Low doses of psilocybin likewise facilitate the extinction of fear memories and may increase hippocampal neurogenesis in rats (Catlow et al. 2013). Furthermore, the spatial tuning of neurons in the prefrontal cortex of rhesus monkeys performing a visual working memory task can be accentuated or attenuated by the delivery of a 5-HT$_{2A}$R agonist or antagonist, respectively (Williams et al. 2002). The 5-HT$_{2A}$R agonist TCB-2 can also improve the working memory of rats with medial forebrain bundle lesions intended to mimic the cognitive effects of Parkinson's Disease (Li et al. 2015). Hippocampal TCB-2 injection during memory consolidation enhances the object memory of mice, and this effect is blocked by pretreatment with a 5-HT$_{2A}$R antagonist, further supporting a key role for 5-HT$_{2A}$R in modulating memory (Zhang et al. 2016).

In addition to animal studies, research in humans has suggested a role for 5-HT$_{2A}$R in regulating mood and memory functions. Genetic research has found that variations of 5-HT$_{2A}$R can influence memory task performance in humans, with carriers of the heterozygous 5-HT$_{2A}$ H452Y polymorphism making more errors during memory tasks (de Quervain et al. 2003), and displaying less right anterior hippocampal activation in response to novel stimuli compared to their homozygous counterparts (Schott et al. 2011). These data suggest a robust influence of 5-HT$_{2A}$R activation in diverse learning and memory processes that are relevant to AD (Zhang and Stackman Jr 2015), raising the possibility that low- or high-dose psychedelic administration may have cognition-enhancing effects in patients with AD. Approaches testing low-dose psychedelics might use a chronic dosing regimen every few days over the course of several weeks to assess pre- and post-treatment performance on validated measures of episodic memory, working memory, visuo-spatial processing, and executive function (e.g., Mini-Mental State Exam, Hopkins Verbal Learning Test, Trail Making Test, Raven's Progressive Matrices, Category Fluency). A similar design could be employed to examine effects of one or more high-dose psychedelic sessions with psychological support, preferably with patients in earlier stages of AD, where there would be less ethical concerns about informed consent regarding study procedures (Kim 2011). Should any signal of cognitive-enhancing effects emerge in initial research, this would pave the way for further study of biological mechanisms using neuroimaging and other biomarker assessment.

3.2 *Psychoplastogenic Effects of Psychedelics and Related Signaling Pathways*

Data from cellular and molecular models additionally suggest classic psychedelics may have potential in treating early-stage AD. A recent study found classic psychedelics to selectively induce structural and functional neuroplasticity in vitro and in vivo in the rat prefrontal cortex at an extent comparable to BDNF, with resulting effects posited as "psychoplastogenic" (Ly et al. 2018). Psychoplastogenic compounds are defined to produce a measurable change in neuroplasticity within 24–72 h of a single administration. Measurable changes in plasticity include changes in neurite growth, dendritic branching, dendritic spine density, synapse number, and intrinsic excitability, among others (Olson 2018). Recently published data consistent with psychoplastogenic effects found a single dose of psilocybin led to significant, rapid increases in the formation, size, and density of dendritic spines in mouse medial frontal cortex occurring within 24 h of dosing, with structural changes persisting up to a month later (Shao et al. 2021). Psilocybin was also found to increase excitatory postsynaptic current frequency and to reduce behavioral signs of learned helplessness in a prolonged stress paradigm in mice (Shao et al. 2021). Psychedelics are thought to induce such psychoplastogenic effects via 5-HT$_{2A}$R receptor stimulation, which upregulates Ania gene expression and affects glutamatergic function (Nichols and Sanders-Bush 2002). Specifically, this activity amplifies α-amino-3-hydroxy-5-methyl-4-isoxazolepropionic acid (AMPA) glutamate receptor signaling, resulting in downstream activation of the mTOR pathway – one of the proposed mechanisms for the neural plasticity – promoting effects of psychoplastogens (Cavalleri et al. 2018). Recent data indicate that 5-HT$_{2A}$R stimulation may impact BDNF (Hutten et al. 2021; Tsybko et al. 2020), and result in upregulated activity-regulated cytoskeleton-associated (Arc) protein expression thought to be involved in cytoskeletal rearrangements for synaptic plasticity (Nichols et al. 2003; Nichols and Sanders-Bush 2002).

Additional preclinical research provides further evidence on relevant mechanisms for classic psychedelics to positively impact biological pathways relevant to AD. In rats, a single LSD administration has been shown to increase expression of immediate early genes (IEGs) implicated in synaptic plasticity in various brain regions including the PFC, midbrain, and hippocampus (Nichols et al. 2003; Nichols and Sanders-Bush 2002). Similarly, psilocybin and other 5-HT$_{2A}$R agonists can induce IEG expression in the mouse cortex (González-Maeso et al. 2007). Psilocybin has also shown dose-dependent and differential alterations in transcriptional regulation in the PFC and hippocampus across multiple plasticity-related genes in rats (Jefsen et al. 2021). Additionally, preliminary data indicate a single administration of the psychedelic 1-(2,5-dimethoxy-4-iodophenyl)-2-aminopropane (DOI) can increase dendritic structural density and plasticity in the PFC, enhance fear extinction, and enhance LTP in mice (Revenga et al. 2021). These effects were mediated by 5-HT$_{2A}$R as evidenced by lack of such effects in 5-HT$_{2A}$R knockout mice. Furthermore, DOI induced lasting changes up to a week post-drug administration in frontal cortex gene expression in mice further suggesting transcriptional and epigenetic mechanisms may mediate lasting effects of serotonergic psychedelics (Revenga et al. 2021). Chronic administration of DOI and other 5-HT$_{2A}$R agonists produced increased proBDNF levels and downregulation of TrkB receptors in mice (Tsybko et al. 2020). Furthermore, the psychedelic 5-MeO-DMT has also been shown to induce neuroplastic changes after a single dose including increased cell growth and maturation in the dentate gyrus of mice (Lima da Cruz et al. 2018).

Finally, a series of recent studies conducted in pigs have also shown lasting changes in PFC gene expression up to a week after a single dose of psilocybin (Donovan et al. 2021). Increased hippocampal synaptic vesicle protein 2A (SV2A) density and decreased hippocampal and PFC 5-HT$_{2A}$R density have been found 24 h post-psilocybin administration, and significant, ongoing increases in SV2A density have been detected at 1 week post-psilocybin administration (Raval et al. 2021). SV2A protein levels are thought to reflect presynaptic density, suggesting psilocybin may increase synaptogenesis up to a week after a single psilocybin exposure in pigs (Raval et al. 2021). Furthermore, novel evidence suggests 5-HT$_{2A}$R inverse agonist administration quickly and significantly reduced brain Aβ levels and improved cognitive function in a mouse model of AD, though this effect was not observed in 5-HT$_{2A}$R knockout mice (Yuede et al. 2021). Taken together, this preclinical evidence suggests classic psychedelics may act via a host of 5-HT$_{2A}$R mediated biological mechanisms to promote rapid changes in genetic expression leading to longer lasting functional and structural brain changes, which may in turn be associated with therapeutic effects observed in human trials (Fig. 1). Although it remains to be seen whether the mechanisms described here lead to clinical improvement in humans, preclinical data on 5-HT$_{2A}$R agonist effects on learning and memory,

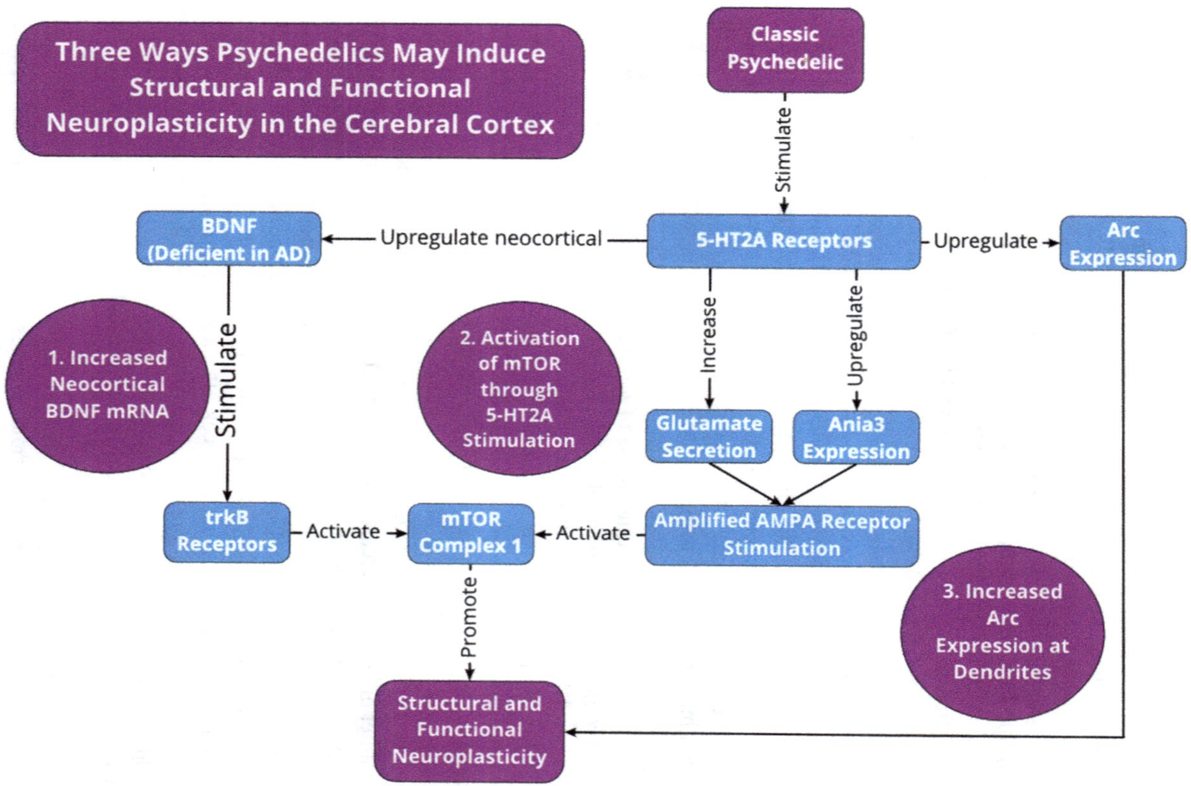

Fig. 1 A diagram of three converging pathways that may be responsible for induced neural plasticity, potentially resulting in lasting beneficial effects following psychedelic administration: 5-HT_{2A}R upregulation of neocortical BDNF, amplification of AMPA receptor activity resulting in downstream activation of mTOR, and upregulated Arc protein expression

combined with the observed neurological, antidepressant, and anxiolytic effects of psychedelics discussed below, present a compelling rationale for targeted investigation of 5-HT_{2A}R agonist effects in patients with AD. Psychoplastogenic effects could present a potential mechanism to slow or reverse atrophy in key brain regions affected by AD and could be studied after chronic low-dose or one or more high-dose psychedelic administration sessions in AD patients using pre- and post-neuroimaging and neuropsychological testing, parallel to research in healthy volunteers described in more detail below (Madsen et al. 2020). Similarly, preclinical findings on 5-HT_{2A}R mediated reductions in Aβ levels (Yuede et al. 2021) could be studied in clinical trials administering classic psychedelics to early-stage AD patients and assessing longitudinal impact on Aβ, cognitive function, and disease progression, providing another possible, complementary therapeutic mechanism for advancing AD treatment.

3.3 *Psychedelics as Anti-inflammatory Agents*

Preclinical studies have shown robust anti-inflammatory effects of classic psychedelics (Flanagan and Nichols 2018). The 5-HT_{2A}R agonist psychedelics (R)-2,4-dimethoxy-4-iodoamphetamine [(R)-DOI] and LSD (among others) have been found to suppress TNF-α induced inflammation in rat aortic smooth muscle cells, with (R)-DOI exhibiting substantial potency in this regard (Yu et al. 2008). These effects were consistent in vivo in mice, showing anti-inflammatory effects of (R)-DOI in aorta, small intestine, and blood at low-dose levels, which were blocked by co-administration of a selective 5-HT_{2A}R antagonist, indicating a central role for 5-HT_{2A}R in anti-inflammatory effects (Nau Jr et al. 2013). In addition to 5-HT_{2A}R mediated anti-inflammatory effects, cellular models suggest some classic psychedelics such as DMT and 5-MeO-DMT may exert additional anti-inflammatory effects via the Sigma-1 receptor, including inhibition of IL-1β, IL-6, and TNF-α (Szabo et al. 2014), proinflammatory cytokines known to be involved in AD pathology (Wang et al. 2015). Furthermore, a recent study demonstrated the DMT containing admixture ayahuasca, but not placebo, significantly reduced levels of the inflammatory biomarker C-reactive protein from pre- to 48 h-post administration, and these reductions were correlated with mood improvements in patients with treatment-resistant depression (Galvão-Coelho et al. 2020). Anti-inflammatory effects of psychedelics have not yet been conclusively studied in human clinical populations but may be observed after repeated low doses of psychedelics or single high doses. Such effects could be studied in AD patients via prospective measurement of cytokines and related inflammatory biomarkers that may also serve as a therapeutic target for treatment during early-stage AD or possibly in later stages using chronic low-dose regimens.

3.4 Psychedelics' Effects in Humans

Recent imaging data provide insight into the activity of the brain during and after acute psychedelic effects. Resting state network connectivity during psilocybin peak effects shows increased between-network functional connectivity and simultaneous decreased within-network connectivity in the DMN, visual networks, and auditory networks (Mason et al. 2020). This altered network connectivity may be due in part to psilocybin's ability to reduce overall activity in both sides of the claustrum, a key structure in the executive control of behavior, while simultaneously modifying the connectivity of the claustrum with different networks such as the frontoparietal task network (Barrett et al. 2020b). These acute effects are time-dependent and may be predicted by baseline global brain connectivity (Preller et al. 2020).

Psilocybin also may mediate glutamate concentration in areas like the hippocampus and medial prefrontal cortex via 5-HT_{2A}R, ultimately leading to the activation of AMPA receptors and the increased expression of BDNF (Hutten et al. 2021; Mason et al. 2020). During acute psilocybin effects, working memory may appear unchanged or impaired, perhaps due to attentional deficits stemming from an impaired ability to ignore irrelevant stimuli (Barrett et al. 2018; Carter et al. 2005). However, although the acute subjective effects of psilocybin last a matter of hours, fMRI research has found longer term effects in the brain such as decreased amygdala response during affective processing tasks a week after administration and increased global functional brain connectivity a full month after administration (Barrett et al. 2020a). Such long-term effects suggest that psilocybin may induce a period of heightened neuroplasticity lasting weeks after initial psilocybin administration.

Volunteers with treatment-resistant depression were found to have decreased amygdala cerebral blood flow and increased DMN integrity 1 day after psilocybin administration, which has been proposed as a potential "reset" mechanism of psychedelics in which networks like the DMN may experience "modular disintegration" acutely and then "re-integration" afterwards associated with therapeutic outcomes (Carhart-Harris et al. 2017). Post-acute changes in functional connectivity have also been found in healthy volunteers 1 day after administration of the classic psychedelic admixture ayahuasca, including increased connectivity within the salience network, decreased connectivity within the DMN, and greater connectivity between the salience network and DMN, with the latter showing association with acute affective changes (Pasquini et al. 2020). Although these findings are not completely consistent with prior post-acute functional connectivity data on psilocybin in depressed patients (Carhart-Harris et al. 2017), they do represent a relevant area for further study using psychedelics and functional neuroimaging in patients with AD, who have shown differential patterns of connectivity alterations related to neuropsychiatric symptoms (Balthazar et al. 2014).

Memory Effects Human studies of psychedelics' effects on memory have largely focused on performance during drug effects, with most finding acute, dose-dependent impairment under the influence of moderate or high doses of psychedelics such as psilocybin (Barrett et al. 2018), LSD (Jarvik et al. 1955; Pokorny et al.

2020), and ayahuasca (Bouso et al. 2013) on various memory and cognitive tasks (Healy 2021). These impairments have been demonstrated across a number of domains such as working memory (Bouso et al. 2013; Wittmann et al. 2007) and word recall (Barrett et al. 2018). However, acute changes in autobiographical memory during psychedelic effects have also been reported, suggesting LSD (Langs 1967) and psilocybin (Carhart-Harris et al. 2012) can facilitate recall and vividness of salient life memories, a potentially relevant mechanism for treatment of AD, which is known to entail episodic memory impairment (Tromp et al. 2015). If psychedelic administration has any long-term effects on human memory, data on persisting brain and mood effects (e.g., Barrett et al. 2020a) suggest that they may not resemble acute effects. However, post-acute effects of psychedelics on cognition and memory in clinical populations have yet to be rigorously studied.

Reducing Depression, Anxiety, and Existential Distress A major focus of recent research has been examining classic psychedelics' effects on mood and anxiety symptoms. Revisiting promising work from the earlier era of research on psychedelics (Grof et al. 1973; Richards et al. 1977), recent double-blind, controlled studies found a single moderate dose of the classic psychedelic psilocybin to produce clinically significant antidepressant effects and reduced anxiety in patients with life-threatening cancer diagnoses (Griffiths et al. 2016; Grob et al. 2011; Ross et al. 2016). In the largest of these contemporary trials, 51 cancer patients were administered a moderate to high dose of psilocybin under supportive conditions, with a majority showing therapeutic reductions in depression and anxiety and improved quality of life that persisted for 6 months (Griffiths et al. 2016). Additional pilot studies have found persisting anxiolytic effects of high-dose LSD treatment in patients with life-threatening illness (Gasser et al. 2014), as well as rapid, sustained antidepressant effects of psilocybin in patients with treatment-resistant major depression lasting 3 months (Carhart-Harris et al. 2016), and rapid antidepressant effects of ayahuasca lasting at least 7 days (Palhano-Fontes et al. 2019). Recent controlled trials have provided further support for antidepressant effects of psilocybin (Carhart-Harris et al. 2021; Davis et al. 2021).

One study of psilocybin in 27 individuals with major depression used a wait-list controlled design, randomizing participants to either an immediate treatment condition in which they received a moderate (20 mg/70 kg) and high (30 mg/70 kg) dose of psilocybin approximately 2 weeks apart with psychological support throughout, or to a control condition in which participants began an identical treatment after an 8-week delay period during which their mood was monitored. Twenty-four individuals completed the study, showing significantly greater decreases in GRID-Hamilton Depression Rating Scale (GRID-HAMD) scores in the immediate treatment group at 1 and 4 weeks after the second psilocybin session compared to the wait-list control group at corresponding timepoints (Davis et al. 2021). After receiving the psilocybin intervention, the wait-list control group also showed statistically significant decreases from baseline in GRID-HAMD and other depression and anxiety measures lasting 4 weeks after the second psilocybin session, with more than half the sample overall (54%) meeting criteria for remission of depression at 4 weeks post-treatment.

Another study used a double-blind, randomized comparative efficacy design to assess effects of two high doses (25 mg) of psilocybin approximately 3 weeks apart compared with 6 weeks of daily oral escitalopram, an approved selective serotonin reuptake inhibitor (SSRI) antidepressant medication in a sample of 59 participants with moderate to severe major depression (Carhart-Harris et al. 2021). Results found significant reductions in depressive symptoms in both groups, with participants who received psilocybin showing greater improvements overall. Although these improvements did not meet statistical significance for superiority of psilocybin in the primary outcome (i.e., Quick Inventory of Depressive Symptomatology-Self-Report), depression remission was found in 57% of participants in the psilocybin condition at the 6 week timepoint compared with 28% in the escitalopram condition, and secondary outcome measures also favored psilocybin, indicating two high doses of psilocybin are at least as effective – if not more so – in treating depression than 6 weeks of daily escitalopram.

Increased Wellbeing and Life Satisfaction A growing body of work has shown sustained well-being benefits after classic psychedelic administration across diverse samples, from healthy volunteers (Griffiths et al. 2008, 2018) and older long-term AIDS survivors (Anderson et al. 2020) to people with a range of health conditions including cancer-related distress (Agin-Liebes et al. 2020; Swift et al. 2017), alcohol dependence (Bogenschutz et al. 2018), nicotine dependence (Noorani et al. 2018), and major depression (Watts et al. 2017). In many cases, such persisting effects are correlated with enduring personality changes such as increased openness, as well as increased life satisfaction and overall well-being (Erritzoe et al. 2018; Griffiths et al. 2008; MacLean et al. 2011; Madsen et al. 2020; Schmid and Liechti 2018; Smigielski et al. 2019a). Although the mechanisms for post-acute alterations in personality, behavior, and well-being are still under investigation, they have been linked to acute psychoactive drug effects that include a sense of insight and meaning (Erritzoe et al. 2018; Griffiths et al. 2008; Smigielski et al. 2019a), spiritual or mystical-type effects characterized by a sense of oneness (Garcia-Romeu et al. 2014; MacLean et al. 2011; Schmid and Liechti 2018), and changes in 5-HT_{2A}R binding (Madsen et al. 2020) and brain network functional connectivity (Barrett et al. 2020a; Sampedro et al. 2017; Smigielski et al. 2019b). That classic psychedelics have shown these persisting benefits across such a wide range of individuals provides good impetus to study them in patients with AD who are known to suffer from substantial decrements to quality of life overall (Karttunen et al. 2011; Shin et al. 2005).

4 Rationale and Approaches for Researching Psychedelics in Patients with AD

The data presented above provide good evidence that for some patients with AD, classic psychedelics may provide potential therapeutic benefits worth exploring further. To this end, we are currently conducting a pilot study to examine the potential of psilocybin to treat neuropsychiatric symptoms (NPS) in patients with early-stage AD and MCI. The trial is the first to our knowledge using moderate (15 mg/70 kg) and high-dose (25 mg/70 kg) psilocybin in patients with early-stage AD or MCI and depressed mood (ClinicalTrials.gov NCT04123314). Because of potential risks in more advanced cases of AD which may include symptoms such as delusions or hallucinations that could be exacerbated by high-dose psychedelic administration (Scarmeas et al. 2005), this research is geared toward earlier phases of the disease, consistent with recommendations that "the field should explore whether the long prodromal phase of AD creates novel possibilities to maintain cellular functionality and brain homeostasis to postpone the phase of irreparable damage and decay" (Sala Frigerio and De Strooper 2016, p. 71).

Clinical research to date has found benefits related largely to higher dose administration of classic psychedelics (Anderson et al. 2020; Bogenschutz et al. 2015; Carhart-Harris et al. 2021; Davis et al. 2021; Griffiths et al. 2016; Johnson et al. 2014). These data also suggest mood, quality of life, and general well-being improvements associated with high-dose psychedelic administration as potential therapeutic targets for patients with AD. Other approaches may consider use of lower repeated dosing of classic psychedelics in this population (Family et al. 2020). Currently available data on psychedelic microdosing (using chronic sub-perceptual doses that are not profoundly psychoactive) have failed to demonstrate consistent benefits in controlled trials (Bershad et al. 2019; Family et al. 2020; Hutten et al. 2020). However, this area remains open for further investigation to expand our understanding of the possible benefits, risks, and mechanisms of psychedelic treatments in AD. Additionally, the potential of classic psychedelics to treat other neurodegenerative disorders represents another compelling direction for future research.

5 Conclusion

Classic psychedelics with psychoplastogenic properties have the potential to be a powerful tool in the treatment of early-stage AD or MCI. Their ability to encourage neuronal growth similar to BDNF, a key protein that MCI patients produce at reduced levels, could possibly slow or even reverse the effects of a disease characterized by neurodegeneration. Agents that selectively induce neural plasticity in the cerebral cortex via direct action on 5-HT_{2A}R, which are highly expressed in layer 5 pyramidal neurons of the cortex, represent an as yet uninvestigated pharmacological class in patients with AD. In sum, three converging biological pathways may be responsible for induced neural plasticity resulting in long-lasting and profound effects following psychedelic administration: 5-$HT_{2A}R$ upregulation of neocortical BDNF, amplification of AMPA receptor activity resulting in downstream activation of mTOR, and upregulated Arc protein expression (Fig. 1). These plasticity-promoting pathways could represent a novel disease modifying treatment approach to treat AD that selectively induces neural plasticity in key cognitive brain regions like the prefrontal cortex, that as a result of the disease are deficient in endogenous plasticity-promoting compounds like BDNF. In addition, classic psychedelics' antidepressant and anxiolytic effects could provide important inroads for promoting psychological benefits in patients struggling with AD and neuropsychiatric comorbidities such as depression and apathy.

Questions remain as to the primary therapeutic mechanisms underlying psychedelic-assisted treatments. Some propose that characteristic mystical-type or ego-dissolving subjective effects of high-dose psychedelics are necessary for psychological benefits, (Yaden and Griffiths 2021) and others posit purely biological mechanisms as necessary and sufficient to achieve lasting positive effects (Olson 2021). It is our contention that there may be truth to both. Neuroplasticity inducing and anti-inflammatory properties of classic psychedelics suggest the potential for purely biological therapeutic activity across several mechanisms, even at doses that would not produce strong psychoactive effects (Flanagan and Nichols 2018; Ly et al. 2018; Shao et al. 2021). Thus, low-dose psychedelic treatments could have specific applications that may not necessitate subjective effects, such as reducing brain atrophy in neurodegenerative conditions, or recent work showing persisting reductions in migraine after a single dose of psilocybin that were not associated with psychoactive effects (Schindler et al. 2021). However, for particular conditions like depression, anxiety, addictions, and existential distress, current evidence suggests the subjective effects of classic psychedelics play a pivotal role, likely driven by their ability to alter core cognitive, emotional, and self-referential processes that can facilitate therapeutic insight, catharsis, and behavior change (Garcia-Romeu et al. 2014; Griffiths et al. 2016; MacLean et al. 2011; Roseman et al. 2018). As such, we recommend continued research of both low- and high-dose psychedelic therapy approaches and to tailor treatments according to clinical target and population.

The present discussion aims to inform the nascent field of clinical psychedelic research in patients with AD (George and Hanson 2019; Vann Jones and O'Kelly 2020). The data presented here, along with ongoing pilot research, set the stage to examine psychedelic treatments as potential avenues to affect disease progression and to enhance well-being and quality of life for patients with AD. We believe this work is both timely and promising, and represents a viable path forward for development of novel therapeutics in AD.

Acknowledgements The authors would like to thank Matthew W. Johnson, Ph.D., Roland R. Griffiths, Ph.D., Frederick S. Barrett, Ph.D., William A. Richards, Ph.D., Mary Cosimano, M. S.W., Annie Umbricht, M.D., and Natalie Gukasyan, M.D. for their support in this work.

Declaration of Interest AGR serves as a scientific advisor to ETHA Natural Botanicals and NeonMind Biosciences.

Funding/Support Funding for this research was provided by the Pineapple Fund, CSP Fund, Heffter Research Institute, and by Tim Ferriss, Matt Mullenweg, Blake Mycoskie, Craig Nerenberg, and the Steven and Alexandra Cohen Foundation.

References

Agin-Liebes GI, Malone T, Yalch MM, Mennenga SE, Ponté KL, Guss J, Bossis AP, Grigsby J, Fischer S, Ross S (2020) Long-term follow-up of psilocybin-assisted psychotherapy for psychiatric and existential distress in patients with life-threatening cancer. J Psychopharmacol 34 (2):155–166

Alzheimer's Association (2021) Alzheimer's disease facts and figures. Alzheimers Dement 17 (3):327–406. https://doi.org/10.1002/alz.12328

Anderson BT, Danforth A, Daroff PR, Stauffer C, Ekman E, Agin-Liebes G, Trope A, Boden MT, Dilley PJ, Mitchell J, Woolley J (2020) Psilocybin-assisted group therapy for demoralized older long-term AIDS survivor men: an open-label safety and feasibility pilot study. EClinicalMedicine 27. https://doi.org/10.1016/j.eclinm.2020.100538

Arancibia S, Silhol M, Moulière F, Meffre J, Höllinger I, Maurice T, Tapia-Arancibia L (2008) Protective effect of BDNF against beta-amyloid induced neurotoxicity in vitro and in vivo in rats. Neurobiol Dis 31(3):316–326. https://doi.org/10.1016/j.nbd.2008.05.012

Ayton S (2021) Brain volume loss due to donanemab. Eur J Neurol 28(9):e67–e68. https://doi.org/10.1111/ene.15007

Aznar S, Hervig ME-S (2016) The 5-HT2A serotonin receptor in executive function: implications for neuropsychiatric and neurodegenerative diseases. Neurosci Biobehav Rev 64:63–82. https://doi.org/10.1016/j.neubiorev.2016.02.008

Balthazar MLF, Pereira FRS, Lopes TM, da Silva EL, Coan AC, Campos BM, Duncan NW, Stella F, Northoff G, Damasceno BP, Cendes F (2014) Neuropsychiatric symptoms in Alzheimer's disease are related to functional connectivity alterations in the salience network. Hum Brain Mapp 35(4):1237–1246. https://doi.org/10.1002/hbm.22248

Banerjee S, Hellier J, Romer R, Dewey M, Knapp M, Ballard C, Baldwin R, Bentham P, Fox C, Holmes C, Katona C, Livingston G, Lawton C, McCrae N, Moniz-Cook E, Murray J, Nurock J, Orrell M, O'Brien J et al (2013) Study of the use of antidepressants for depression in dementia: the HTA -SADD trial – a multicentre, randomised, double-blind, placebo-controlled trial of the clinical effectiveness and cost-effectiveness of sertraline and mirtazapine. Health Technol Assess 17(7):Article 7. http://www.hta.ac.uk/research/HTAjournal.shtml

Banning LCP, Ramakers IHGB, Rosenberg PB, Lyketsos CG, Leoutsakos J-MS (2021) Alzheimer's disease biomarkers as predictors of trajectories of depression and apathy in cognitively normal individuals, mild cognitive impairment, and Alzheimer's disease dementia. Int J Geriatr Psychiatry 36(1):224–234. https://doi.org/10.1002/gps.5418

Barrett FS, Carbonaro TM, Hurwitz E, Johnson MW, Griffiths RR (2018) Double-blind comparison of the two hallucinogens psilocybin and dextromethorphan: effects on cognition. Psychopharmacology 235(10):2915–2927. https://doi.org/10.1007/s00213-018-4981-x

Barrett FS, Doss MK, Sepeda ND, Pekar JJ, Griffiths RR (2020a) Emotions and brain function are altered up to one month after a single high dose of psilocybin. Sci Rep 10(1):1–14. https://doi.org/10.1038/s41598-020-59282-y

Barrett FS, Krimmel SR, Griffiths RR, Seminowicz DA, Mathur BN (2020b) Psilocybin acutely alters the functional connectivity of the claustrum with brain networks that support perception, memory, and attention. NeuroImage 218:116980. https://doi.org/10.1016/j.neuroimage.2020.116980

Bateman RJ, Xiong C, Benzinger TLS, Fagan AM, Goate A, Fox NC, Marcus DS, Cairns NJ, Xie X, Blazey TM, Holtzman DM, Santacruz A, Buckles V, Oliver A, Moulder K, Aisen PS, Ghetti B, Klunk WE, McDade E et al (2012) Clinical and biomarker changes in dominantly inherited Alzheimer's disease. N Engl J Med 367(9):795–804. https://doi.org/10.1056/NEJMoa1202753

Bershad AK, Schepers ST, Bremmer MP, Lee R, de Wit H (2019) Acute subjective and behavioral effects of microdoses of lysergic acid diethylamide in healthy human volunteers. Biol Psychiatry 86(10):792–800. https://doi.org/10.1016/j.biopsych.2019.05.019

Bloom GS (2014) Amyloid-β and tau: the trigger and bullet in Alzheimer disease pathogenesis. JAMA Neurol 71(4):505–508. https://doi.org/10.1001/jamaneurol.2013.5847

Bogenschutz MP, Forcehimes AA, Pommy JA, Wilcox CE, Barbosa PCR, Strassman RJ (2015) Psilocybin-assisted treatment for alcohol dependence: a proof-of-concept study. J Psychopharmacol 29(3):289–299

Bogenschutz MP, Podrebarac SK, Duane JH, Amegadzie SS, Malone TC, Owens LT, Ross S, Mennenga SE (2018) Clinical interpretations of patient experience in a trial of psilocybin-assisted psychotherapy for alcohol use disorder. Front Pharmacol 9. https://doi.org/10.3389/fphar.2018.00100

Boulougouris V, Glennon JC, Robbins TW (2008) Dissociable effects of selective 5-HT 2A and 5-HT 2C receptor antagonists on serial spatial reversal learning in rats. Neuropsychopharmacology 33(8):2007–2019. https://doi.org/10.1038/sj.npp.1301584

Bouso JC, Fábregas JM, Antonijoan RM, Rodríguez-Fornells A, Riba J (2013) Acute effects of ayahuasca on neuropsychological performance: differences in executive function between experienced and occasional users. Psychopharmacology 230(3):415–424. https://doi.org/10.1007/s00213-013-3167-9

Braak H, Braak E (1991) Neuropathological stageing of Alzheimer-related changes. Acta Neuropathol 82(4):239–259. https://doi.org/10.1007/BF00308809

Brier MR, Gordon B, Friedrichsen K, McCarthy J, Stern A, Christensen J, Owen C, Aldea P, Su Y, Hassenstab J, Cairns NJ, Holtzman DM, Fagan AM, Morris JC, Benzinger TLS, Ances BM (2016) Tau and Aβ imaging, CSF measures, and cognition in Alzheimer's disease. Sci Transl Med 8(338):338ra66. https://doi.org/10.1126/scitranslmed.aaf2362

Buckner RL, Snyder AZ, Shannon BJ, LaRossa G, Sachs R, Fotenos AF, Sheline YI, Klunk WE, Mathis CA, Morris JC, Mintun MA (2005) Molecular, structural, and functional characterization of Alzheimer's disease: evidence for a relationship between default activity, amyloid, and memory. J Neurosci 25(34):7709–7717. https://doi.org/10.1523/JNEUROSCI.2177-05.2005

Butzlaff M, Ponimaskin E (2016) The role of serotonin receptors in Alzheimer's disease. Opera Med Physiol 2(1):77–86

Caccamo A, Magrì A, Medina DX, Wisely EV, López-Aranda MF, Silva AJ, Oddo S (2013) mTOR regulates tau phosphorylation and degradation: implications for Alzheimer's disease and other tauopathies. Aging Cell 12(3):370–380. https://doi.org/10.1111/acel.12057

Cammalleri M, Lütjens R, Berton F, King AR, Simpson C, Francesconi W, Sanna PP (2003) Time-restricted role for dendritic activation of the mTOR-p70S6K pathway in the induction of late-phase long-term potentiation in the CA1. Proc Natl Acad Sci 100(24):14368–14373. https://doi.org/10.1073/pnas.2336098100

Carhart-Harris RL, Leech R, Williams TM, Erritzoe D, Abbasi N, Bargiotas T, Hobden P, Sharp DJ, Evans J, Feilding A, Wise RG, Nutt DJ (2012) Implications for psychedelic-assisted psychotherapy: functional magnetic resonance imaging study with psilocybin. Br J Psychiatry 200(3):238–244. https://doi.org/10.1192/bjp.bp.111.103309

Carhart-Harris RL, Bolstridge M, Rucker J, Day CMJ, Erritzoe D, Kaelen M, Bloomfield M, Rickard JA, Forbes B, Feilding A, Taylor D, Pilling S, Curran VH, Nutt DJ (2016) Psilocybin with psychological support for treatment-resistant depression: an open-label feasibility study. Lancet Psychiatry 3(7):619–627. https://doi.org/10.1016/S2215-0366(16)30065-7

Carhart-Harris RL, Roseman L, Bolstridge M, Demetriou L, Pannekoek JN, Wall MB, Tanner M, Kaelen M, McGonigle J, Murphy K, Leech R, Curran HV, Nutt DJ (2017) Psilocybin for treatment-resistant depression: FMRI-measured brain mechanisms. Sci Rep 7(1):13187. https://doi.org/10.1038/s41598-017-13282-7

Carhart-Harris R, Giribaldi B, Watts R, Baker-Jones M, Murphy-Beiner A, Murphy R, Martell J, Blemings A, Erritzoe D, Nutt DJ (2021) Trial of psilocybin versus escitalopram for depression. N Engl J Med 384(15):1402–1411

Carter OL, Burr DC, Pettigrew JD, Wallis GM, Hasler F, Vollenweider FX (2005) Using psilocybin to investigate the relationship between attention, working memory, and the serotonin 1A and 2A receptors. J Cogn Neurosci 17(10):1497–1508. https://doi.org/10.1162/089892905774597191

Catlow BJ, Song S, Paredes DA, Kirstein CL, Sanchez-Ramos J (2013) Effects of psilocybin on hippocampal neurogenesis and extinction of trace fear conditioning. Exp Brain Res 228 (4):481–491. https://doi.org/10.1007/s00221-013-3579-0

Cavalleri L, Merlo Pich E, Millan MJ, Chiamulera C, Kunath T, Spano PF, Collo G (2018) Ketamine enhances structural plasticity in mouse mesencephalic and human iPSC-derived dopaminergic neurons via AMPAR-driven BDNF and mTOR signaling. Mol Psychiatry 23 (4):812–823. https://doi.org/10.1038/mp.2017.241

Chakraborty S, Lennon JC, Malkaram SA, Zeng Y, Fisher DW, Dong H (2019) Serotonergic system, cognition, and BPSD in Alzheimer's disease. Neurosci Lett 704:36–44. https://doi.org/10.1016/j.neulet.2019.03.050

Chen M-K, Mecca AP, Naganawa M, Finnema SJ, Toyonaga T, Lin S, Najafzadeh S, Ropchan J, Lu Y, McDonald JW, Michalak HR, Nabulsi NB, Arnsten AFT, Huang Y, Carson RE, van Dyck CH (2018) Assessing synaptic density in Alzheimer disease with synaptic vesicle glycoprotein 2A positron emission tomographic imaging. JAMA Neurol 75(10):1215. https://doi.org/10.1001/jamaneurol.2018.1836

Ciaramella A, Salani F, Bizzoni F, Orfei MD, Langella R, Angelucci F, Spalletta G, Taddei AR, Caltagirone C, Bossù P (2013) The stimulation of dendritic cells by amyloid beta 1–42 reduces BDNF production in Alzheimer's disease patients. Brain Behav Immun 32:29–32. https://doi.org/10.1016/j.bbi.2013.04.001

Davis AK, Barrett FS, May DG, Cosimano MP, Sepeda ND, Johnson MW, Finan PH, Griffiths RR (2021) Effects of psilocybin-assisted therapy on major depressive disorder: a randomized clinical trial. JAMA Psychiat 78(5):481–489

de Quervain DJ-F, Henke K, Aerni A, Coluccia D, Wollmer MA, Hock C, Nitsch RM, Papassotiropoulos A (2003) A functional genetic variation of the 5-HT2a receptor affects human memory. Nat Neurosci 6(11):1141–1142. https://doi.org/10.1038/nn1146

Dennis EL, Thompson PM (2014) Functional brain connectivity using fMRI in aging and Alzheimer's disease. Neuropsychol Rev 24(1):49–62. https://doi.org/10.1007/s11065-014-9249-6

Doggrell SA (2021) Still grasping at straws: donanemab in Alzheimer's disease. Expert Opin Investig Drugs 30(8):797–801. https://doi.org/10.1080/13543784.2021.1948010

Donovan LL, Johansen JV, Ros NF, Jaberi E, Linnet K, Johansen SS, Ozenne B, Issazadeh-Navikas S, Hansen HD, Knudsen GM (2021) Effects of a single dose of psilocybin on behaviour, brain 5-HT2A receptor occupancy and gene expression in the pig. Eur Neuropsychopharmacol 42:1–11. https://doi.org/10.1016/j.euroneuro.2020.11.013

Dotson VM, Beydoun MA, Zonderman AB (2010) Recurrent depressive symptoms and the incidence of dementia and mild cognitive impairment. Neurology 75(1):27–34. https://doi.org/10.1212/WNL.0b013e3181e62124

Dubois B, Feldman HH, Jacova C, Hampel H, Molinuevo JL, Blennow K, DeKosky ST, Gauthier S, Selkoe D, Bateman R, Cappa S, Crutch S, Engelborghs S, Frisoni GB, Fox NC, Galasko D, Habert M-O, Jicha GA, Nordberg A et al (2014) Advancing research diagnostic criteria for Alzheimer's disease: the IWG-2 criteria. Lancet Neurol 13(6):614–629. https://doi.org/10.1016/S1474-4422(14)70090-0

Eide FF, Vining ER, Eide BL, Zang K, Wang X-Y, Reichardt LF (1996) Naturally occurring truncated trkB receptors have dominant inhibitory effects on brain-derived neurotrophic factor signaling. J Neurosci 16(10):3123–3129. https://doi.org/10.1523/JNEUROSCI.16-10-03123.1996

Erritzoe D, Roseman L, Nour MM, MacLean K, Kaelen M, Nutt DJ, Carhart-Harris RL (2018) Effects of psilocybin therapy on personality structure. Acta Psychiatr Scand 138(5):368–378. https://doi.org/10.1111/acps.12904

Family N, Maillet EL, Williams LTJ, Krediet E, Carhart-Harris RL, Williams TM, Nichols CD, Goble DJ, Raz S (2020) Safety, tolerability, pharmacokinetics, and pharmacodynamics of low dose lysergic acid diethylamide (LSD) in healthy older volunteers. Psychopharmacology 237 (3):841–853. https://doi.org/10.1007/s00213-019-05417-7

Flanagan TW, Nichols CD (2018) Psychedelics as anti-inflammatory agents. Int Rev Psychiatry 30 (4):363–375. https://doi.org/10.1080/09540261.2018.1481827

Frick LR, Bernardez-Vidal M, Hocht C, Zanutto BS, Rapanelli M (2015) Dual role of serotonin in the acquisition and extinction of reward-driven learning: involvement of 5-HT1A, 5-HT2A and 5-HT3 receptors. Behav Brain Res 277:193–203. https://doi.org/10.1016/j.bbr.2014.06.025

Furr A, Lapiz-Bluhm MD, Morilak DA (2012) 5-HT2A receptors in the orbitofrontal cortex facilitate reversal learning and contribute to the beneficial cognitive effects of chronic citalopram treatment in rats. Int J Neuropsychopharmacol 15(9):1295–1305. https://doi.org/10.1017/S1461145711001441

Galvão-Coelho NL, de Menezes Galvão AC, de Almeida RN, Palhano-Fontes F, Campos Braga I, Lobão Soares B, Maia-de-Oliveira JP, Perkins D, Sarris J, de Araujo DB (2020) Changes in inflammatory biomarkers are related to the antidepressant effects of Ayahuasca. J Psychopharmacol 34(10):1125–1133

Garcia-Romeu A, Griffiths RR, Johnson MW (2014) Psilocybin-occasioned mystical experiences in the treatment of tobacco addiction. Curr Drug Abuse Rev 7(3):157–164

Garcia-Romeu A, Kersgaard B, Addy PH (2016) Clinical applications of hallucinogens: a review. Exp Clin Psychopharmacol 24(4):229

Garwood CJ, Pooler AM, Atherton J, Hanger DP, Noble W (2011) Astrocytes are important mediators of a β-induced neurotoxicity and tau phosphorylation in primary culture. Cell Death Dis 2(6):e167–e167

Gasser P, Holstein D, Michel Y, Doblin R, Yazar-Klosinski B, Passie T, Brenneisen R (2014) Safety and efficacy of lysergic acid diethylamide-assisted psychotherapy for anxiety associated with life-threatening diseases. J Nerv Ment Dis 202(7):513

George DR, Hanson R (2019) Imagining a role for psychedelics in dementia care. Am J Geriatr Psychiatry 27(9):1028–1030. https://doi.org/10.1016/j.jagp.2019.03.008

González-Maeso J, Weisstaub NV, Zhou M, Chan P, Ivic L, Ang R, Lira A, Bradley-Moore M, Ge Y, Zhou Q, Sealfon SC, Gingrich JA (2007) Hallucinogens recruit specific cortical 5-HT2A receptor-mediated signaling pathways to affect behavior. Neuron 53(3):439–452. https://doi.org/10.1016/j.neuron.2007.01.008

Graff-Radford J, Yong KXX, Apostolova LG, Bouwman FH, Carrillo M, Dickerson BC, Rabinovici GD, Schott JM, Jones DT, Murray ME (2021) New insights into atypical Alzheimer's disease in the era of biomarkers. Lancet Neurol 20(3):222–234. https://doi.org/10.1016/S1474-4422(20)30440-3

Griffiths RR, Richards WA, Johnson MW, McCann UD, Jesse R (2008) Mystical-type experiences occasioned by psilocybin mediate the attribution of personal meaning and spiritual significance 14 months later. J Psychopharmacol 22(6):621–632

Griffiths RR, Johnson MW, Carducci MA, Umbricht A, Richards WA, Richards BD, Cosimano MP, Klinedinst MA (2016) Psilocybin produces substantial and sustained decreases in depression and anxiety in patients with life-threatening cancer: a randomized double-blind trial. J Psychopharmacol 30(12):1181–1197. https://doi.org/10.1177/0269881116675513

Griffiths RR, Johnson MW, Richards WA, Richards BD, Jesse R, MacLean KA, Barrett FS, Cosimano MP, Klinedinst MA (2018) Psilocybin-occasioned mystical-type experience in combination with meditation and other spiritual practices produces enduring positive changes in psychological functioning and in trait measures of prosocial attitudes and behaviors. J Psychopharmacol 32(1):49–69

Grob CS, Danforth AL, Chopra GS, Hagerty M, McKay CR, Halberstadt AL, Greer GR (2011) Pilot study of psilocybin treatment for anxiety in patients with advanced-stage cancer. Arch Gen Psychiatry 68(1):71. https://doi.org/10.1001/archgenpsychiatry.2010.116

Grof S, Goodman LE, Richards WA, Kurland AA (1973) LSD-assisted psychotherapy in patients with terminal cancer. Int Pharmacopsychiatry 8:129–144. https://doi.org/10.1159/000467984

Grothe M, Heinsen H, Teipel SJ (2012) Atrophy of the cholinergic basal forebrain over the adult age range and in early stages of Alzheimer's disease. Biol Psychiatry 71(9):805–813. https://doi.org/10.1016/j.biopsych.2011.06.019

Hafkemeijer A, van der Grond J, Rombouts SARB (2012) Imaging the default mode network in aging and dementia. Biochim Biophys Acta 1822(3):431–441. https://doi.org/10.1016/j.bbadis.2011.07.008

Harvey JA, Quinn JL, Liu R, Aloyo VJ, Romano AG (2004) Selective remodeling of rabbit frontal cortex: relationship between 5-HT2A receptor density and associative learning. Psychopharmacology 172(4):435–442. https://doi.org/10.1007/s00213-003-1687-4

Hasselbalch SG, Madsen K, Svarer C, Pinborg LH, Holm S, Paulson OB, Waldemar G, Knudsen GM (2008) Reduced 5-HT2A receptor binding in patients with mild cognitive impairment. Neurobiol Aging 29(12):1830–1838. https://doi.org/10.1016/j.neurobiolaging.2007.04.011

He X, Qin W, Liu Y, Zhang X, Duan Y, Song J, Li K, Jiang T, Yu C (2014) Abnormal salience network in normal aging and in amnestic mild cognitive impairment and Alzheimer's disease. Hum Brain Mapp 35(7):3446–3464. https://doi.org/10.1002/hbm.22414

He W, Goodkind D, Kowal P (2016) International population reports, P95/16-1, An aging world: 2015, U.S. Census Bureau. U.S. Government Publishing Office, Washington, DC

Healy CJ (2021) The acute effects of classic psychedelics on memory in humans. Psychopharmacology 238(3):639–653. https://doi.org/10.1007/s00213-020-05756-w

Herbert J, Lucassen PJ (2016) Depression as a risk factor for Alzheimer's disease: genes, steroids, cytokines and neurogenesis – what do we need to know? Front Neuroendocrinol 41:153–171. https://doi.org/10.1016/j.yfrne.2015.12.001

Hoeffer CA, Klann E (2010) mTOR signaling: at the crossroads of plasticity, memory and disease. Trends Neurosci 33(2):67–75. https://doi.org/10.1016/j.tins.2009.11.003

Holm P, Ettrup A, Klein AB, Santini MA, El-Sayed M, Elvang AB, Stensbøl TB, Mikkelsen JD, Knudsen GM, Aznar S (2010) Plaque deposition dependent decrease in 5-HT 2A serotonin receptor in AβPPswe/PS1dE9 amyloid overexpressing mice. J Alzheimers Dis 20 (4):1201–1213. https://doi.org/10.3233/JAD-2010-100117

Hutten NRPW, Mason NL, Dolder PC, Theunissen EL, Holze F, Liechti ME, Feilding A, Ramaekers JG, Kuypers KPC (2020) Mood and cognition after administration of low LSD doses in healthy volunteers: a placebo controlled dose-effect finding study. Eur Neuropsychopharmacol 41:81–91. https://doi.org/10.1016/j.euroneuro.2020.10.002

Hutten NRPW, Mason NL, Dolder PC, Theunissen EL, Holze F, Liechti ME, Varghese N, Eckert A, Feilding A, Ramaekers JG, Kuypers KPC (2021) Low doses of LSD acutely increase BDNF blood plasma levels in healthy volunteers. ACS Pharmacol Transl Sci 4(2):461–466. https://doi.org/10.1021/acsptsci.0c00099

Ismail Z, Agüera-Ortiz L, Brodaty H, Cieslak A, Cummings J, Fischer CE, Gauthier S, Geda YE, Herrmann N, Kanji J, Lanctôt KL, Miller DS, Mortby ME, Onyike CU, Rosenberg PB, Smith EE, Smith GS, Sultzer DL, Lyketsos C, for the N. P. I. A. of the I. S. of to A. A. R. and T. (NPS-P. of ISTAART) (2017) The mild behavioral impairment checklist (MBI-C): a rating scale for neuropsychiatric symptoms in pre-dementia populations. J Alzheimers Dis 56 (3):929–938. https://doi.org/10.3233/JAD-160979

Jack CR Jr, Wiste HJ, Botha H, Weigand SD, Therneau TM, Knopman DS, Graff-Radford J, Jones DT, Ferman TJ, Boeve BF (2019) The bivariate distribution of amyloid-β and tau: relationship with established neurocognitive clinical syndromes. Brain 142(10):3230–3242

Jarvik ME, Abramson HA, Hirsch MW (1955) Lysergic acid diethylamide (LSD-25): VI. Effect upon recall and recognition of various stimuli. J Psychol 39(2):443–454. https://doi.org/10.1080/00223980.1955.9916194

Jaworski J, Sheng M (2006) The growing role of mTOR in neuronal development and plasticity. Mol Neurobiol 34(3):205–219. https://doi.org/10.1385/MN:34:3:205

Jefsen OH, Elfving B, Wegener G, Müller HK (2021) Transcriptional regulation in the rat prefrontal cortex and hippocampus after a single administration of psilocybin. J Psychopharmacol 35 (4):483–493. https://doi.org/10.1177/0269881120959614

Johnson MW, Richards WA, Griffiths RR (2008) Human hallucinogen research: guidelines for safety. J Psychopharmacol 22(6):603–620

Johnson MW, Garcia-Romeu A, Cosimano MP, Griffiths RR (2014) Pilot study of the 5-HT2AR agonist psilocybin in the treatment of tobacco addiction. J Psychopharmacol 28(11):983–992

Johnson MW, Griffiths RR, Hendricks PS, Henningfield JE (2018) The abuse potential of medical psilocybin according to the 8 factors of the controlled substances act. Neuropharmacology 142:143–166

Joie RL, Perrotin A, Barré L, Hommet C, Mézenge F, Ibazizene M, Camus V, Abbas A, Landeau B, Guilloteau D, Sayette VL, Eustache F, Desgranges B, Chételat G (2012) Region-specific hierarchy between atrophy, hypometabolism, and β-amyloid (Aβ) load in Alzheimer's disease dementia. J Neurosci 32(46):16265–16273. https://doi.org/10.1523/JNEUROSCI.2170-12.2012

Karran E, Hardy J (2014) A critique of the drug discovery and phase 3 clinical programs targeting the amyloid hypothesis for Alzheimer disease. Ann Neurol 76(2):185–205. https://doi.org/10.1002/ana.24188

Karran E, Mercken M, Strooper BD (2011) The amyloid cascade hypothesis for Alzheimer's disease: an appraisal for the development of therapeutics. Nat Rev Drug Discov 10 (9):698–712. https://doi.org/10.1038/nrd3505

Karttunen K, Karppi P, Hiltunen A, Vanhanen M, Välimäki T, Martikainen J, Valtonen H, Sivenius J, Soininen H, Hartikainen S, Suhonen J, Pirttilä T (2011) Neuropsychiatric symptoms and quality of life in patients with very mild and mild Alzheimer's disease. Int J Geriatr Psychiatry 26(5):473–482. https://doi.org/10.1002/gps.2550

Killin LOJ, Starr JM, Shiue IJ, Russ TC (2016) Environmental risk factors for dementia: a systematic review. BMC Geriatr 16(1):175. https://doi.org/10.1186/s12877-016-0342-y

Kim SYH (2011) The ethics of informed consent in Alzheimer disease research. Nat Rev Neurol 7 (7):410–414. https://doi.org/10.1038/nrneurol.2011.76

King AR, Martin IL, Seymour KA (1972) Reversal learning facilitated by a single injection of lysergic acid diethylamide (LSD 25) in the rat. Br J Pharmacol 45(1):161P–162P

Kinney JW, Bemiller SM, Murtishaw AS, Leisgang AM, Salazar AM, Lamb BT (2018) Inflammation as a central mechanism in Alzheimer's disease. Alzheimer's Dementia Transl Res Clin Interv 4:575–590. https://doi.org/10.1016/j.trci.2018.06.014

Klaassens BL, van Gerven J, van der Grond J, de Vos F, Möller C, Rombouts SA (2017) Diminished posterior precuneus connectivity with the default mode network differentiates normal aging from Alzheimer's disease. Front Aging Neurosci 9:97

Knopman DS, Jones DT, Greicius MD (2021) Failure to demonstrate efficacy of aducanumab: an analysis of the EMERGE and ENGAGE trials as reported by Biogen, December 2019. Alzheimers Dement 17(4):696–701. https://doi.org/10.1002/alz.12213

Kunkle BW, Grenier-Boley B, Sims R, Bis JC, Damotte V, Naj AC, Boland A, Vronskaya M, van der Lee SJ, Amlie-Wolf A, Bellenguez C, Frizatti A, Chouraki V, Martin ER, Sleegers K, Badarinarayan N, Jakobsdottir J, Hamilton-Nelson KL, Moreno-Grau S et al (2019) Genetic meta-analysis of diagnosed Alzheimer's disease identifies new risk loci and implicates Aβ, tau, immunity and lipid processing. Nat Genet 51(3):414–430. https://doi.org/10.1038/s41588-019-0358-2

Lai MK, Tsang SW, Alder JT, Keene J, Hope T, Esiri MM, Francis PT, Chen CP (2005) Loss of serotonin 5-HT2A receptors in the postmortem temporal cortex correlates with rate of cognitive decline in Alzheimer's disease. Psychopharmacology 179(3):673–677. https://doi.org/10.1007/s00213-004-2077-2

Lanctôt KL, Amatniek J, Ancoli-Israel S, Arnold SE, Ballard C, Cohen-Mansfield J, Ismail Z, Lyketsos C, Miller DS, Musiek E, Osorio RS, Rosenberg PB, Satlin A, Steffens D, Tariot P, Bain LJ, Carrillo MC, Hendrix JA, Jurgens H, Boot B (2017) Neuropsychiatric signs and symptoms of Alzheimer's disease: new treatment paradigms. Alzheimer's Dementia Transl Res Clin Interv 3(3):440–449. https://doi.org/10.1016/j.trci.2017.07.001

Langs RJ (1967) Stability of earliest memories under LSD-25 AND PLACEBO. J Nerv Ment Dis 144(3):171–184

Leoutsakos J-MS, Forrester SN, Lyketsos CG, Smith GS (2015) Latent classes of neuropsychiatric symptoms in NACC controls and conversion to mild cognitive impairment or dementia. J Alzheimers Dis 48(2):483–493. https://doi.org/10.3233/JAD-150421

Li L-B, Zhang L, Sun Y-N, Han L-N, Wu Z-H, Zhang Q-J, Liu J (2015) Activation of serotonin2A receptors in the medial septum-diagonal band of Broca complex enhanced working memory in the hemiparkinsonian rats. Neuropharmacology 91:23–33. https://doi.org/10.1016/j.neuropharm.2014.11.025

Lima da Cruz RV, Moulin TC, Petiz LL, Leão RN (2018) A single dose of 5-MeO-DMT stimulates cell proliferation, neuronal survivability, morphological and functional changes in adult mice ventral dentate gyrus. Front Mol Neurosci 11. https://doi.org/10.3389/fnmol.2018.00312

Liu Y, Yoo M-J, Savonenko A, Stirling W, Price DL, Borchelt DR, Mamounas L, Lyons WE, Blue ME, Lee MK (2008) Amyloid pathology is associated with progressive monoaminergic neurodegeneration in a transgenic mouse model of Alzheimer's disease. J Neurosci 28 (51):13805–13814. https://doi.org/10.1523/JNEUROSCI.4218-08.2008

Ly C, Greb AC, Cameron LP, Wong JM, Barragan EV, Wilson PC, Burbach KF, Soltanzadeh Zarandi S, Sood A, Paddy MR, Duim WC, Dennis MY, McAllister AK, Ori-McKenney KM, Gray JA, Olson DE (2018) Psychedelics promote structural and functional neural plasticity. Cell Rep 23(11):3170–3182. https://doi.org/10.1016/j.celrep.2018.05.022

Lyketsos CG, Lopez O, Jones B, Fitzpatrick AL, Breitner J, DeKosky S (2002) Prevalence of neuropsychiatric symptoms in dementia and mild cognitive impairment: results from the cardiovascular health study. JAMA 288(12):1475. https://doi.org/10.1001/jama.288.12.1475

Lyketsos CG, Colenda CC, Beck C, Blank K, Doraiswamy MP, Kalunian DA, Yaffe K (2006) Position statement of the American Association for Geriatric Psychiatry regarding principles of care for patients with dementia resulting from Alzheimer Disease. Am J Geriatr Psychiatry 14 (7):561–573. https://doi.org/10.1097/01.JGP.0000221334.65330.55

Lyketsos CG, Carrillo MC, Ryan JM, Khachaturian AS, Trzepacz P, Amatniek J, Cedarbaum J, Brashear R, Miller DS (2011) Neuropsychiatric symptoms in Alzheimer's disease. Alzheimers Dement 7(5):532–539. https://doi.org/10.1016/j.jalz.2011.05.2410

MacLean KA, Johnson MW, Griffiths RR (2011) Mystical experiences occasioned by the hallucinogen psilocybin lead to increases in the personality domain of openness. J Psychopharmacol 25 (11):1453–1461

Madsen MK, Fisher PM, Stenbæk DS, Kristiansen S, Burmester D, Lehel S, Páleníček T, Kuchař M, Svarer C, Ozenne B, Knudsen GM (2020) A single psilocybin dose is associated with long-term increased mindfulness, preceded by a proportional change in neocortical 5-HT2A receptor binding. Eur Neuropsychopharmacol 33:71–80. https://doi.org/10.1016/j.euroneuro.2020.02.001

Marner L, Frokjaer VG, Kalbitzer J, Lehel S, Madsen K, Baaré WF, Knudsen GM, Hasselbalch SG (2012) Loss of serotonin 2A receptors exceeds loss of serotonergic projections in early Alzheimer's disease: a combined [11C] DASB and [18F] altanserin-PET study. Neurobiol Aging 33(3):479–487

Marucci G, Buccioni M, Ben DD, Lambertucci C, Volpini R, Amenta F (2021) Efficacy of acetylcholinesterase inhibitors in Alzheimer's disease. Neuropharmacology 190:108352. https://doi.org/10.1016/j.neuropharm.2020.108352

Masliah E, Mallory M, Alford M, DeTeresa R, Hansen LA, McKeel DW, Morris JC (2001) Altered expression of synaptic proteins occurs early during progression of Alzheimer's disease. Neurology 56(1):127–129. https://doi.org/10.1212/WNL.56.1.127

Mason NL, Kuypers KPC, Müller F, Reckweg J, Tse DHY, Toennes SW, Hutten NRPW, Jansen JFA, Stiers P, Feilding A, Ramaekers JG (2020) Me, myself, bye: regional alterations in glutamate and the experience of ego dissolution with psilocybin. Neuropsychopharmacology 45(12):2003–2011. https://doi.org/10.1038/s41386-020-0718-8

Mecca AP (2019) 1ciii - AD molecular: molecular imaging of Alzheimer's disease: PET imaging of neurotransmitter systems. In: Becker JT, Cohen AD (eds) Progress in molecular biology and translational science, vol vol 165. Academic Press, pp 139–165. https://doi.org/10.1016/bs.pmbts.2019.04.003

Mevel K, Chételat G, Eustache F, Desgranges B (2011) The default mode network in healthy aging and Alzheimer's disease. Int J Alzheimers Dis 2011:e535816. https://doi.org/10.4061/2011/535816

Michaelsen K, Zagrebelsky M, Berndt-Huch J, Polack M, Buschler A, Sendtner M, Korte M (2010) Neurotrophin receptors TrkB.T1 and p75NTR cooperate in modulating both functional and structural plasticity in mature hippocampal neurons. Eur J Neurosci 32(11):1854–1865. https://doi.org/10.1111/j.1460-9568.2010.07460.x

Miguel-Álvarez M, Santos-Lozano A, Sanchis-Gomar F, Fiuza-Luces C, Pareja-Galeano H, Garatachea N, Lucia A (2015) Non-steroidal anti-inflammatory drugs as a treatment for Alzheimer's disease: a systematic review and meta-analysis of treatment effect. Drugs Aging 32(2):139–147

Minichiello L (2009) TrkB signalling pathways in LTP and learning. Nat Rev Neurosci 10 (12):850–860. https://doi.org/10.1038/nrn2738

Mintun MA, Lo AC, Duggan Evans C, Wessels AM, Ardayfio PA, Andersen SW, Shcherbinin S, Sparks J, Sims JR, Brys M, Apostolova LG, Salloway SP, Skovronsky DM (2021) Donanemab in early Alzheimer's disease. N Engl J Med 384(18):1691–1704. https://doi.org/10.1056/NEJMoa2100708

Nau F Jr, Yu B, Martin D, Nichols CD (2013) Serotonin 5-HT 2A receptor activation blocks TNF-α mediated inflammation in vivo. PLoS One 8(10):e75426

Nelson JC, Devanand DP (2011) A systematic review and meta-analysis of placebo-controlled antidepressant studies in people with depression and dementia. J Am Geriatr Soc 59 (4):577–585. https://doi.org/10.1111/j.1532-5415.2011.03355.x

Nichols DE (2016) Psychedelics. Pharmacol Rev 68(2):264–355. https://doi.org/10.1124/pr.115.011478

Nichols DE (2020) Psilocybin: from ancient magic to modern medicine. J Antibiot 73(10):679–686. https://doi.org/10.1038/s41429-020-0311-8

Nichols CD, Sanders-Bush E (2002) A single dose of lysergic acid diethylamide influences gene expression patterns within the mammalian brain. Neuropsychopharmacology 26(5):634–642. https://doi.org/10.1016/S0893-133X(01)00405-5

Nichols CD, Garcia EE, Sanders-Bush E (2003) Dynamic changes in prefrontal cortex gene expression following lysergic acid diethylamide administration. Mol Brain Res 111 (1):182–188. https://doi.org/10.1016/S0169-328X(03)00029-9

Noorani T, Garcia-Romeu A, Swift TC, Griffiths RR, Johnson MW (2018) Psychedelic therapy for smoking cessation: qualitative analysis of participant accounts. J Psychopharmacol 32 (7):756–769. https://doi.org/10.1177/0269881118780612

Olson DE (2018) Psychoplastogens: a promising class of plasticity-promoting neurotherapeutics. J Exp Neurosci 12:1179069518800508. https://doi.org/10.1177/1179069518800508

Olson DE (2021) The subjective effects of psychedelics may not be necessary for their enduring therapeutic effects. ACS Pharmacol Transl Sci 4(2):563–567. https://doi.org/10.1021/acsptsci.0c00192

Ossenkoppele R, Schonhaut DR, Schöll M, Lockhart SN, Ayakta N, Baker SL, O'Neil JP, Janabi M, Lazaris A, Cantwell A, Vogel J, Santos M, Miller ZA, Bettcher BM, Vossel KA,

Kramer JH, Gorno-Tempini ML, Miller BL, Jagust WJ, Rabinovici GD (2016) Tau PET patterns mirror clinical and neuroanatomical variability in Alzheimer's disease. Brain 139 (5):1551–1567. https://doi.org/10.1093/brain/aww027

Ozben T, Ozben S (2019) Neuro-inflammation and anti-inflammatory treatment options for Alzheimer's disease. Clin Biochem 72:87–89. https://doi.org/10.1016/j.clinbiochem.2019.04.001

Palhano-Fontes F, Barreto D, Onias H, Andrade KC, Novaes MM, Pessoa JA, Mota-Rolim SA, Osório FL, Sanches R, dos Santos RG, Tófoli LF, Silveira GO, Yonamine M, Riba J, Santos FR, Silva-Junior AA, Alchieri JC, Galvão-Coelho NL, Lobão-Soares B et al (2019) Rapid antidepressant effects of the psychedelic ayahuasca in treatment-resistant depression: a randomized placebo-controlled trial. Psychol Med 49(4):655–663. https://doi.org/10.1017/S0033291718001356

Palmqvist S, Schöll M, Strandberg O, Mattsson N, Stomrud E, Zetterberg H, Blennow K, Landau S, Jagust W, Hansson O (2017) Earliest accumulation of β-amyloid occurs within the default-mode network and concurrently affects brain connectivity. Nat Commun 8(1):1214. https://doi.org/10.1038/s41467-017-01150-x

Palop JJ, Mucke L (2016) Network abnormalities and interneuron dysfunction in Alzheimer disease. Nat Rev Neurosci 17(12):777–792. https://doi.org/10.1038/nrn.2016.141

Pasquini L, Rahmani F, Maleki-Balajoo S, La Joie R, Zarei M, Sorg C, Drzezga A, Tahmasian M (2019) Medial temporal lobe disconnection and hyperexcitability across Alzheimer's disease stages. J Alzheimer's Dis Rep 3(1):103–112. https://doi.org/10.3233/ADR-190121

Pasquini L, Palhano-Fontes F, Araujo DB (2020) Subacute effects of the psychedelic ayahuasca on the salience and default mode networks. J Psychopharmacol 34(6):623–635. https://doi.org/10.1177/0269881120909409

Peng S, Wuu J, Mufson EJ, Fahnestock M (2005) Precursor form of brain-derived neurotrophic factor and mature brain-derived neurotrophic factor are decreased in the pre-clinical stages of Alzheimer's disease. J Neurochem 93(6):1412–1421. https://doi.org/10.1111/j.1471-4159.2005.03135.x

Peters ME, Schwartz S, Han D, Rabins PV, Steinberg M, Tschanz JT, Lyketsos CG (2015) Neuropsychiatric symptoms as predictors of progression to severe Alzheimer's dementia and death: the cache county dementia progression study. Am J Psychiatr 172(5):460–465. https://doi.org/10.1176/appi.ajp.2014.14040480

Pokorny T, Duerler P, Seifritz E, Vollenweider FX, Preller KH (2020) LSD acutely impairs working memory, executive functions, and cognitive flexibility, but not risk-based decision-making. Psychol Med 50(13):2255–2264. https://doi.org/10.1017/S0033291719002393

Preller KH, Duerler P, Burt JB, Ji JL, Adkinson B, Stämpfli P, Seifritz E, Repovš G, Krystal JH, Murray JD, Anticevic A, Vollenweider FX (2020) Psilocybin induces time-dependent changes in global functional connectivity. Biol Psychiatry 88(2):197–207. https://doi.org/10.1016/j.biopsych.2019.12.027

Putcha D, Brickhouse M, O'Keefe K, Sullivan C, Rentz D, Marshall G, Dickerson B, Sperling R (2011) Hippocampal hyperactivation associated with cortical thinning in Alzheimer's disease signature regions in non-demented elderly adults. J Neurosci 31(48):17680–17688. https://doi.org/10.1523/JNEUROSCI.4740-11.2011

Raval NR, Johansen A, Donovan LL, Ros NF, Ozenne B, Hansen HD, Knudsen GM (2021) A single dose of psilocybin increases synaptic density and decreases 5-HT2A receptor density in the pig brain. Int J Mol Sci 22(2):835. https://doi.org/10.3390/ijms22020835

Reiff CM, Richman EE, Nemeroff CB, Carpenter LL, Widge AS, Rodriguez CI, Kalin NH, McDonald WM (2020) Psychedelics and psychedelic-assisted psychotherapy. Am J Psychiatr 177(5):391–410. https://doi.org/10.1176/appi.ajp.2019.19010035

Revenga MF, Zhu B, Guevara CA, Naler LB, Saunders JM, Zhou Z, Toneatti R, Sierra S, Wolstenholme JT, Beardsley PM, Huntley GW, Lu C, González-Maeso J (2021) Prolonged epigenetic and synaptic plasticity alterations following single exposure to a psychedelic in mice. BioRxiv 2021(02):24.432725. https://doi.org/10.1101/2021.02.24.432725

Rice L, Bisdas S (2017) The diagnostic value of FDG and amyloid PET in Alzheimer's disease—a systematic review. Eur J Radiol 94:16–24. https://doi.org/10.1016/j.ejrad.2017.07.014

Rich JB, Rasmusson DX, Folstein MF, Carson KA, Kawas C, Brandt J (1995) Nonsteroidal anti-inflammatory drugs in Alzheimer's disease. Neurology 45(1):51–55. https://doi.org/10.1212/WNL.45.1.51

Richards WA, Rhead JC, DiLeo FB, Yensen R, Kurland AA (1977) The peak experience variable in DPT-assisted psychotherapy with cancer patients. J Psychedelic Drugs 9(1):1–10

Romano AG, Quinn JL, Li L, Dave KD, Schindler EA, Aloyo VJ, Harvey JA (2010) Intrahippocampal LSD accelerates learning and desensitizes the 5-HT2A receptor in the rabbit, Romano et al. Psychopharmacology 212(3):441–448. https://doi.org/10.1007/s00213-010-2004-7

Roseman L, Nutt DJ, Carhart-Harris RL (2018) Quality of acute psychedelic experience predicts therapeutic efficacy of psilocybin for treatment-resistant depression. Front Pharmacol 8:974

Rosenberg PB, Martin BK, Frangakis C, Mintzer JE, Weintraub D, Porsteinsson AP, Schneider LS, Rabins PV, Munro CA, Meinert CL, Lyketsos CG, Drye LT (2010) Sertraline for the treatment of depression in Alzheimer disease. Am J Geriatr Psychiatry 18(2):136–145. https://doi.org/10.1097/JGP.0b013e3181c796eb

Ross S, Bossis A, Guss J, Agin-Liebes G, Malone T, Cohen B, Mennenga SE, Belser A, Kalliontzi K, Babb J, Su Z, Corby P, Schmidt BL (2016) Rapid and sustained symptom reduction following psilocybin treatment for anxiety and depression in patients with life-threatening cancer: a randomized controlled trial. J Psychopharmacol 30(12):1165–1180. https://doi.org/10.1177/0269881116675512

Ruthirakuhan M, Herrmann N, Vieira D, Gallagher D, Lanctôt KL (2019) The roles of apathy and depression in predicting Alzheimer disease: a longitudinal analysis in older adults with mild cognitive impairment. Am J Geriatr Psychiatry 27(8):873–882. https://doi.org/10.1016/j.jagp.2019.02.003

Sala Frigerio C, De Strooper B (2016) Alzheimer's disease mechanisms and emerging roads to novel therapeutics. Annu Rev Neurosci 39(1):57–79. https://doi.org/10.1146/annurev-neuro-070815-014015

Sampedro F, de la Fuente Revenga M, Valle M, Roberto N, Domínguez-Clavé E, Elices M, Luna LE, Crippa JAS, Hallak JEC, de Araujo DB, Friedlander P, Barker SA, Álvarez E, Soler J, Pascual JC, Feilding A, Riba J (2017) Assessing the psychedelic "after-glow" in Ayahuasca users: post-acute neurometabolic and functional connectivity changes are associated with enhanced mindfulness capacities. Int J Neuropsychopharmacol 20(9):698–711. https://doi.org/10.1093/ijnp/pyx036

Scarmeas N, Brandt J, Albert M, Hadjigeorgiou G, Papadimitriou A, Dubois B, Sarazin M, Devanand D, Honig L, Marder K, Bell K, Wegesin D, Blacker D, Stern Y (2005) Delusions and hallucinations are associated with worse outcome in Alzheimer disease. Arch Neurol 62 (10). https://doi.org/10.1001/archneur.62.10.1601

Schindler EAD, Sewell RA, Gottschalk CH, Luddy C, Flynn LT, Lindsey H, Pittman BP, Cozzi NV, D'Souza DC (2021) Exploratory controlled study of the migraine-suppressing effects of psilocybin. Neurotherapeutics 18(1):534–543. https://doi.org/10.1007/s13311-020-00962-y

Schmid Y, Liechti ME (2018) Long-lasting subjective effects of LSD in normal subjects. Psychopharmacology 235(2):535–545. https://doi.org/10.1007/s00213-017-4733-3

Schott BH, Seidenbecher CI, Richter S, Wüstenberg T, Debska-Vielhaber G, Schubert H, Heinze H-J, Richardson-Klavehn A, Düzel E (2011) Genetic variation of the serotonin 2a receptor affects hippocampal novelty processing in humans. PLoS One 6(1):e15984. https://doi.org/10.1371/journal.pone.0015984

Schwindt GC, Chaudhary S, Crane D, Ganda A, Masellis M, Grady CL, Stefanovic B, Black SE (2013) Modulation of the default-mode network between rest and task in Alzheimer's disease. Cereb Cortex 23(7):1685–1694. https://doi.org/10.1093/cercor/bhs160

Selkoe DJ (2002) Alzheimer's disease is a synaptic failure. Science 298(5594):789–791. https://doi.org/10.1126/science.1074069

Serrano-Pozo A, Frosch MP, Masliah E, Hyman BT (2011a) Neuropathological alterations in Alzheimer disease. Cold Spring Harb Perspect Med 1(1):a006189. https://doi.org/10.1101/cshperspect.a006189

Serrano-Pozo A, Mielke ML, Gómez-Isla T, Betensky RA, Growdon JH, Frosch MP, Hyman BT (2011b) Reactive glia not only associates with plaques but also parallels tangles in Alzheimer's disease. Am J Pathol 179(3):1373–1384. https://doi.org/10.1016/j.ajpath.2011.05.047

Sestieri C, Corbetta M, Romani GL, Shulman GL (2011) Episodic memory retrieval, parietal cortex, and the default mode network: functional and topographic analyses. J Neurosci 31 (12):4407–4420. https://doi.org/10.1523/JNEUROSCI.3335-10.2011

Sevigny J, Chiao P, Bussière T, Weinreb PH, Williams L, Maier M, Dunstan R, Salloway S, Chen T, Ling Y, O'Gorman J, Qian F, Arastu M, Li M, Chollate S, Brennan MS, Quintero-Monzon O, Scannevin RH, Arnold HM et al (2016) The antibody aducanumab reduces Aβ plaques in Alzheimer's disease. Nature 537(7618):50–56. https://doi.org/10.1038/nature19323

Shao L-X, Liao C, Gregg I, Davoudian PA, Savalia NK, Delagarza K, Kwan AC (2021) Psilocybin induces rapid and persistent growth of dendritic spines in frontal cortex in vivo. Neuron. https://doi.org/10.1016/j.neuron.2021.06.008

Shin I-S, Carter M, Masterman D, Fairbanks L, Cummings JL (2005) Neuropsychiatric symptoms and quality of life in Alzheimer disease. Am J Geriatr Psychiatry 13(6):469–474. https://doi.org/10.1097/00019442-200506000-00005

Shoghi-Jadid K, Small GW, Agdeppa ED, Kepe V, Ercoli LM, Siddarth P, Read S, Satyamurthy N, Petric A, Huang S-C, Barrio JR (2002) Localization of neurofibrillary tangles and beta-amyloid plaques in the brains of living patients with Alzheimer disease. Am J Geriatr Psychiatry 10 (1):24–35. https://doi.org/10.1097/00019442-200201000-00004

Smigielski L, Kometer M, Scheidegger M, Krähenmann R, Huber T, Vollenweider FX (2019a) Characterization and prediction of acute and sustained response to psychedelic psilocybin in a mindfulness group retreat. Sci Rep 9(1):14914. https://doi.org/10.1038/s41598-019-50612-3

Smigielski L, Scheidegger M, Kometer M, Vollenweider FX (2019b) Psilocybin-assisted mindfulness training modulates self-consciousness and brain default mode network connectivity with lasting effects. NeuroImage 196:207–215. https://doi.org/10.1016/j.neuroimage.2019.04.009

Smith GS, Barrett FS, Joo JH, Nassery N, Savonenko A, Sodums DJ, Marano CM, Munro CA, Brandt J, Kraut MA, Zhou Y, Wong DF, Workman CI (2017) Molecular imaging of serotonin degeneration in mild cognitive impairment. Neurobiol Dis 105:33–41. https://doi.org/10.1016/j.nbd.2017.05.007

Soto M, Andrieu S, Nourhashemi F, Ousset PJ, Ballard C, Robert P, Vellas B, Lyketsos CG, Rosenberg PB (2015) Medication development for agitation and aggression in Alzheimer disease: review and discussion of recent randomized clinical trial design. Int Psychogeriatr 27 (2):181–197. https://doi.org/10.1017/S1041610214001720

Sperling RA, Aisen PS, Beckett LA, Bennett DA, Craft S, Fagan AM, Iwatsubo T, Jack CR, Kaye J, Montine TJ, Park DC, Reiman EM, Rowe CC, Siemers E, Stern Y, Yaffe K, Carrillo MC, Thies B, Morrison-Bogorad M et al (2011) Toward defining the preclinical stages of Alzheimer's disease: recommendations from the National Institute on Aging-Alzheimer's association workgroups on diagnostic guidelines for Alzheimer's disease. Alzheimers Dement 7(3):280–292. https://doi.org/10.1016/j.jalz.2011.03.003

Steinberg M, Shao H, Zandi P, Lyketsos CG, Welsh-Bohmer KA, Norton MC, Breitner JCS, Steffens DC, Tschanz JT (2008) Point and 5-year period prevalence of neuropsychiatric symptoms in dementia: the Cache County Study. Int J Geriatr Psychiatry 23(2):170–177. https://doi.org/10.1002/gps.1858

Stewart WF, Kawas C, Corrada M, Metter EJ (1997) Risk of Alzheimer's disease and duration of NSAID use. Neurology 48(3):626–632. https://doi.org/10.1212/WNL.48.3.626

Stopford CL, Snowden JS, Thompson JC, Neary D (2008) Variability in cognitive presentation of Alzheimer's disease. Cortex 44(2):185–195. https://doi.org/10.1016/j.cortex.2005.11.002

Swift TC, Belser AB, Agin-Liebes G, Devenot N, Terrana S, Friedman HL, Guss J, Bossis AP, Ross S (2017) Cancer at the dinner table: experiences of psilocybin-assisted psychotherapy for the treatment of cancer-related distress. J Humanist Psychol 57(5):488–519

Szabo A, Kovacs A, Frecska E, Rajnavolgyi E (2014) Psychedelic N, N-dimethyltryptamine and 5-methoxy-N, N-dimethyltryptamine modulate innate and adaptive inflammatory responses through the sigma-1 receptor of human monocyte-derived dendritic cells. PLoS One 9(8): e106533

Sze C-I, Troncoso JC, Kawas C, Mouton P, Price DL, Martin LJ (1997) Loss of the presynaptic vesicle protein synaptophysin in hippocampus correlates with cognitive decline in Alzheimer disease. J Neuropathol Exp Neurol 56(8):933–944. https://doi.org/10.1097/00005072-199708000-00011

Tang L, Wang Y, Chen Y, Chen L, Zheng S, Bao M, Xiang J, Luo H, Li J, Li Y (2017) The association between 5HT2A T102C and behavioral and psychological symptoms of dementia in Alzheimer's disease: a meta-analysis. Biomed Res Int 2017:e5320135. https://doi.org/10.1155/2017/5320135

Tanila H (2017) The role of BDNF in Alzheimer's disease. Neurobiol Dis 97:114–118. https://doi.org/10.1016/j.nbd.2016.05.008

Tanzi RE (2012) The genetics of Alzheimer disease. Cold Spring Harb Perspect Med 2(10): a006296. https://doi.org/10.1101/cshperspect.a006296

Tayeb HO, Yang HD, Price BH, Tarazi FI (2012) Pharmacotherapies for Alzheimer's disease: beyond cholinesterase inhibitors. Pharmacol Ther 134(1):8–25. https://doi.org/10.1016/j.pharmthera.2011.12.002

Theofilas P, Ehrenberg AJ, Dunlop S, Di Lorenzo Alho AT, Nguy A, Leite REP, Rodriguez RD, Mejia MB, Suemoto CK, Ferretti-Rebustini REDL, Polichiso L, Nascimento CF, Seeley WW, Nitrini R, Pasqualucci CA, Jacob Filho W, Rueb U, Neuhaus J, Heinsen H, Grinberg LT (2017) Locus coeruleus volume and cell population changes during Alzheimer's disease progression: a stereological study in human postmortem brains with potential implication for early-stage biomarker discovery. Alzheimers Dement 13(3):236–246. https://doi.org/10.1016/j.jalz.2016.06.2362

Tomasi D, Volkow ND (2012) Aging and functional brain networks. Mol Psychiatry 17 (5):549–558. https://doi.org/10.1038/mp.2011.81

Tramutola A, Lanzillotta C, Domenico FD (2017) Targeting mTOR to reduce Alzheimer-related cognitive decline: from current hits to future therapies. Expert Rev Neurother 17(1):33–45. https://doi.org/10.1080/14737175.2017.1244482

Tromp D, Dufour A, Lithfous S, Pebayle T, Després O (2015) Episodic memory in normal aging and Alzheimer disease: insights from imaging and behavioral studies. Ageing Res Rev 24:232–262. https://doi.org/10.1016/j.arr.2015.08.006

Tsybko AS, Ilchibaeva TV, Filimonova EA, Eremin DV, Popova NK, Naumenko VS (2020) The chronic treatment with 5-HT2A receptor agonists affects the behavior and the BDNF system in mice. Neurochem Res 45(12):3059–3075. https://doi.org/10.1007/s11064-020-03153-5

Vann Jones SA, O'Kelly A (2020) Psychedelics as a treatment for Alzheimer's disease dementia. Front Synapt Neurosci 12. https://doi.org/10.3389/fnsyn.2020.00034

Versijpt J, Van Laere KJ, Dumont F, Decoo D, Vandecapelle M, Santens P, Goethals I, Audenaert K, Slegers G, Dierckx RA, Korf J (2003) Imaging of the 5-HT2A system: age-, gender-, and Alzheimer's disease-related findings. Neurobiol Aging 24(4):553–561. https://doi.org/10.1016/S0197-4580(02)00137-9

Wang W-Y, Tan M-S, Yu J-T, Tan L (2015) Role of pro-inflammatory cytokines released from microglia in Alzheimer's disease. Ann Transl Med 3(10). https://doi.org/10.3978/j.issn.2305-5839.2015.03.49

Watts R, Day C, Krzanowski J, Nutt D, Carhart-Harris R (2017) Patients' accounts of increased "connectedness" and "acceptance" after psilocybin for treatment-resistant depression. J Humanist Psychol 57(5):520–564. https://doi.org/10.1177/0022167817709585

Weintraub S, Wicklund AH, Salmon DP (2012) The neuropsychological profile of Alzheimer disease. Cold Spring Harb Perspect Med 2(4):a006171. https://doi.org/10.1101/cshperspect.a006171

Williams GV, Rao SG, Goldman-Rakic PS (2002) The physiological role of 5-HT2A receptors in working memory. J Neurosci 22(7):2843–2854. https://doi.org/10.1523/JNEUROSCI.22-07-02843.2002

Wimo A, Ballard C, Brayne C, Gauthier S, Handels R, Jones RW, Jonsson L, Khachaturian AS, Kramberger M (2014) Health economic evaluation of treatments for Alzheimer's disease: impact of new diagnostic criteria. J Intern Med 275(3):304–316. https://doi.org/10.1111/joim.12167

Wittmann M, Carter O, Hasler F, Cahn BR, Grimberg U, Spring P, Hell D, Flohr H, Vollenweider FX (2007) Effects of psilocybin on time perception and temporal control of behaviour in humans. J Psychopharmacol 21(1):50–64. https://doi.org/10.1177/0269881106065859

Yaden DB, Griffiths RR (2021) The subjective effects of psychedelics are necessary for their enduring therapeutic effects. ACS Pharmacol Transl Sci 4(2):568–572. https://doi.org/10.1021/acsptsci.0c00194

Yu B, Becnel J, Zerfaoui M, Rohatgi R, Boulares AH, Nichols CD (2008) Serotonin 5-HT2A receptor activation suppresses TNF-α-induced inflammation with extraordinary potency. J Pharmacol Exp Ther. https://doi.org/10.1124/jpet.108.143461

Yuede CM, Wallace CE, Davis TA, Gardiner WD, Hettinger JC, Edwards HM, Hendrix RD, Doherty BM, Yuede KM, Burstein ES, Cirrito JR (2021) Pimavanserin, a 5HT2A receptor inverse agonist, rapidly suppresses Aβ production and related pathology in a mouse model of Alzheimer's disease. J Neurochem 156(5):658–673. https://doi.org/10.1111/jnc.15260

Yuki D, Sugiura Y, Zaima N, Akatsu H, Takei S, Yao I, Maesako M, Kinoshita A, Yamamoto T, Kon R, Sugiyama K, Setou M (2014) DHA-PC and PSD-95 decrease after loss of synaptophysin and before neuronal loss in patients with Alzheimer's disease. Sci Rep 4 (1):7130. https://doi.org/10.1038/srep07130

Zhang G, Stackman RW Jr (2015) The role of serotonin 5-HT2A receptors in memory and cognition. Front Pharmacol 6:225

Zhang G, Ásgeirsdóttir HN, Cohen SJ, Munchow AH, Barrera MP, Stackman RW (2013) Stimulation of serotonin 2A receptors facilitates consolidation and extinction of fear memory in C57BL/6J mice. Neuropharmacology 64:403–413. https://doi.org/10.1016/j.neuropharm.2012.06.007

Zhang G, Cinalli D, Cohen SJ, Knapp KD, Rios LM, Martínez-Hernández J, Luján R, Stackman RW (2016) Examination of the hippocampal contribution to serotonin 5-HT2A receptor-mediated facilitation of object memory in C57BL/6J mice. Neuropharmacology 109:332–340. https://doi.org/10.1016/j.neuropharm.2016.04.033

Zhao Q-F, Tan L, Wang H-F, Jiang T, Tan M-S, Tan L, Xu W, Li J-Q, Wang J, Lai T-J, Yu J-T (2016) The prevalence of neuropsychiatric symptoms in Alzheimer's disease: systematic review and meta-analysis. J Affect Disord 190:264–271. https://doi.org/10.1016/j.jad.2015.09.069

Zhou J, Seeley WW (2014) Network dysfunction in Alzheimer's disease and frontotemporal dementia: implications for psychiatry. Biol Psychiatry 75(7):565–573. https://doi.org/10.1016/j.biopsych.2014.01.020

Zotova E, Nicoll JA, Kalaria R, Holmes C, Boche D (2010) Inflammation in Alzheimer's disease: relevance to pathogenesis and therapy. Alzheimers Res Ther 2(1):1. https://doi.org/10.1186/alzrt24

Psilocybin for Trauma-Related Disorders

Amanda J. Khan, Ellen Bradley, Aoife O'Donovan, and Joshua Woolley

Contents

Abstract Posttraumatic stress disorder (PTSD) is a debilitating, chronic disorder and efficacy rates of current PTSD treatments are underwhelming. There is a critical need for innovative approaches. We provide an overview of trauma and PTSD and cite literature providing converging evidence of the therapeutic potential of psilocybin for PTSD. No study to date has investigated psilocybin or psilocybin-assisted psychotherapy (PAP) as treatments for PTSD. An open-label study in traumatized AIDS survivors found that PAP reduced PTSD symptoms, attachment anxiety, and demoralization. Several PAP trials show preliminary efficacy in facilitating confronting traumatic memories, decreasing emotional avoidance, depression, anxiety, pessimism, and disconnection from others, and increasing acceptance, self-compassion, and forgiveness of abusers, all of which are relevant to PTSD recovery. There is also early evidence that other classic psychedelics may produce large reductions in PTSD symptoms in combat veterans. However, this body of literature is small, mechanisms are not yet well understood, and the risks of using psychedelic

---

A. J. Khan, E. Bradley, A. O'Donovan, and J. Woolley (✉)
Department of Psychiatry, University of California, San Francisco, CA, USA
e-mail: amanda.khan@ucsf.edu; ellen.bradley@ucsf.edu; aoife.odonovan@ucsf.edu; josh.woolley@ucsf.edu

Curr Topics Behav Neurosci (2022) 56: 319–332
https://doi.org/10.1007/7854_2022_366
Published Online: 17 June 2022

compounds for trauma-related disorders need further study. In sum, evidence supports further investigation of PAP as a radically new approach for treating PTSD.

Keywords Psilocybin · Posttraumtic Stress Disorder · Psychedelics · Trauma · Treatment

1 Introduction

The recent resurgence of interest in using psychedelics in mainstream healthcare has led to numerous studies investigating their therapeutic potential for multiple psychiatric disorders (e.g., Galvão-Coelho et al. 2021; Mithoefer et al. 2018). Trauma-related disorders such as posttraumatic stress disorder (PTSD) have notoriously chronic or treatment-resistant trajectories and there is an urgent need for novel interventions (Steenkamp et al. 2020). The majority of psychedelic-assisted clinical trials for PTSD have focused on using MDMA (3,4-Methylenedioxymethamphetamine), an entactogen now designated as a Breakthrough Therapy for PTSD by the U.S. Food and Drug Administration (FDA) (Mithoefer et al. 2018). However, a different group of psychoactive compounds, "classic psychedelics," which includes psilocybin, lysergic acid diethylamide (LSD), and N,N-dimethyltryptamine (DMT), may provide important and unique therapeutic benefit for PTSD. In this chapter, we examine the theoretical framework for using psilocybin and other classic psychedelics to treat PTSD and trauma-related disorders and review relevant findings.

2 PTSD and Trauma-Related Disorders

Over 70% of the population will experience a traumatic event in their lifetime (e.g., Benjet et al. 2016). This number is likely even higher given that the current psychiatric diagnostic framework of trauma (American Psychiatric Association 2013) fails to capture experiences of racial trauma, emotional abuse and neglect, perpetration-based traumas (e.g., killing in combat), and institutional betrayal (e.g., Litz et al. 2009; Williams et al. 2018). Both the cumulative amount and type of trauma are important predictors of subsequent mental and physical health sequelae (Cloitre et al. 2009; Karam et al. 2014). Trauma exposure, particularly in childhood, can cause neurobiological, physiological, psychological, immunological, and epigenetic changes (e.g., Nemeroff 2004; Nöthling et al. 2020). The sequelae of trauma are numerous and include PTSD, depression, anxiety, suicidal thoughts and behaviors, personality disorders, somatic complaints, eating disorders, moral injury, substance use disorders, and psychotic disorder, cardiovascular, and autoimmune disorders (e.g., Chen et al. 2010; Cloitre et al. 2009; Litz et al. 2009; O'Donovan et al. 2015; Shalev et al. 1998). In this chapter, we primarily focus on the specific psychological sequelae of PTSD, although trauma very likely plays a role in the development and maintenance of other disorders.

PTSD is a debilitating psychiatric disorder that involves re-experiencing, avoidance, negative alterations in cognition and mood, and hyperarousal (American Psychiatric Association 2013). Stemming from exposure to traumatic event(s), PTSD is characterized by emotion dysregulation and can cause long-lasting impairments in social and occupational functioning (Ehring and Quack 2010; Rodriguez et al. 2012). Lifetime prevalence rates range widely (7–50%) depending on the population, with higher rates in veterans than civilians, and amongst those with interpersonal trauma (Fulton et al. 2015; Kessler et al. 2017). PTSD is associated with chronic symptom trajectories, increased psychiatric and medical morbidity, and decreased quality of life (Sareen et al. 2007).

Prevailing theoretical models of factors involved in the etiology and maintenance of PTSD are largely rooted in principles of (1) fear conditioning, with a focus on increased fear acquisition and impaired extinction, and (2) cognitive theory, emphasizing negative appraisals of the trauma and its meaning (Ehlers and Clark 2008; Foa and Kozak 1986). Beyond these models, individuals with PTSD are more likely to have insecure attachment styles, use avoidance-related emotion regulation strategies like suppression, and have strongly negative self- and other-concepts (e.g., Seligowski et al. 2015; Woodhouse et al. 2015). In addition, the extent to which a person perceives the trauma(s) as central to their identity and life story is a robust predictor of risk for and severity of PTSD (for review, see Gehrt et al. 2018). Indeed, how a person organizes their knowledge of self and consequently, others in relation to self appears central to PTSD. These self-other concepts can subsequently serve as reference points for interpreting everyday occurrences, which then contribute to the developing and/or strengthening of stable, global beliefs (Berntsen and Rubin 2006). Overall, these models highlight the breadth of important treatment targets robustly involved in PTSD.

Currently available evidence-based treatments (EBTs) for PTSD are inadequate (Steenkamp et al. 2020). Existing pharmacotherapies (e.g., selective serotonin reuptake inhibitors (SSRIs)) provide limited symptom relief and likely lack the specificity to address the unique neurobiology of PTSD (see Ostacher and Cifu 2019). Critically, up to 60% of patients do not respond adequately to medications (Watts et al. 2013), and latest clinical guidelines even advise against most available pharmacotherapies (Ostacher and Cifu 2019). Rooted in fear conditioning and cognitive models of PTSD, front-line psychotherapies include Prolonged Exposure (PE) and Cognitive Processing Therapy (CPT). Both treatments show moderate efficacy in reducing PTSD symptoms compared to waitlist and treatment as usual (e.g., Merz et al. 2019). However, as many as 60–72% of patients retain their PTSD diagnoses (see Steenkamp et al. 2015; Watts et al. 2013) and dropout rates reach up to 40% (e.g., Goetter et al. 2015). Some have theorized that fear conditioning models, while effective, may be too narrow to explain the breadth of symptoms associated with PTSD (e.g., Krystal et al. 2017; Markowitz et al. 2015). Indeed, trauma exposure and PTSD are both linked to structural, functional, and connectivity alterations in brain networks responsible for not only fear conditioning, but also self-concept, emotion regulation, and memory (see review Liberzon and Abelson 2016; Akiki et al. 2017). In sum, development of alternatives to current pharmacologic and psychotherapeutic EBTs is essential for improving outcomes for this patient population.

3 Brief Review of Psilocybin

Psilocybin is the naturally occurring prodrug of psilocin (4-hydroxy-dimethyltryptamine) and has been used by Indigenous peoples of Central and South America for centuries (Wasson 1980). Psilocin is a serotonergic (5-hydroxytryptamine, 5-HT) agonist, primarily exerting its psychedelic effects through the 5HT2A receptors but also binding to 5HT2C, 5HT1A, and 5HT1B receptors (Halberstadt and Geyer 2011). Acutely, psilocybin can cause profound, dose-dependent changes in sensory perception and cognition including auditory and visual hallucinations and derealization (Barrett et al. 2015; Kometer and Vollenweider 2018) that last from 3 to 6 h after oral ingestion. Studies examining the clinical safety of oral dosages ranging from 0.29 to 0.43 mg/kg indicate psilocybin is generally well tolerated (Bogenschutz and Ross 2016), has low physiological toxicity, and is not associated with compulsive drug seeking (e.g., Tylš et al. 2014; van Amsterdam et al. 2011). Evidence from preclinical models, neuroimaging work, and clinical trials (Baumeister et al. 2014; Carhart-Harris et al. 2012; Catlow et al. 2013; Herzog et al. 2020; Kraehenmann et al. 2015; Kringelbach et al. 2020; Ly et al. 2018; Petri et al. 2014; Raval et al. 2021; Schindler et al. 2018) suggest that psilocybin and related compounds have the potential to alleviate PTSD symptoms via multiple mechanisms (see review by Vollenweider and Preller 2020). Further investigation is critical to understanding the effects of classic psychedelics for trauma-related disorders.

4 Review of the Empirical Literature of Psilocybin for Treating PTSD

To date, no studies have examined the efficacy of psilocybin as a PTSD treatment (see review Krediet et al. 2020) or have reported quantitative information on trauma exposure rates in participants. One study examined PTSD symptoms as a secondary outcome in a single-arm, open-label, trial of psilocybin-assisted group psychotherapy (Anderson et al. 2020). In a sample of 18 older, gay-identified, long-term AIDS survivors, PTSD severity declined from baseline to the end of treatment with gains maintained at 3-month follow-up with moderate effect sizes. However, only 3 of the 18 participants had a baseline PTSD severity score above clinical cut-off, limiting the conclusions that can be drawn. Attachment anxiety also significantly decreased from baseline to 3-month follow-up in the same study (Stauffer et al. 2020), which is in line with research in healthy volunteers showing psilocybin acutely reduces rejection sensitivity (Preller et al. 2015).

Studies focused on depression and anxiety also point to psilocybin's ability to improve negative cognitions, avoidance, anxiety, and disconnectedness, overlapping symptoms between PTSD and other disorders. Four studies reported that psilocybin-assisted therapy reduces depression symptoms in patients with treatment-resistant depression (TRD) and in cancer patients (see meta-analysis Galvão-Coelho et al. 2021). Across all four studies, depression remission rates remained high (60–80%) at 3- or 6-month follow-up and patients reported being less pessimistic about their future (Watts et al. 2017). In follow-up qualitative interviews from one study (Watts et al. 2017), participants reported post-treatment shifts from avoiding traumatic memories and painful emotions to confronting and accepting them, increased understanding and compassion for past abusers, access to a fuller range of autobiographical material, and a sense of reconnection with self, others, and the world. In the three studies in cancer patients, psilocybin reduced trait anxiety related to having a life-threatening illness and some participants discussed unearthing and processing childhood traumas that were realized to be unhealed (Malone et al. 2018). Additionally, a prospective self-report study found psilocybin (and related compounds) reduced experiential avoidance, which in turn correlated with decreases in depression and suicide ideation (SI) (Zeifman et al. 2020). Finally, a cross-sectional self-report study in Black, Indigenous, and People of Color (BIPOC) who experienced racial trauma reported that increases in psychological flexibility following ingestion of psilocybin (and related compounds) were associated with decreases in posttraumatic stress symptoms (Davis et al. 2021; Williams et al. 2021). Although additional research focused on PTSD specifically is clearly needed, these findings provide some insight into how psilocybin may help trauma-related disorders across a range of trauma types.

5 Review of the Empirical Literature of Other Classic Psychedelics for Treating PTSD

To the best of our knowledge, no studies have investigated the potential of LSD to treat PTSD (see review Krediet et al. 2020). Given that it is also a 5HT2A agonist and facilitates associative learning (Romano et al. 2010), it may have similar therapeutic effects for PTSD as those theorized for psilocybin, but clinical evidence is needed. Ayahuasca (which includes the psychedelic tryptamine, DMT) has also been proposed as a candidate treatment for PTSD (e.g., Nielson and Megler 2014). Studies in rodents and healthy volunteers show that ayahuasca increases serotonin and induces alterations in activity in regions involved in episodic memory, contextual associations of emotional events, and reactivity (de Castro-Neto et al. 2013; Riba et al. 2006). Another study found DMT with pharmahuasca (Harmaline) normalized trauma-induced reactive oxygen species production and PTSD associated gene expression that overlapped with human PTSD (Kelley et al. 2022). A recent open-label study examined the effectiveness of using 5-MeO-DMT (another psychedelic tryptamine found in several plants and certain desert toads) and ibogaine to treat PTSD in combat veterans ($N = 65$) (Davis et al. 2020). PTSD symptoms, suicidal ideation (SI), depression, anxiety, and psychological flexibility all showed significant reductions from 1 month pre- to 1 month post-treatment, with large effect sizes. Notably, the dual action on serotonergic (5-MeO-DMT) and Kappa opioid and NMDA (ibogaine) receptors may have provided unique, additive benefits; it is unclear whether using only one psychedelic agent would have yielded the same results. No study has empirically investigated 5-MeO-DMT or ibogaine as monotherapies for PTSD. In one large survey study assessing 5-MEO-DMT use and effects, 21% reported having PTSD and of those, 79% reported symptom improvement (18% no change, 3% worsening) (Davis et al. 2018). These findings suggest the potential of 5-MeO-DMT and ayahuasca for PTSD, but intervention studies evaluating each of these compounds are essential. Additionally, in an online survey study, participants reported improvements in PTSD symptoms following use of mescaline, another primarily serotonergic psychoactive alkaloid (Agin-Liebes et al. 2021).

LSD and ayahuasca also have demonstrated anxiolytic and antidepressant effects, facilitated engagement with salient memories, and improved positive self-other concepts that may be relevant for treating PTSD. In a placebo-controlled pilot study, LSD-assisted psychotherapy significantly reduced anxiety related to having a life-threatening illness (Gasser et al. 2014). Of the eight participants in the LSD condition, one did have PTSD, but no data regarding PTSD symptom changes were reported. In qualitative interviews a year later, participants reported treatment helped them confront memories and access emotion, which they viewed as helpful in restructuring their beliefs about trust and the world (Gasser et al. 2015). There is also evidence a single dose of LSD produces lasting reductions in anxiety in healthy volunteers (Schmid and Liechti 2018) and increases feelings of trust and closeness (Dolder et al. 2016), effects likely to benefit individuals with PTSD. In open-label and placebo-controlled trials in depressed participants, ayahuasca significantly reduced depression severity (Osório et al. 2015; Palhano-Fontes et al. 2019). These findings provide preliminary evidence that classic psychedelics can reduce anxiety, improve mood, and facilitate trauma processing.

6 Considerations for Other Trauma-Related Disorders

6.1 Complicated Grief

Complicated Grief (CG) is a distinct syndrome characterized by protracted and impairing grief in response to losing someone, causing more impairment and significant distress than traditional grief trajectories (American Psychiatric Association 2013). Core symptoms include pervasive yearning for the deceased, persistent preoccupation with the deceased, and avoiding grief-related stimuli. There is preliminary evidence that psilocybin and ayahuasca reduce grief. Although no study has investigated psilocybin to treat CG, the psilocybin group therapy for demoralization trial did find significant decreases in CG symptoms, with improvement maintained at 3-month follow-up (Anderson et al. 2020). Two additional studies examined the effectiveness of ayahuasca for bereavement. In a small survey study, ayahuasca reduced grief and preoccupations and improved self-other forgiveness significantly more than peer-support grief groups (González et al. 2017). In a small sample of 50 bereaved individuals attending ayahuasca ceremonies, symptoms of grief significantly decreased from baseline to end of treatment and up to 12 months after (González et al. 2020). Acceptance and defusion mediated the improvement in grief symptoms. These findings are encouraging and future research should prioritize investigating the therapeutic potential of psilocybin and ayahuasca for CG.

6.2 *Borderline Personality Disorder*

Borderline personality disorder (BPD) is theorized as a trauma-related disorder with high rates of comorbidity with PTSD and notable symptom overlap (e.g., Cloitre et al. 2013). To date, no study has investigated the efficacy of psilocybin or classic psychedelics for treating BPD. Notably, three participants in the psilocybin group therapy trial in AIDS survivors met criteria for BPD (Anderson et al. 2020), but no data on BPD symptom changes were reported. Similarly, in a clinical trial of ayahuasca for TRD, 76% of participants had a cluster B personality disorder, but no data on personality disorder changes were reported (Palhano-Fontes et al. 2019). Of most relevance to BPD, two studies show psilocybin reduces anxious attachment and rejection sensitivity (Stauffer et al. 2020; Preller et al. 2015) and another study found psilocybin helped people re-connect with close others who had wronged them (Watts et al. 2017). Psilocybin also increases feelings of empathy, and trait openness while decreasing trait neuroticism (Carhart et al. 2018; Roseman et al. 2018). In conjunction with results showing psilocybin induces decreased amygdala reactivity to negative faces (Roseman et al. 2018), these findings raise the possibility that psilocybin could influence mechanisms central to the relational symptoms of BPD such as fear of abandonment, which might have downstream effects on behavioral efforts to avoid this outcome. Finally, psilocybin and related compounds have been shown to actually decrease suicide-related behaviors (see review Zeifman et al. 2021; Ross et al. 2021), hinting at a therapeutic potential for this common BPD symptom. Given the prevalence of self-directed violence in BPD, a critical future consideration is whether psilocybin and related compounds should be integrated with existing EBTs like Dialectical Behavior Therapy (Zeifman and Wagner 2020).

7 Precautions and Possible Contraindications

Given the absence of psilocybin therapy trials for PTSD, the safety and tolerability of this treatment for this population remains unknown. Three of the most obvious potential adverse effects and contraindications include dissociative episodes, concomitant medications, and self-directed violence. Dissociative features are a subtype of PTSD symptoms and are either common or are central to other trauma-related disorders (e.g., BPD, dissociative identity disorder). In the scant studies examining classic psychedelics, no data on baseline or subsequent dissociative episodes has been reported. In the small psilocybin trial by Anderson et al. (2020), one person experienced a flashback a few days after the dosing session. Interestingly, participants in some other psilocybin trials (e.g., Watts et al. 2017) have reported *more* embodiment, not less, possibly suggesting psilocybin could help trauma-related dissociation. Critical lines of future research include the general safety profile of psilocybin-assisted therapy for dissociative subtypes of PTSD and related phenomenology (i.e., flashbacks), BPD, and dissociative disorders. We also need to determine whether people with dissociative features are more likely to experience dissociation during psilocybin sessions, whether that is associated with adverse or *positive* acute and long-term outcomes (e.g., Roseman et al. 2018), and whether psilocybin increases subsequent dissociation frequency.

An alternative approach may be to consider using psycholytic doses, or low doses that provide that serve to lubricate the therapeutic process and socialize users to the effects (Garcia-Romeu and Richards 2018). In theory, psycholytic doses may provide an opportunity for mild relaxation of ego defenses, which could be leveraged to deepen psychotherapeutic processes (Majic et al. 2015). Those with a high propensity for dissociation or re-experiencing symptoms, or who are more apprehensive about psychedelics may also particularly benefit from such a graded approach. Similar to starting any medication, however, using the "start low, go slow" approach will likely be safest. Research comparing the efficacy of macrodosing (i.e., full dose), psycholytic, and microdosing will greatly improve our understanding of how best to use these compounds to treat PTSD. The basic components of preparation sessions for PTSD patients need not necessarily be altered from modern psychedelic protocols (i.e., review safety procedures, what ifs, agreed upon touch, rescue procedures), but clinical judgment should always be used. Regarding concerns about potentially intense exposure experiences, further study is needed to understand tolerability of psilocybin in PTSD. Thus far, psychedelic therapies for PTSD are associated with high ratings of acceptability, satisfaction, and openness to further therapy and low dropout (e.g., Barone et al. 2019; Davis et al. 2020; Feder et al. 2021). However, it is critical to recognize the potential for bias in these findings particularly given that participants in early studies may have unusually positive expectations for treatment.

Another concern is that little is known about possible interactions between psilocybin and psychotropics typically used to treat PTSD. Anecdotal evidence suggests that chronic use of SSRIs, serotonin-norepinephrine reuptake inhibitors

(SNRIs), and antipsychotics all attenuate the psychedelic effects of psilocybin and related compounds (e.g., Bonson et al. 1996). There is also theoretical risk of inducing serotonin syndrome by combining classic psychedelics with other serotonergic agents, which can be life-threatening. However, high quality evidence is lacking. Because these risks are not well-understood, all modern clinical trials have required patients to abstain from most psychotropics and other medications that may influence psilocybin metabolism and/or precipitate adverse effects. Future trials should explore which whether SSRIs, the most common PTSD pharmacotherapy, and other concomitant medications are safe and appropriate to continue during psilocybin-therapy. Regarding risk of self-directed violence, two large population studies reported psilocybin use is associated with *reduced* SI, planning, and attempt (Johansen and Krebs 2015; Krebs and Johansen 2013). This same pattern of findings was also found in a longitudinal study of women sex workers with high trauma exposure (Argento et al. 2017). Psilocybin and ayahuasca have also been associated with reduced SI in clinical trials for depression (e.g., Carhart-Harris et al. 2018; Zeifman et al. 2021). Though this early evidence suggests psilocybin may confer a protective effect (for review Zeifman et al. 2021), research examining changes in self-directed violence in people with trauma-related disorders specifically is essential.

8 Conclusions

Trauma exposure is ubiquitous and the downstream effects on mental and physical health, overall functioning, and quality of life can be devastating. Psilocybin and other classic psychedelics may offer unique value for healing trauma-related disorders through dynamic neuronal and neuromodulation changes across large-scale networks throughout the whole brain. Although there is preliminary support, there is a clear need for rigorous clinical studies that specifically test the efficacy of these compounds in trauma-related disorder samples. Regardless of clinical population, future psychedelic-assisted therapy studies would greatly benefit from measuring trauma exposure, including those falling outside the DSM Criterion A (e.g., racial trauma), PTSD symptoms, and pre- and post-treatment SI and related behaviors. Progress in this field will also require careful investigation of potential adverse effects and contraindications. Future studies that compare the efficacy of different psychedelic-assisted therapy approaches, evaluate dose and sequencing options, determine personalized medicine guidelines, and assess scalability are needed. Notwithstanding the early stage of this work, evidence to date supports further investigation of PAP as a radically new approach for treating PTSD.

References

Agin-Liebes G, Haas TF, Lancelotta R, Uthaug MV, Davis JGRAK (2021) Naturalistic use of mescaline is associated with self-reported psychiatric improvements and enduring positive life changes. ACS Pharmacol Trans Sci 4(2):543–552. https://doi.org/10.1021/acsptsci.1c00018

Akiki TJ, Averill CL, Abdallah CG (2017) A network-based neurobiological model of PTSD: evidence from structural and functional neuroimaging studies. Curr Psychiatry Rep 19(11):81

American Psychiatric Association (2013) Diagnostic and statistical manual of mental disorders: DSM-5, 5th edn. American Psychiatric Publishing, Washington

Anderson BT, Danforth A, Daroff R et al (2020) Psilocybin-assisted group therapy for demoralized older long-term AIDS survivor men: an open-label safety and feasibility pilot study. EClinicalMedicine 27:100538

Argento E, Strathdee SA, Tupper K et al (2017) Does psychedelic drug use reduce risk of suicidality? Evidence from a longitudinal community-based cohort of marginalised women in a Canadian setting. BMJ Open 7(9):e016025

Barone W, Beck J, Mitsunaga-Whitten M et al (2019) Perceived benefits of MDMA-assisted psychotherapy beyond symptom reduction: qualitative follow-up study of a clinical trial for individuals with treatment-resistant PTSD. J Psychoactive Drugs 51(2):199–208

Barrett FS, Johnson MW, Griffiths RR (2015) Validation of the revised mystical experience questionnaire in experimental sessions with psilocybin. J Psychopharmacol 29:1182–1190

Baumeister D, Barnes G, Giaroli G, Tracy D (2014) Classical hallucinogens as antidepressants? a review of pharmacodynamics and putative clinical roles. Ther Adv Psychopharmacol 4(4):156–169. https://doi.org/10.1177/2045125314527985

Benjet C, Bromet E, Karam EG et al (2016) The epidemiology of traumatic event exposure worldwide: results from the world mental health survey consortium. Psychol Med 46(2): 327–343

Berntsen D, Rubin DC (2006) The centrality of event scale: a measure of integrating a trauma into one's identity and its relation to post-traumatic stress disorder symptoms. Behav Res Ther 44(2): 219–231

Bogenschutz MP, Ross S (2016) Therapeutic applications of classic hallucinogens. In: Halberstadt AL, Vollenweider FX, Nichols DE (eds) Behavioral neurobiology of psychedelic drugs, vol 36. Springer, Berlin, pp 361–391

Bonson K, Buckholtz JW, Murphy DL (1996) Chronic administration of serotonergic antidepressants attenuates the subjective effects of LSD in humans. Neuropsychopharmacology 14(6): 425–436

Carhart-Harris RL, Bolstridge M, Day CMJ, Rucker J, Watts R, Erritzoe DE, Kaelen M, Giribaldi B, Bloomfield M, Pilling S, Rickard JA, Forbes B, Feilding A, Taylor D, Curran HV, Nutt DJ (2018) Psilocybin with psychological support for treatment-resistant depression: six-month follow-up. Psychopharmacology 235(2):399–408. https://doi.org/10.1007/s00213-017-4771-x

Carhart-Harris RL, Erritzoe D, Williams T et al (2012) Neural correlates of the psychedelic state as determined by fMRI studies with psilocybin. Proc Natl Acad Sci U S A 109:2138–2143

Catlow BJ, Song S, Paredes DA et al (2013) Effects of psilocybin on hippocampal neurogenesis and extinction of trace fear conditioning. Exp Brain Res 228(4):481–491

Chen LP, Murad MH, Paras ML et al (2010) Sexual abuse and lifetime diagnosis of psychiatric disorders: systematic review and meta-analysis. Mayo Clin Proc 85(7):618–629

Cloitre M, Stolbach BC, Herman JL et al (2009) A developmental approach to complex PTSD: childhood and adult cumulative trauma as predictors of symptom complexity. J Trauma Stress 22(5):399–408

Cloitre M, Garvert DW, Brewin CR et al (2013) Evidence for proposed ICD-11 PTSD and complex PTSD: a latent profile analysis. Eur J Psychotraumatol 4:1–12

de Castro-Neto EF, da Cunha RH, da Silveira DX, Yonamine M, Gouveia TLF, Cavalheiro EA, Amado D, da Graça Naffah-Mazzacoratti M (2013) Changes in aminoacidergic and

monoaminergic neurotransmission in the hippocampus and amygdala of rats after ayahuasca ingestion. World J Biol Chem 4(4):141. https://doi.org/10.4331/wjbc.v4.i4.141
Davis AK, Barsuglia JP, Lancelotta R, Grant RM, Renn E (2018) The epidemiology of 5-methoxy-N, N-dimethyltryptamine (5-MeO-DMT) use: Benefits, consequences, patterns of use, subjective effects, and reasons for consumption. J Psychopharmacol 32(7):779–792. https://doi.org/10.1177/0269881118769063
Davis AK, Averill LA, Sepeda ND et al (2020) Psychedelic treatment for trauma-related psychological and cognitive impairment among US special operations forces veterans. Chron Stress 4: 2470547020939564
Davis AK, Xin Y, Sepeda ND, Garcia-Romeu A, Williams MT (2021) Increases in psychological flexibility mediate relationship between acute psychedelic effects and decreases in racial trauma symptoms among people of color. Chron Stress 5:247054702110356. https://doi.org/10.1177/24705470211035607
Dolder PC, Schmid Y, Müller F et al (2016) LSD acutely impairs fear recognition and enhances emotional empathy and sociality. Neuropsychopharmacology 41:2638–2646
Ehlers A, Clark DM (2008) Post-traumatic stress disorder: the development of effective psychological treatments. Nord J Psychiatry 62(Suppl 47):11–18
Ehring T, Quack D (2010) Emotion regulation difficulties in trauma survivors: the role of trauma type and PTSD symptom severity. Behav Ther 41:587–598
Feder A, Costi S, Rutter SB et al (2021) A randomized controlled trial of repeated ketamine administration for chronic posttraumatic stress disorder. Am J Psychiatry 178(2):193–202
Foa EB, Kozak MJ (1986) Emotional processing of fear: exposure to corrective information. Psychol Bull 99:20–35
Fulton JJ, Calhoun PS, Wagner HR et al (2015) The prevalence of posttraumatic stress disorder in operation enduring freedom/operation Iraqi freedom (OEF/OIF) veterans: a meta-analysis. J Anxiety Disord 31:98–107
Galvão-Coelho NL, Marx W, Gonzalez M et al (2021) Classic serotonergic psychedelics for mood and depressive symptoms: a meta-analysis of mood disorder patients and healthy participants. Psychopharmacology (Berl) 238:341–354
Garcia-Romeu A, Richards WA (2018) Current perspectives on psychedelic therapy: use of serotonergic hallucinogens in clinical interventions. Int Rev Psychiatry 30(4):291–316. https://doi.org/10.1080/09540261.2018.1486289
Gasser P, Holstein D, Michel Y et al (2014) Safety and efficacy of lysergic acid diethylamide-assisted psychotherapy for anxiety associated with life-threatening diseases. J Nerv Ment Dis 202:513–520
Gasser P, Kirchner K, Passie T (2015) LSD-assisted psychotherapy for anxiety associated with a life-threatening disease: a qualitative study of acute and sustained subjective effects. J Psychopharmacol 29(1):57–68. https://doi.org/10.1177/0269881114555249
Gehrt TB, Berntsen D, Hoyle RH et al (2018) Psychological and clinical correlates of the centrality of event scale: a systematic review. Clin Psychol Rev 65:57–80
Goetter EM, Bui E, Ojserkis RA et al (2015) A systematic review of dropout from psychotherapy for posttraumatic stress disorder among Iraq and Afghanistan combat veterans. J Trauma Stress 28(5):401–409
González D, Carvalho M, Cantillo J et al (2017) Potential use of ayahuasca in grief therapy. Omega (Westport) 1:30222817710879
González D, Cantillo J, Pérez I et al (2020) Therapeutic potential of ayahuasca in grief: a prospective, observational study. Psychopharmacology (Berl) 237:1171–1182
Halberstadt AL, Geyer MA (2011) Multiple receptors contribute to the behavioral effects of indoleamine hallucinogens. Neuropharmacology 61(3):364–381
Herzog R, Mediano PAM, Rosas FE et al (2020) A mechanistic model of the neural entropy increase elicited by psychedelic drugs. Sci Rep 10:17725
Johansen PØ, Krebs TS (2015) Psychedelics not linked to mental health problems or suicidal behavior: a population study. J Psychopharmacol 29(3):270–279

Karam EG, Friedman MG, Hill ED et al (2014) Cumulative traumas and risk thresholds: 12 month PTSD in the world mental health (WMH) surveys. Depress Anxiety 31(2):130–142

Kelley DP, Venable K, Destouni A, Bilac G, Ebenezer P, Stadler K, Nichols C, Barker S, Francis J (2022) Pharmahuasca and DMT rescue ROS production and differentially expressed genes observed after predator and psychosocial stress: relevance to human PTSD. ACS Chem Neurosci 13(2):257–274. https://doi.org/10.1021/acschemneuro.1c00660

Kessler RC, Aguilar-Gaxiola S, Alonso J et al (2017) Trauma and PTSD in the WHO world mental health surveys. Eur J Psychotraumatol 8(sup5):1353383

Kometer M, Vollenweider FX (2018) Serotonergic hallucinogen-induced visual perceptual alterations. Curr Top Behav Neurosci 36:257–282

Kraehenmann R, Preller KH, Scheidegger M, Pokorny T, Bosch OG, Seifritz E, Vollenweider FX (2015) Psilocybin-induced decrease in amygdala reactivity correlates with enhanced positive mood in healthy volunteers. Biol Psychiatry 78(8):572–581. https://doi.org/10.1016/j.biopsych.2014.04.010

Krebs TS, Johansen PØ (2013) Psychedelics and mental health: a population study. PLoS One 8: e63972

Krediet E, Bostoen T, Breeksema J, van Schagen A, Passie T, Vermetten E (2020) Reviewing the potential of psychedelics for the treatment of PTSD. Int J Neuropsychopharmacol 23(6): 385–400

Krystal JH, Abdallah CG, Averill LA et al (2017) Synaptic loss and the pathophysiology of PTSD: implications for ketamine as a prototype novel therapeutic. Curr Psychiatry Rep 19(10):74

Liberzon I, Abelson JL (2016) Context processing and the neurobiology of post-traumatic stress disorder. Neuron 92(1):14–30. https://doi.org/10.1016/j.neuron.2016.09.039

Litz BT, Stein N, Delaney E et al (2009) Moral injury and moral repair in war veterans: a preliminary model and intervention strategy. Clin Psychol Rev 29(8):695–706

Ly C, Greb AC, Cameron LP et al (2018) Psychedelics promote structural and functional neural plasticity. Cell Rep 23(11):3170–3182

Majic T, Schmidt TT, Gallinat J (2015) Peak experiences and the afterglow phenomenon: when and how do therapeutic effects of hallucinogens depend on psychedelic experiences? J Psychopharmacol 29(3):241–253

Malone TC, Mennenga SE, Guss J, Podrebarac SK, Owens LT, Bossis AP, Belser AB, Agin-Liebes G, Bogenschutz MP, Ross S (2018) Individual experiences in four cancer patients following psilocybin-assisted psychotherapy. Front Pharmacol 9:256. https://doi.org/10.3389/fphar.2018.00256

Markowitz JC, Petkova E, Neria Y et al (2015) Is exposure necessary? A randomized clinical trial of interpersonal psychotherapy for PTSD. Am J Psychiatry 172(5):430–440

Merz J, Schwarzer G, Gerger H (2019) Comparative efficacy and acceptability of pharmacological, psychotherapeutic, and combination treatments in adults with posttraumatic stress disorder: a network meta-analysis. JAMA Psychiat 76(9):904–913

Mithoefer MC, Mithoefer AT, Feduccia AA et al (2018) 3,4-methylenedioxymethamphetamine (MDMA)-assisted psychotherapy for post-traumatic stress disorder in military veterans, firefighters, and police officers: a randomised, double-blind, dose-response, phase 2 clinical trial. Lancet Psychiatry 5:486–497

Nardou R, Lewis EM, Rothhaas R et al (2019) Oxytocin-dependent reopening of a social reward learning critical period with MDMA. Nature 569:116–120

Nemeroff CB (2004) Neurobiological consequences of childhood trauma. J Clin Psychiatry 65:18–28

Nielson JL, Megler JD (2014) Ayahuasca as a candidate therapy for PTSD. In: The therapeutic use of ayahuasca, vol 9783642404269. Springer, pp 41–58

Nöthling J, Malan-Müller S, Abrahams N et al (2020) Epigenetic alterations associated with childhood trauma and adult mental health outcomes: a systematic review. World J Biol Psychiatry 21(7):493–512

O'Donovan A, Cohen BE, Seal KH et al (2015) Elevated risk for autoimmune disorders in Iraq and Afghanistan veterans with posttraumatic stress disorder. Biol Psychiatry 77(4):365–374
Osório FL, Sanches RF, Macedo LR et al (2015) Antidepressant effects of a single dose of ayahuasca in patients with recurrent depression: a preliminary report. Br J Psychiatry 37:13–20
Ostacher MJ, Cifu AS (2019) Management of posttraumatic stress disorder. JAMA 321(2):200–201
Palhano-Fontes F, Barreto D, Onias H et al (2019) Rapid antidepressant effects of the psychedelic ayahuasca in treatment-resistant depression: a randomized placebo-controlled trial. Psychol Med 49(4):655–663
Petri G, Expert P, Turkheimer F et al (2014) Homological scaffolds of brain functional networks. J R Soc Interface 11:20140873
Preller K, Pokorny T, Krähenmann R et al (2015) The effect of 5-HT2A/1a agonist treatment on social cognition, empathy, and social decision-making. Eur Psychiatry 30(S1):1–1
Raval NR, Johansen A, Donovan LL et al (2021) A single dose of psilocybin increases synaptic density and decreases 5-HT_{2A} receptor density in the pig brain. Int J Mol Sci 22:835
Riba J, Romero S, Grasa E et al (2006) Increased frontal and paralimbic activation following ayahuasca, the pan-amazonian inebriant. Psychopharmacology (Berl) 186(1):93–98
Rodriguez P, Holowka DW, Marx BP (2012) Assessment of posttraumatic stress disorder-related functional impairment: a review. J Rehabil Res Dev 49:649–666
Romano AG, Quinn JL, Li L, Dave KD, Schindler EA, Harvey VJAJA (2010) Intrahippocampal LSD accelerates learning and desensitizes the 5-HT2A receptor in the rabbit Romano et al. Psychopharmacology 212(3):441–448. https://doi.org/10.1007/s00213-010-2004-7
Roseman L, Nutt DJ, Carhart-Harris RL (2018) Quality of acute psychedelic experience predicts therapeutic efficacy of psilocybin for treatment-resistant depression. Front Pharmacol 8:974
Ross S, Agin-Liebes G, Lo S, Zeifman RJ, Ghazal L, Benville J, Corso SF, Real CB, Guss J, Bossis A, Mennenga SE (2021) Acute and sustained reductions in loss of meaning and suicidal ideation following psilocybin-assisted psychotherapy for psychiatric and existential distress in life-threatening cancer. ACS Pharmacol Trans Sci 4(2):553–562. https://doi.org/10.1021/acsptsci.1c00020
Sareen J, Cox BJ, Stein MB et al (2007) Physical and mental comorbidity, disability, and suicidal behavior associated with posttraumatic stress disorder in a large community sample. Psychosom Med 69(3):242–248
Schindler EAD, Wallace RM, Sloshower JA et al (2018) Neuroendocrine associations underlying the persistent therapeutic effects of classic serotonergic psychedelics. Front Pharmacol 1(9):177
Schmid Y, Liechti ME (2018) Long-lasting subjective effects of LSD in normal subjects. Psychopharmacology (Berl) 235(2):535–545
Seligowski AV, Lee DJ, Bardeen JR et al (2015) Emotion regulation and posttraumatic stress symptoms: a meta-analysis. Cogn BehavTher 44(2):87–102
Shalev AY, Freedman S, Peri T et al (1998) Prospective study of posttraumatic stress disorder and depression following trauma. Am J Psychiatry 155(5):630–637
Stauffer CS, Anderson BT, Ortigo KM et al (2020) Psilocybin-assisted group therapy and attachment: observed reduction in attachment anxiety and influences of attachment insecurity on the psilocybin experience. ACS Pharmacol Transl Sci 4(2):526–532
Steenkamp MM, Litz BT, Hoge CW et al (2015) Psychotherapy for military-related PTSD: a review of randomized clinical trials. JAMA 314(5):489–500
Steenkamp MM, Litz BT, Marmar CR (2020) First-line psychotherapies for military-related PTSD. JAMA 323(7):656–657
Tylš F, Páleníček T, Horáček J (2014) Psilocybin – summary of knowledge and new perspectives. Eur Neuropsychopharmacol 24(3):342–356
van Amsterdam J, Opperhuizen A, van den Brink W (2011) Harm potential of magic mushroom use: a review. Regul Toxicol Pharmacol 59(3):423–429
Vollenweider FX, Preller KH (2020) Psychedelic drugs: neurobiology and potential for treatment of psychiatric disorders. Nat Rev Neurosci:1–14

Wasson RG (1980) The wondrous mushroom. Mycolatry in Mesoamerica. Ethnomycoloical studies no. 7. McGraw-Hill Book Company, New York

Watts BV, Schnurr PP, Mayo L et al (2013) Meta-analysis of the efficacy of treatments for posttraumatic stress disorder. J Clin Psychiatry 74(6):e541–e550

Watts R, Day C, Krzanowski J et al (2017) Patients' accounts of increased "connectedness" and "acceptance" after psilocybin for treatment-resistant depression. J Humanist Psychol 57(5): 520–564

Williams MT, Metzger IW, Leins C et al (2018) Assessing racial trauma within a DSM–5 framework: the UConn racial/ethnic stress & trauma survey. Pract Innov 3(4):242–260

Williams MT, Davis AK, Xin Y et al (2021) People of color in North America report improvements in racial trauma and mental health symptoms following psychedelic experiences. Drugs 28(3): 215–226

Woodhouse S, Ayers S, Field AP (2015) The relationship between adult attachment style and post-traumatic stress symptoms: a meta-analysis. J Anxiety Disord 35:103–117

Zeifman RJ, Wagner AC (2020) Exploring the case for research on incorporating psychedelics within interventions for borderline personality disorder. J Contextual Behav Sci 15:1–11

Zeifman RJ, Wagner AC, Watts R et al (2020) Post-psychedelic reductions in experiential avoidance are associated with decreases in depression severity and suicidal ideation. Front Psych 11: 782

Zeifman RJ, Singhal B et al (2021) On the relationship between classic psychedelics and suicidality: a systematic review. ACS Pharmacol Transl Sci 4(2):436–451

Printed in the USA
CPSIA information can be obtained
at www.ICGtesting.com
LVHW011252130823
755086LV00005B/225

9 783031 121838